Machinery Vibration

Machinery Vibration

Measurement and Analysis

Victor Wowk

McGraw-Hill, Inc.

New York St. Louis San Francisco Auckland Bogotá
Caracas Lisbon London Madrid Mexico Milan
Montreal New Delhi Paris San Juan São Paulo
Singapore Sydney Tokyo Toronto

Library of Congress Cataloging-in-Publication Data

Wowk, Victor.
Machinery vibration : measurement and analysis / Victor Wowk.
p. cm.
Includes bibliographical references and index.
ISBN 0-07-071936-5
1. Machinery—Vibration. I. Title.
TJ177.W68 1991
621.8'11—dc20 91-4034
CIP

1 2 3 4 5 6 7 8 9 0 DOC/DOC 9 7 6 5 4 3 2 1

ISBN 0-07-071936-5

The sponsoring editor for this book was Robert Hauserman, the editing supervisor was Peggy Lamb, and the production supervisor was Donald F. Schmidt. This book was set in Century Schoolbook. It was composed by McGraw-Hill's Professional Book Group composition unit.

Printed and bound by R. R. Donnelley & Sons, Company.

Contents

Preface

Machinery Vibration: Measurement and Analysis is intended to be a companion student text to a three-day vibration workshop. The workshop utilizes many physical models in the classroom to demonstrate concepts and practice techniques; however, this text can stand alone as an introduction to the field of machinery vibration.

This book contains basic information and fundamental knowledge about machinery vibration. The physical phenomenon of vibration is explained in practical terms and related to machines. The measurement of vibration and the interpretation of the measured data are nontrivial skills, but these data hold the key to understanding dynamic forces within machines.

The objective of the book is to train an engineer or technician to take vibration measurements and to interpret the results. The interpretation can be for several reasons—to diagnose condition, to judge severity, to predict time to failure, to extend life, or to just understand the cause of a noise or vibration problem. This process of understanding is the first step of two general steps in any problem resolution. This first step is a two-part process, to gather data and diagnose the cause, or measurement and analysis, and is the subject matter of this book. The second general step, to perform remedial action, is usually straightforward after the cause is understood. This second step, corrective methods, is the subject of a subsequent book.

This book focuses on the conventional instruments and methods to measure machinery vibrations that a field engineer or technician is likely to encounter. There are other instruments of a more sophisticated nature for unique applications. The author has chosen to limit the coverage of these other instruments and focus instead on the conventional, cost-effective methods of data acquisition.

The subject of *Machinery Vibration: Measurement and Analysis* is presented in small bite-size pieces. This makes it digestible. The book also starts with the premise that the reader has zero knowledge and background in vibrations. This is hardly true of anyone in the dynamic world that we live in, but it allows a fresh perspective to be developed in the reader. My hope is to take a complicated subject and present it as a natural experience that everyone has been exposed to.

Victor Wowk

ABOUT THE AUTHOR

Victor Wowk, a registered professional engineer and member of the Vibration Institute, is the owner of Machine Dynamics, a consulting firm specializing in machinery vibrations, based in Albuquerque, New Mexico. He has extensive field experience in vibration analysis which includes machinery diagnostics and applying corrective techniques, such as dynamic balancing and precision alignment. He is an instructor for both in-plant and public Machinery Vibration Workshops. He was previously a senior manufacturing engineer at Honeywell, Inc., Defense Avionics Systems Division, where he was involved in the vibration testing of military avionics. Prior to that, he was a mechanical engineer at Hewlett-Packard, Loveland Instrument Division, where he was responsible for production of the spectrum analyzer product line as well as plant engineering and maintenance.

Chapter

1

Introduction

There are generally two situations in which vibration measurements are taken. One is in a surveillance mode to check the health of machinery on a routine basis. The second situation is during an analysis process where the ultimate goal is to fix a problem. In the latter case, vibration measurements are taken to understand the cause, so that an appropriate fix can be undertaken. In either situation, surveillance or analysis, there are several types of instruments available to take the measurements, and at least as many ways to acquire the data. Going one step further, the analysis of that data is a high-level reasoning function that introduces the human variable. The instruments, however, faithfully and consistently measure the physical quantities of vibratory motion, albeit they display it differently.

The instruments used to measure vibration have advanced tremendously in the past 15 years. Paramount among all these modern instruments is the fast Fourier transform (FFT) spectrum analyzer which represents the state of the art in vibration measurements. Its cost has decreased to levels below older tuneable filter analyzers. The FFT spectrum analyzer provides more information of higher quality at a lower cost, and it is the instrument of choice for the vast majority of vibration problems. It can provide the "big picture" of the dynamic motion in real time and store the data for later analysis. For these reasons, the FFT spectrum analyzer is used almost exclusively throughout this book. It is the most powerful instrument to use for vibration analysis, but it is generally not the easiest to use. The front panel has approximately 40 buttons, some of which have multiple functions, and hidden menus. A large part of learning modern vibration analysis is overcoming the button phobia of current instrumentation.

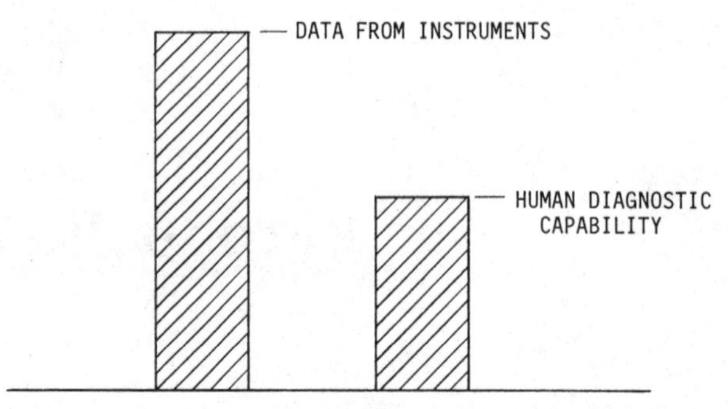

Figure 1.1 Human diagnostic capability lags behind instrument development.

Modern instruments have greatly advanced our ability to acquire dynamic data. The interpretation of that data is the weakest link in the analysis chain. The sophistication of modern instruments exceeds our diagnostic capability. There is generally more information available than we can interpret. If this concept were presented in a bar graph, it would look something like Fig. 1.1.

Certainly, the acquisition of good data is the most important step in any analysis situation. There are some hazards to acquiring good data with modern, complicated instruments, and the reader will be guided around some of these obstacles. The important point to be made with this illustration is that today's instruments can produce good data and a lot of it. What has lagged behind is the human ability to absorb and digest it. The purpose for this book is to raise the level of human diagnostic capability, and fill this gap.

This book focuses on the diagnosis of machinery faults using the vibration signals emanating from the machine itself. Mathematics and the theory of vibration are only briefly introduced to explain concepts. The major emphasis of the book is on hands-on techniques using available instrumentation to solve real vibration problems. Case histories will be presented of electric motors, pumps, fans, and piping problems.

This book is intended for operating and maintenance personnel with responsibility for rotating equipment. It is useful to engineers, managers, and technical tradespeople. Some background knowledge in electronics and machine repair is helpful. The formal mathematics of algebra and trigonometry are a necessary prerequisite. Some calculus is presented, but it is not necessary in the application of vibration technology. This book is designed for those practicing technical professionals who need solutions to real vibration problems and who would rather dispense with long analytical and theoretical explanations.

Objectives

After studying this book, the reader will:

- Understand the physical phenomenon of vibration in machines.
- Understand why vibration monitoring is important.
- Have a working knowledge of the terms used in vibration.
- Understand the relationship of machinery vibration analysis to similar fields such as acoustics and testing.
- Have a working knowledge of the instruments used to measure vibration.
- Be able to recognize typical vibration problems.
- Be able to recognize imbalance and misalignment in rotating equipment.
- Be able to apply state-of-the-art techniques to analyze bearings, gears, couplings, and looseness.
- Be able to diagnose internal defects from vibration signatures and judge their severity against industry standards.
- Be able to apply trending and set up a predictive maintenance program using vibration signatures.
- Understand resonance.
- Understand narrowband frequency analysis and FFT spectrum analyzers.
- Understand structural vibration problems.

Why Vibrations?

Vibration of mechanical equipment is generally not good. It causes excessive wear of bearings, it causes cracking, it causes fasteners to come loose, it causes electric relays to malfunction, it causes electronic malfunctions through the fracture of solder joints, it abrades insulation around electrical conductors causing shorts, it causes noise, and it is generally uncomfortable for humans.

Motorcycles are a good case in point (Fig. 1.2). Some older motorcycles had reputations for losing parts on the road. I remember, as a teenager, seeing the headlight vibrate off of a British motorcycle and hang suspended from its wires while idling at a stoplight. The rider reached over the handlebars, ripped the wires, and stuffed the headlight in his jacket.

Figure 1.2 A vibration machine.

Not all vibration is harmful. Some is benign. It is the task for the analyst to determine benign from harmful. Vibrations that will result in future failure are the ones that need to be identified and corrected. These harmful vibrations are the symptoms of significant forces that cause wear at load points and structural cracking. Harmful symptoms might be serious imbalance, misalignment, resonance, and shock pulses. Benign vibrations are those that are due to the normal functioning of a machine. Benign effects might be a motor hum, vane passing frequency, and broadband turbulence due to fluid motion. These will not usually cause failure, although they could be a perceptual problem for coexisting humans.

You have analyzed vibration if you have ever ridden in any type of vehicle. You are continuously exposed to the sound and feel of its vibrations. Specifically, you are very interested in any change that you perceive. Imagine that you couldn't hear while driving a car. How would you know that something was wrong mechanically? You would have the first indication when it vibrated the structure so badly that you could feel it, or the car just stopped. By listening, you have a very sensitive indicator of mechanical condition.

A common criteria for quality judgment is lack of vibration. Vibration is directly correlated to machine longevity, in two ways:

1. A low vibration level when new generally indicates that it will last a long time.
2. The vibration level increases when a machine is heading for a breakdown.

Maintenance typically represents 15 to 40 percent of the total cost of production. Assume that yours is an operation that generates $10 million in annual sales, and the cost of production (cost of goods sold) is $6 million. Twenty-eight percent (average maintenance cost) of $6 million is $1,680,000. If maintenance costs could be reduced by 10 percent, then you could add $168,000 to your bottom-line profit. Your operation may need to generate another $1 million in sales to equal a

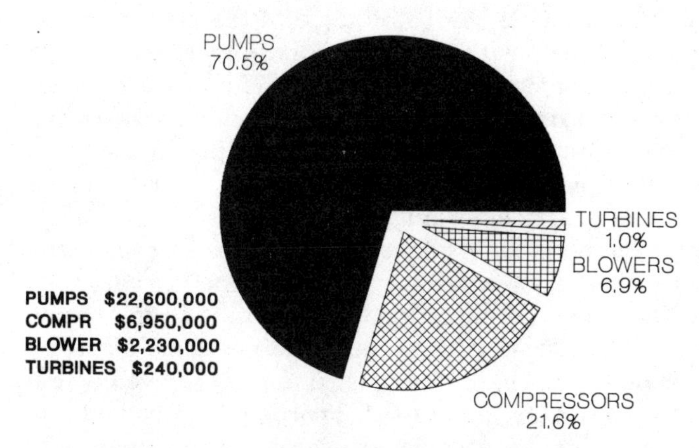

Figure 1.3 Rotating equipment costs at a petrochemical operation, 1973 to 1982. (*From Charles Jackson.*)

$168,000 profit. By improving maintenance, it is possible to obtain the same profit gain without any increase in sales.

You may ask if a 10 percent decrease in maintenance cost is achievable. The answer is yes. Improvements of 20 to 50 percent are achievable in all manufacturing and processing plants. Some higher returns have been realized in some operations.

The data in Fig. 1.3 came from a typical petrochemical operation over a 10-year period. Notice that the total maintenance costs are over $32 million in 1982 dollars. These were true costs, including drivers, in cases, and costs include labor and parts overhead. There are two other important points to be observed from this data.

First, pumps represent more than 70 percent of the total cost. Turbines are the most expensive machines, but the large number of common pumps that must be maintained consume far more dollars over the years to keep running. The second point is the typical breakdown of costs which follow the Pareto principle.[1] One type of machine accounts for more than two-thirds of the problems. If any significant impact is to be made in reducing maintenance costs, then changes must

[1]Vilfredo Pareto, an Italian economist (1848–1923), studied the unequal distribution of wealth. Various other researchers, especially in quality control, have observed the phenomenon of the vital few and the trivial many. The term *Pareto principle* was first identified and applied as a universal principle by J. M. Juran in the *Quality Control Handbook,* 3d ed., McGraw-Hill, New York, 1979 (now J. M. Juran and F. M. Gryna, *Juran's Quality Control Handbook,* 4th ed., 1988). Today, in the context of quality improvement, the principle can be loosely defined as a hierarchy of contributors. There is always a Number 1 contributor that accounts for the bulk of the data, a Number 2 contributor with a smaller percentage, and so forth. For any quality-control program to be effective, the Number 1 contributor must be identified, and improvement efforts must be directed first toward that contributor.

be directed to the pumps. This chart is unique to the petrochemical industry. Other operations, with different mixtures of machinery types, may have a different machine that is the Number 1 sink for maintenance costs. For example, in the semiconductor industry the movement of air is the primary need for rotating equipment, and fans may occupy the largest piece of the pie. The important concept here is to first classify the machinery according to dollars consumed in maintenance to identify the Number 1 offender. Corrective measures can then be directed to where they will have the most benefit.

Why vibration? Because there is a hidden gold mine in improving maintenance operations. These improvements do not come about by decreasing staff or budgets. These improvements are achieved by doing things smarter—by not buying unneeded parts, by easily doubling the life of machinery, and by decreasing energy consumption as a result of not generating noise and vibration.

A case history to illustrate this point is about a 25-year-old government building. It has four quadrant fans, two chillers, two cooling towers with gearboxes, and about a dozen pumps for hot and cold water circulation. This structure and its equipment is typical of a 50,000-ft^2 office building. The building was to be refurbished because of asbestos used in the original construction. The engineers thought that this would also be a good time to replace most of the mechanical equipment. The estimate just to replace the pumps was $30,000. A vibration survey was commissioned to assess the mechanical condition of all the rotating equipment. One quadrant fan was found to be out of balance, a deteriorating bearing was found on another fan, a fan motor had a bad rear bearing, and one pump was slightly misaligned. The remainder of the rotating equipment was found to be in good condition—better than new in some cases. The cooling-tower fans and gearboxes were especially smooth. The maintenance supervisor claimed that the only maintenance these cooling towers had seen in the 16 years he had been there was an occasional greasing. Unfortunately, that's where most of the asbestos was, and the cooling towers had to be removed. The rest of the rotating equipment was repaired as necessary and left in place. This equipment had obviously been well maintained over the years, and the owner was now reaping the benefits by applying $30,000 of unnecessary mechanical equipment replacement to other parts of the project. However, the real gold mine was the cash outflow that did *not* occur because the cooling towers did *not* require rebuilding. If two cooling towers can operate trouble-free for 25 years, then *all* cooling towers can operate trouble-free for 25 years. It is possible.

This vibration survey took one day to accomplish. This story illustrates one of the benefits of knowledge. One day's worth of vibration analysis was worth $30,000 of benefit to the owner. This type of payback is typical. It is not unusual to have return on investments for vibration analysis of 1000 to 100,000 percent. This is hard to beat at any financial institution.

Statistically, the majority of maintenance work is on common pumps, fans, and motors. Even industries that have high-speed, expensive machinery like turbines and compressors, spend most of their maintenance effort on common slow-speed machinery.

Vibration itself is usually a bad situation, but it is also a symptom of an internal defect. Specifically, it is a very sensitive and early predictor of developing defects. Catastrophic failures are usually preceded by a change in vibration, sometimes months before the actual failure.

The operating machine is talking to us in its own language telling us all about its internal condition. We should listen and interpret what it is saying. This is where instruments help. A machine's vibration signature is like a heartbeat. Figure 1.4 depicts a doctor with a motor on the examining table. A vibration analyst performs the same diagnostic service on machinery that a medical doctor performs on a human patient.

Why vibration? Because vibration is the best indicator of overall mechanical condition and the earliest indicator of defects developing. There are other indicators, e.g., temperature, pressure, flow, and oil analysis.

As Fig. 1.5 indicates, these other parameters are useful and valuable. At times, this additional data should be sought and correlated with vibration data to obtain a complete analysis. If only one indicator

Figure 1.4 A vibration analyst is analogous to a machine doctor.

WHY VIBRATION?

	TEMPERATURE	PRESSURE	FLOW	OIL ANALYSIS	VIBRATION
OUT OF BALANCE					X
MISALIGNMENT BENT SHAFT	X				X
DAMAGED ROLLING ELEMENT BEARINGS	X			X	X
DAMAGED JOURNAL BEARINGS	X	X	X	X	X
DAMAGED OR WORN GEARS				X	X
MECHANICAL LOOSENESS					X
NOISE					X
CRACKING					X

Figure 1.5 Vibration is the best indicator of machine health.

is to be used to monitor machine health, then vibration is usually the best choice.

Demonstration of Floor Vibrations

Everything in the universe vibrates. Heavenly bodies revolving are vibrations; tides are vibrations, molecular Brownian movements are vibrations; atoms vibrate. There are no static forces because superimposed on all "constant" forces are microvibrations. These microvibrations cause wear at joints and interfaces. In time, with the assistance of gravity, they will cause everything to crumble to the ground.

To prove that a building is vibrating, attach a transducer to the floor, column, electrical outlet, or air grille, and display its output on a spectrum analyzer. Some of the steady-state vibration frequencies that you may observe could be:

<5 Hz	Possibly building sway
10–30 Hz	Fans
~30 Hz	1800-rpm (revolutions per minute) motors
~60 Hz	3600-rpm motors or compressors
120 Hz	Transformers

There may also be vibrations that come and go at specific frequencies (amplitude rises and falls). These are most likely resonances. Tap or walk on the floor. Ask someone far away to drop something on the floor. Disconnect the input briefly to show the "instrument noise floor" and verify that these vibrations are really in the building even though we cannot sense them.

This illustrates the sensitivity of modern instruments. Some industries are sensitive to minute floor vibrations, like semiconductor and medical facilities, where optics are used. The sources of these vibrations are typical building mechanical and electrical equipment, i.e., motors, pumps, fans, and transformers. Just because the vibration is below the level of human perception does not mean that it is to be ignored. It could very well be affecting sensitive instruments or processes.

Chapter

2

Overview of the Entire Field and a Brief History

Earthquakes and Wind

Earthquakes and wind are naturally occurring vibrations of direct consequence to humans. Ground motions are monitored by seismic recording stations around the world. The frequency and severity of earthquakes vary geographically. Figure 2.1 is a current risk map for ground motion activity in the United States. Darker areas represent greater risk.

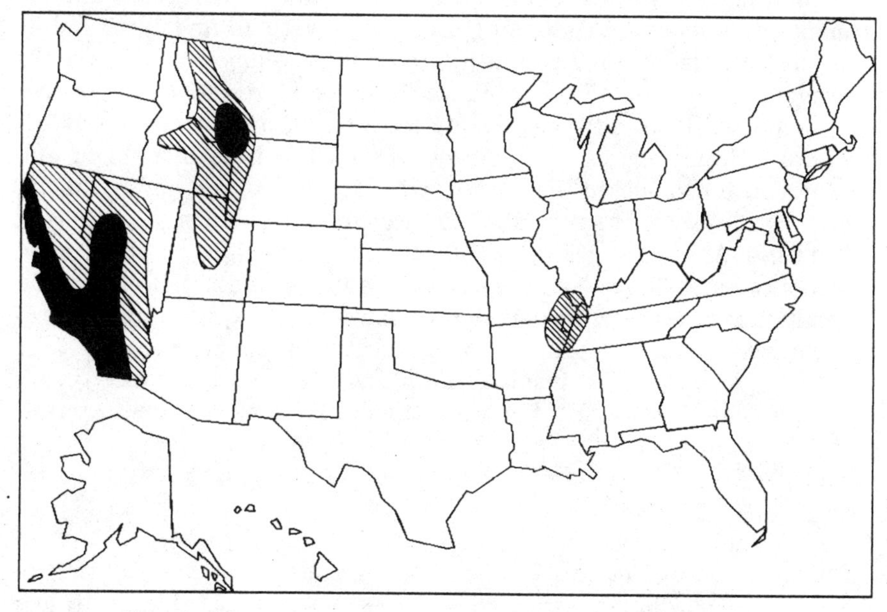

Figure 2.1 Seismic risk map.

It must be pointed out that this map and others like it represent statistical data accumulated since recording began. Events prior to this are not recorded on such maps. There is evidence that the east coast of the United States experienced a "very large" earthquake during the past 100,000 years. Figure 2.1 represents data observed only during the past 450 years. Therefore, maps such as this depict the probability of frequent earth tremors based on recent historical data. They are mostly ignorant of infrequent events that are probably massive in destructive capability. Any place on the surface of this planet can experience earth movements in terms of geologic time.

There are special considerations for mechanical equipment in seismically active areas. As the building sways horizontally, the mechanical equipment will sway on its spring isolators. This movement can cause it to fall off its springs and tear its electrical supply conduit. Piping and attached ductwork can be damaged. To limit such damage, mechanical equipment is restrained from excessive movement. This restraint can be in the form of cables, snubbers, or hard-bolting the equipment to the floor. These restraints complicate the vibration isolation task, and in some cases allow machinery vibrations to transmit to the building. The use of seismic restraints is further discussed in the isolation chapter of *Machinery Vibration: Corrective Methods* (1992).

Wind is a naturally occurring random vibration. Random vibrations will be discussed later. For now, *random vibrations* can be described as those that do not repeat regularly. Random vibrations can excite structures to vibrate at their natural resonant frequencies. Some examples are power line galloping, Aeolian vibrations, and duct rumble.

The Tacoma Narrows Bridge collapsed in November 1940 (4 months after opening) due to wind-induced structural resonances. Vortices around the I-beam roadway during a steady 42-mi/h wind caused torsional twisting of the span. It swayed as much as 8 ft (steel and concrete) for about an hour until it collapsed into the gorge. Future structural designs had to take into account the random vibrations of the wind. Any structure more than one story in height of typical steel construction will have some sway due to wind. These motions can be easily measured with modern instruments used for machinery vibrations, and a competent analyst will recognize them as such.

Fans generate wind inside of buildings. They also produce turbulence inside ducts which have the capability to excite the duct wall into resonance, i.e., duct rumble.

Music

Musical instruments have produced vibrations that are pleasing to the human ear for centuries. These vibrations are caused by exciting a string, a reed, a column of air, or a membrane to vibrate at its natural frequency. These vibrations are sinusoidal, and they repeat regularly with a tone.

Vibration analysis and music are closely related technologies. Listening for a flute from an orchestra performance is similar to listening for a bearing from a fan.

Acoustics

Sound is air in vibration. Excessive amplitude (or volume) is damaging to the ears. Some interior spaces (like concert halls and recording studios) need to be especially quiet. This is the field of architectural acoustics. The significant factors here are reverberation time and the reduction of background noise. This field is mostly science but still retains some art to produce the desired effect.

The Occupational Safety and Health Administration (OSHA) has placed limits on the maximum noise exposure a worker can receive in an 8-h workday to avoid permanent hearing loss. These are enforced in industrial environments but routinely exceeded at entertainment and sporting events. The scientific field concerned with hearing damage is *audiology*.

Unwanted sound is defined as *noise,* and noise is now perceived as a form of environmental pollution. People are becoming more sensitive to noise pollution, especially around airports, traffic corridors, industrial areas, and commercial buildings. Some communities have instituted ordinances defining the maximum sound level permissible at residential property lines. By exchanging the vibration transducer for a microphone, the same instruments used in machinery vibration analysis can be used for acoustical analysis. A separate chapter later in this book is devoted to acoustics as it relates to machinery vibration.

Two other related fields are:

1. Acoustic emissions
2. Ultrasonics

Acoustic emissions are the sounds generated in materials when they are strained. The sounds are usually very minute and are actually the shock waves generated when crystal boundaries deform along their slip planes. Theoretically, these are the same as the low-frequency "groans" heard from the earth just prior to and during earthquakes. Some high-performance fighter aircraft are now equipped with acoustic emission sensors on their skin and other structural parts to detect excessive strains.

Ultrasonics is the generation of vibrations above the human range of hearing (i.e., greater than 20 kHz). These vibrations are used for testing materials, welding plastics, and imaging of interior soft body tissues.

Finite Element Analysis

Using the finite element analysis (FEA) technique, a structure is divided into small pieces that have mass, elasticity, and damping. The pieces are all connected together, and some forces are input to see what happens to the whole structure. This is a design technique done on paper (or computer) *before* any hardware exists to predict how it will perform. Figure 2.2 represents this division on a section of pipe.

FEA is mentioned here because it is a dynamic engineering design tool. It attempts to predict motions due to temperature changes or force inputs. It uses the same concepts of mass, spring, and damper as vibration analysis does. Vibration analysis is the study of how a structure responds *after* it is built.

Modal Analysis

Modal analysis is the exciting of a structure with shakers in order to measure how it bends. This analysis is routinely done in the aerospace industry. The purpose for doing this is to place stiffeners in the proper places where they will be most effective.

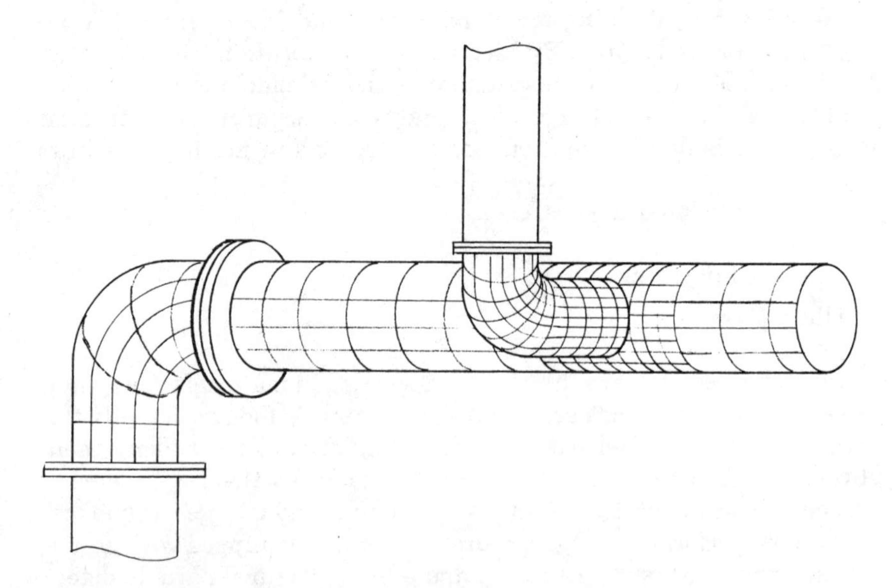

Figure 2.2 Finite element analysis of a pipe section.

When "modal analysis" is mentioned, most engineers think of a very complicated test with many shakers, a hundred or so transducers mounted on the structure, and a very large and fast computer to process all this data. This is not always necessary, and later in this book a simple modal analysis technique using a hammer and some sand will be presented. Another method using a simple hand-held vibration meter will also be described.

Vibration Testing

Vibration testing is generally of three kinds. One is to see how a structure will respond. Modal analysis falls into this category. A second type tests production assemblies, especially electronics. The assembled product is intentionally exposed to vibration before shipment to find the weak connections. This wrings out the infant mortalities. It is more cost-effective to repair it while it is still in the factory. A third type of vibration testing measures acceptance. Vibration limits are sometimes specified on the purchase documents of new equipment. There are industry standards for various equipment and how much it should be allowed to shake.

Shock

Shock is an impulse characterized by suddenness and severity. It is a single-impulse event of short duration. Shock usually causes significant relative displacement. This large initial displacement is usually the cause of failure as materials plastically deform or fracture when material strengths are exceeded. The impact generates a shock pulse that travels through the material, reflecting at the boundaries. The energy of the shock pulse is absorbed by damping. Metal-to-metal impacts are characterized by short-duration (several milliseconds) shock pulses that can clearly be seen on an oscilloscope using an appropriate transducer, such as an accelerometer. These can usually be seen coming from most bearings that have begun to wear. Shock pulses also have the capability to set up vibrations at the system's natural frequency.

Many machinery vibrations are actually shock pulses or transients that repeat. Modern vibration instruments have the capability to capture and store these transients for analysis at "people" speed.

Generating Vibrations

To keep things in perspective, vibrations are sometimes used profitably in industrial applications. In addition to generating vibrations for testing, some production processes use vibrations. These are vibratory finishers, conveyors, and hoppers to move bulk materials, compactors for soil, and pile driving. Vibrations are also used to improve the efficiency of casting operations.

A new application of generating vibrations is the active cancellation of unwanted noise and vibration. Figure 2.3 shows two waves of equal magnitude and frequency, but 180° out of phase. When these two waves are mixed, the result is zero. In active cancellation, waves are generated that are 180° out of phase with the objectionable noise or vibration. These are amplified and fed to an appropriate speaker or other transducer. When the source vibration and the generated vibration mix, they cancel each other and the result is quiet. This same concept of generating 180° out-of-phase signals has been applied to small

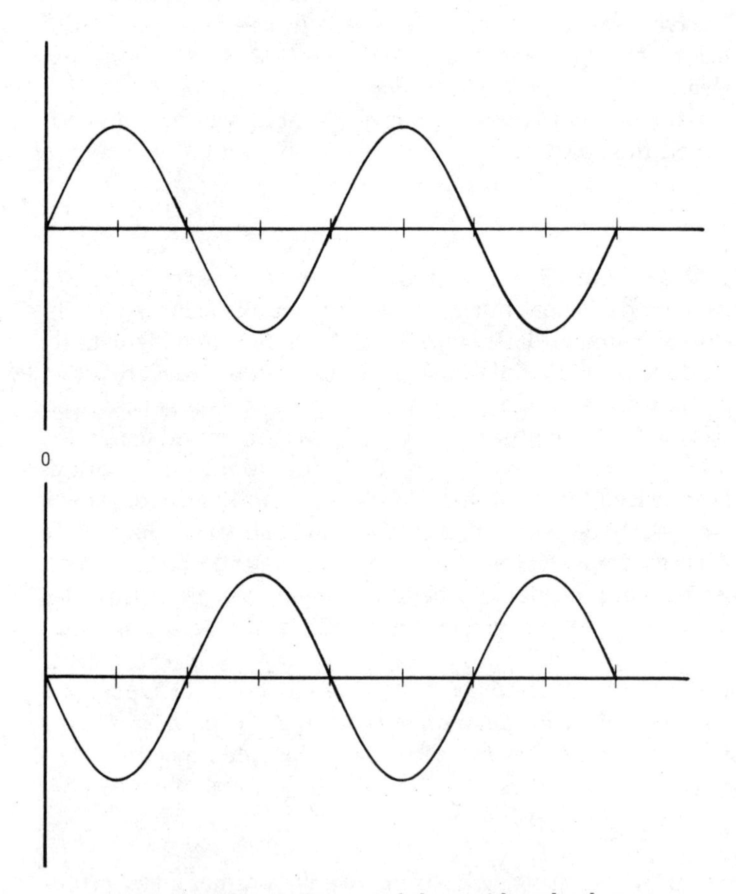

Figure 2.3 Two vibrations 180° out of phase with each other.

tables. In this way, vibrations from the floor are canceled, and the tabletop is stable.

Machinery Monitoring

The focus of this book is on the understanding of machinery vibrations in a maintenance environment. In the last 5 years, there has been an explosion in the sales of FFT spectrum analyzers. Part of the reason for this has been a decrease in cost, but a bigger reason is the tailoring of these instruments for machine monitoring. The FFT spectrum analyzer, since its introduction in the 1970s, has been a superior instrument for machine monitoring. It was, however, not an easy instrument to use. It was large and heavy and had many front panel buttons that were foreign to even an electronic technician. Some had hidden menus, and there seemed to be no standardization among manufacturers concerning its front panel layout. The adaptations that tailored this instrument to machine monitoring have been to allow a maintenance person to use it. These adaptations were portability, small size, light weight, battery power, shoulder straps so the maintenance person could use both hands, fewer buttons, and larger memories.

The power of the FFT spectrum analyzer in a machine monitoring environment is its capability to display vibration data in the frequency domain as a spectrum. Figure 2.4 is a vibration spectrum of a 10-hp motor-pump combination. The vertical axis is amplitude, and the horizontal axis is frequency. There are three features of this display that make it so useful for machine monitoring. First, each peak is associated with a specific defect. Second, this spectrum display will not change very much if the machine does not change. Of greater significance is that when the machine changes because of a developing internal defect, the spectrum changes. Third, an overall number can be assigned to this spectrum display, and this overall can be easily tracked. Virtually all modern instruments used today for machine monitoring can measure the overall and produce a spectrum. The machine and problem condition associated with the spectrum of Fig. 2.4 is further discussed in Chapter 5 under "Piping."

There are two general types of machinery monitoring using vibration:

1. Machine diagnostics
2. Predictive maintenance

Machine diagnostics is examining the vibration signals reaching the outer casting of a machine and from these determine its internal condition. This is analysis of an existing problem. *Predictive maintenance* (*PM*) is letting a machine run as long as it runs smoothly. The vibra-

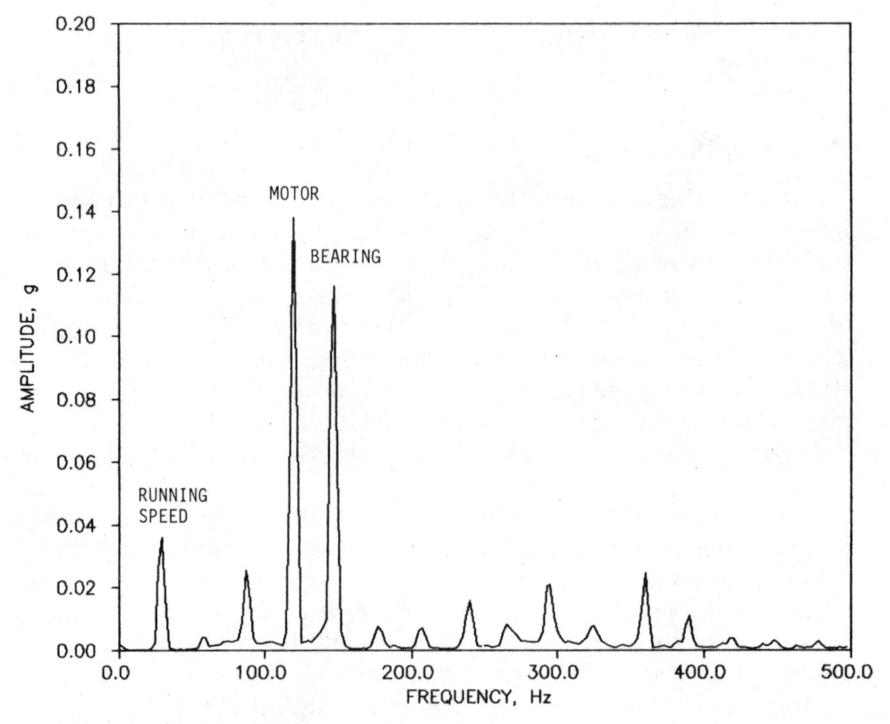

Figure 2.4 A vibration spectrum.

tion signals are periodically measured and plotted. This is a health checkup. When defects develop and the vibration signature changes, then PM raises a warning and predicts the remaining running time before a breakdown occurs.

History

Prior to the industrial revolution

Let's discuss the historical events that have occurred that brought us to the present state of technology in machine monitoring. Prior to the industrial revolution, when machines turned slowly, the frequency of vibrations was also very slow and of no consequence. The displacements, however, were quite large because of the manufacturing tolerances of the day. When the deflections exceeded the tensile strength at some extreme fiber is when failures occurred. The engineering disciplines of strength of materials and maximum deflections arose to handle the failures of that time. Figure 2.5 is an illustration of a slow-turning machine prior to the industrial revolution. Today, manufacturing processes produce accurate geometries and less dimensional variability. Rarely do machines fail

Figure 2.5 Pump for extracting water from mines c. 1550. A = shaft, B = treadmill, C and D = gears, E = pulley, F = unknown purpose, G = padded pistons. (*From Georgius Agricola,* 'De Re Metallica.')

because of excessive strain. Rather, failures occur as cracking due to excessive cyclical stresses. This fatigue failure mode is dependent on the number of cycles, hence the speed, rather than the tensile strength of materials. Failures occur at stress levels far below the ultimate material strength. The appropriate science for designing long-life machinery today is no longer strength of materials but vibration.

The field of vibration (and dynamic analysis) has its roots in the art of music. The ancient musical instrument makers understood some basic concepts of length of a string and its tone, about resonant cavities, and about harmonics.

Galileo wrote the first treatise on modern dynamics in 1590. He established the relationship between length and frequency. His works on the oscillation of a simple pendulum and the vibration of strings are of fundamental significance in the theory of vibrations. Other famous scientists laid the analytical foundations for vibration analysis by attempting to understand the natural phenomena and developing mathematical theories. Fourier introduced the expansion of a periodic function in terms of harmonics. More will be said about Fourier later. Bernoulli studied vibrating beams and strings. Lagrange developed the equations that describe the exchange of energy in elastic systems that vibrate. These are very useful in the study of flexible structures, like spacecraft. Lord Rayleigh published the theory of sound in 1877, which is still considered a standard reference today. He also developed the method of computing approximate natural frequencies of vibrating bodies using an energy method.

Most of the analytical background mathematics for vibrations were conceived from 100 to 500 years ago. An interesting observation can be made concerning these famous scientists. Most of them "tinkered" with vibrations as a sideline. They earned a living with some other endeavor and only "played" with vibration theory either for amusement or for the challenge. It seems that there was a higher calling to understanding dynamics. In this intellectual environment, the mathematics and theory of vibration were conceived. The mathematics were very cumbersome, and analysis was done only on the simplest of structures.

Machines and technology

During the latter half of the nineteenth century, electrical machines proliferated, and their speeds were higher, from 1800 to 3600 rpm. Figure 2.6 illustrates a typical rotor with a mass unbalance as a single heavy spot of mass m. By the centrifugal force formula,

$$F = mr\omega^2$$

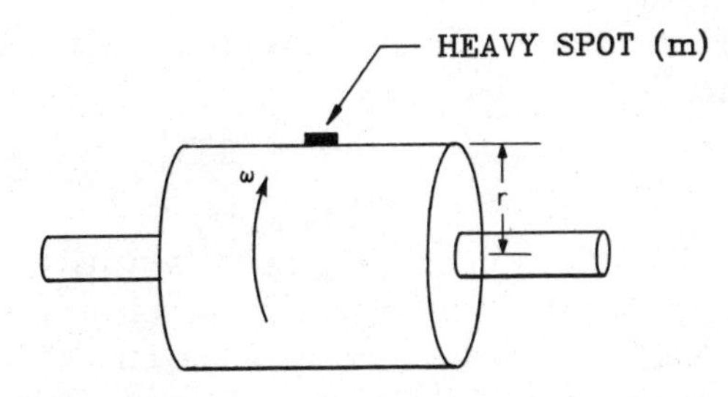

Figure 2.6 A rotor with an imaginary heavy spot.

The force due to this heavy spot is proportional to the product of the mass times the radius times the square of the speed. The source of many machine vibrations is mass unbalance. As machine speeds increased, the residual mass unbalances, multiplied by large speeds, became very large forces.

In 1934, Thearle, an engineer at General Electric, published "Dynamic Balancing of Rotating Machinery in the Field," discussing the influence coefficient method of two-plane balancing in an ASME transaction. This two-plane balancing method was necessary to achieve a better balance condition of long machines, such as generators, and is still used today. About the same time, Professor Den Hartog published his first edition of a book entitled *Mechanical Vibrations*. This book was the first thorough treatment of machine vibrations.

Since the start of the industrial revolution, the world has seen a shift from manual processes to machines and technology. Technically, machines convert energy between chemical, electrical, and mechanical forms. Practically for civilization, machines generate power and move fluids. The trend has been to faster machines and lighter weight. Both contribute to unwanted vibrations.

Years ago, "unwanted vibrations" was mostly a subjective judgment. It was either uncomfortable to be around or too noisy for good communication. Travel in vehicles without spring isolators was a bone-jarring experience. During World War II it was realized that "unwanted vibrations" can destroy valuable equipment. Radios with tubes in World War II aircraft routinely failed in flight.

It was disheartening to bring a new turbine on-line only to see it fly apart in a short time. The financial loss to the owners was staggering. The energy utilities put people and money to work solving the problems of high-speed rotating equipment, and today they have a handle on it.

Three things happened in the past 50 years to move the study of vibrations from an interesting pastime to a serious scientific discipline. The first was necessity: Things were falling apart at higher speeds. The second was sensitive transducers. The third was the advent of electronics and the digital computer that could handle the tedious calculations very fast and error free.

Post-World War II

Advances in electronics after World War II made possible sensitive transducers and instruments to measure small vibrations. It was possible to identify an enemy submarine miles away by its vibration signature. It was not only possible to identify a specific submarine but the listener could tell the condition of its motors by the sounds. This same technology is used today to detect and analyze internal machine condition.

Early missiles shook themselves apart from resonances due to random vibrations from the engines and wind turbulence. The next time you are in a jet aircraft, put your ear against the cabin wall. You can hear the random vibrations of the boundary layer flow over the skin of the aircraft.

In the early 1950s, the British Overseas Aircraft Corporation introduced the first commercial jet transport, the *de Havilland Comet.* Figure 2.7 is an illustration of a *Comet.* By 1954, there were five unexplained crashes of the *de Havilland Comet.* Two crashed in the Atlantic with no survivors. Investigators found cracking in the metal skin caused by flexing of the airframe during takeoffs. The cracks passed through the rectangular windows on the sides of the fuselage. Every time the aircraft took off, the nose flexed upward. This was one cycle of vibration. With each takeoff, and flexing cycle, the cracks progressed a little farther until the final fatal flexing accelerated the crack propagation to its limits. Today, jet aircraft have oval windows, which was part of the solution to the problem with the *de Havilland Comet.* With the grounding of the *Comet,* the British lost their lead in the jet transport market.

In the 1960s, five Lockheed *Electras* crashed. Two disintegrated in midair. The cause was self-destructive wing vibrations set up by abnormal propeller rotation. Lockheed fixed the problem, but the air-

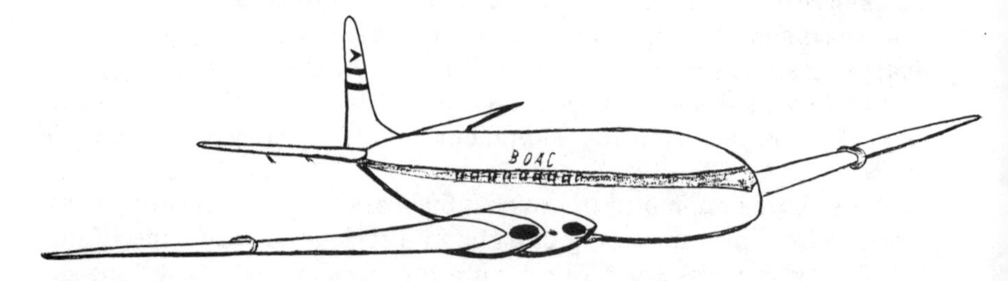

Figure 2.7 *de Havilland Comet.*

lines would not buy the plane anymore. The U.S. Navy purchased the modified *Electras,* and they are still flying as the *P-3 Orion.*

The submarine and the aerospace industry have advanced the state of the art of vibrations. Industry is now reaping the benefits by applying similar technology to machine health monitoring.

The history of vibrations can be summed up in a few words:

Music and mathematics

Faster and lighter machines

Submarines and aircraft

Electronic instruments

Chapter

3

Basic Concepts and Theory

This chapter presents the fundamental knowledge required to understand what vibration is. The physics of the phenomena are described in relation to machines. Mathematics is kept to a minimum. Only those mathematics necessary for machine monitoring are introduced.

Two Kinds of Vibration

There are two basic kinds of vibration: whole body motion and pressure waves. Modern accelerometers are sensitive to both, whereas older velocity transducers detected only whole body motion. Indeed, in the past, the theory of vibration has been presented from the perspective of whole body motion only. The theory presented in this book is an extension of the whole body motion concepts to include pressure waves. The emphasis will be on machinery.

A significant amount of vibration measurements are being made today from sources originating as contact point forces within machines. Decisions for machinery repair are being made based on these signals. Therefore, a clear understanding of how forces transmit as pressure waves is necessary for a vibration analyst.

In whole body motion, the body is considered rigid and its center of gravity moves (Fig. 3.1*a*). That is, the entire body moves as a rigid mass, and the motion of every particle in it tracks precisely with the motion of its center. This is an ideal situation and never precisely true, but it is closely approximated with machines made of metal. Most of the theory in this chapter is based on whole body motion.

In pressure wave vibration, the body is not rigid, and waves move through the body (Fig. 3.1*b*). The center of gravity is more or less stationary. Pressure waves originate at a contact point and radiate outward from there traveling at the speed of sound through that material (approximately 16,410 ft/sec in iron). "Radiating outward" is not a

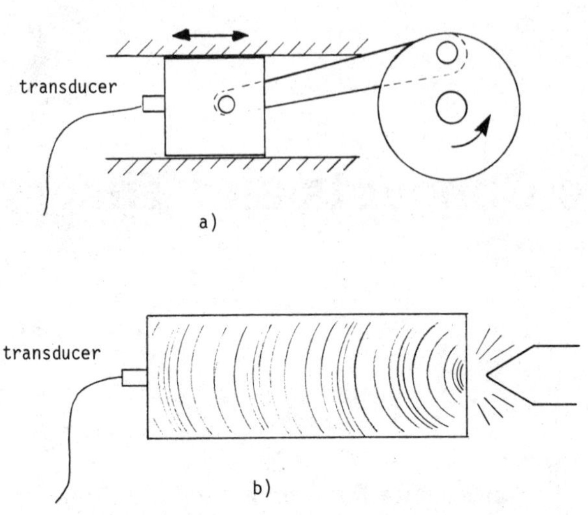

Figure 3.1 (*a*) Whole body motion. (*b*) Pressure wave.

good expression for this motion, since the waves are constrained by the surface of the metal. Unlike acoustic waves in air that radiate outward as a sphere in all directions, the pressure waves travel along the structural members of the machine, that is, along its shaft, through bearings, its housing, and into the concrete foundation. Some of that energy couples to the air and produces sound. All along its path, this pressure wave encounters discontinuities such as joints, boundaries, and material changes. At each discontinuity, some of the wave energy is transmitted through, some is reflected, and some is absorbed. Eventually, the pressure wave dies away just as a sound fades away in air.

These pressure waves are actually the mechanism by which forces are transmitted through materials. The origin of the pressure wave is an impact (usually metal-to-metal impact in machinery) where two parts make physical contact. The contact force is then transmitted through both contacting materials as pressure waves in each, at their respective speeds. This is how impact forces are transmitted through materials—as a longitudinal pressure wave. If an accelerometer were attached to the surface of a machine, an approaching pressure wave will cross the boundaries of attachment, enter the accelerometer's housing, and excite its sensing crystal, before being reflected back. This is how pressure waves are detected with an appropriate transducer, and it will be seen that the coupling method to the surface is critical. Further discussion of this, and transducers, are in later chapters.

An analogy can be made between whole body motion and pressure waves. Assume that I was standing on the boardwalk in Atlantic City,

New Jersey, looking out over the ocean. Through the day a "sea breeze" would be blowing in from the ocean. During the night, a "land breeze" would be blowing out from land to the water. This pattern repeats on a daily cycle, and large volumes of air move as a whole. This is whole body motion. Now imagine that someone fired a canon from a ship 1 mi out at sea. The sound from this canon blast would travel as a pressure wave at approximately 1100 ft/sec. I would hear the canon blast about 5 sec after seeing its flash. No significant air motion took place. This is a pressure wave.

I will return to pressure wave vibrations in this book where appropriate. For the remainder of this chapter, most of the theory deals with whole body motion so that the concepts of sine waves can be presented.

Oscillation

Whole body vibration is an *oscillation,* that is, a movement back and forth as time passes. An example is a swinging pendulum. The source of all vibrations are forces. A force causes the initial movement, and forces sustain the continued motion. A heavy spot on a rotor causes a centrifugal force as it rotates. This force, going around during rotation, creates a strain on the shaft which transmits through the bearings to the housing. Mass imbalance is just one force causing vibration in machinery. There are other forces that can set machinery into oscillatory motion. A rotating machine can be thought of as a mechanical oscillator.

Even under steady-state operating conditions, the vibration generated is rarely perfectly repetitive in amplitude. It must be recognized that rotating machinery is a dynamic condition and that other factors add into the total response, such as resonances, random effects due to turbulence, and interference from other sources. Nevertheless, many machine vibrations are repetitive, and they can be analyzed as oscillatory motions.

Mass, Spring, and Damper

Referring to Fig. 3.2, a mass is the object that moves. A spring is what the mass strains against when it moves. A spring can be thought of as a restraining force. When the mass is displaced, the spring supplies a force to return it. A damper is a device to convert energy of motion into heat. An example is an automotive shock absorber.

A vibration system generally includes a means for storing potential energy in a spring, a means for storing kinetic energy in a mass, and a means by which energy is lost in a damper. Vibration involves the

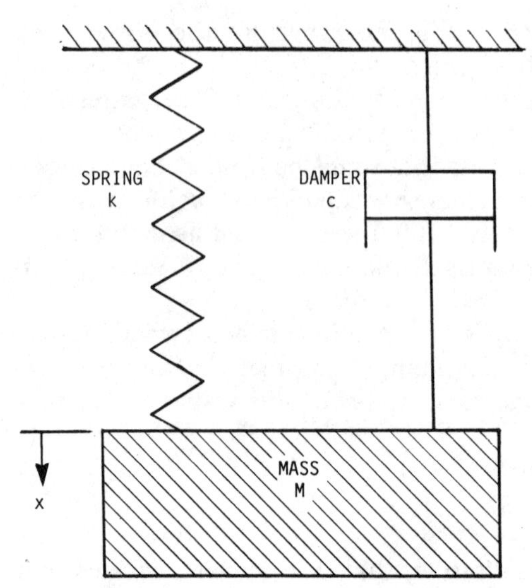

Figure 3.2 Single-degree-of-freedom system.

transfer of energy from its potential energy to its kinetic energy, alternately.

In a rotating machine, the mass that moves is generally the rotor with some residual imbalance. This residual imbalance generates a centrifugal force that strains against the shaft, bearings, and the support system. These latter items all have some stiffness and become the springs in our simplified model. They also have some mass, and they move. This movement is what allows us to detect the vibration by sensitive transducers on the outside. Damping is provided by internal hysteresis, friction, and viscous drag. There is no such thing as a pure mass, a pure spring, or a pure damper. These are idealized concepts. A mass has some stiffness, and a spring, being composed of matter, must have some mass. In reality, these three properties of mass, spring, and damper reside to varying degrees in all mechanical components.

Let's put aside the damper for the time being and concentrate on the mass and spring. This is valid because mechanical systems made entirely from metal have very little damping. The concept of damping will be covered more later.

Referring to Fig. 3.3, the mass and spring form a single-degree-of-freedom model because there is only one mass (if we assume the spring to have negligible weight), and it moves in only one direction, up and down. This is an idealized model and is generally not true for

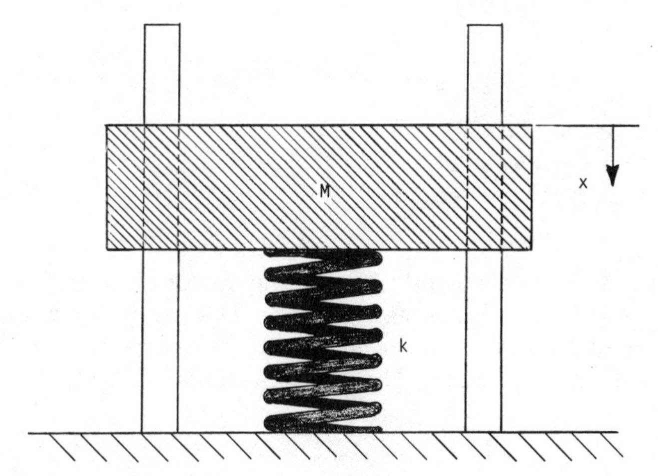

Figure 3.3 Simple spring mass system.

machinery. Machines usually vibrate in all directions at once, but their vibration is sometimes stronger in a specific direction. The single-degree-of-freedom model is useful to help visualize what is going on.

If the mass were displaced some distance x and released, the mass would oscillate up and down. The force moving the mass is, by Newton's second law,

$$F = ma = m\ddot{x}$$

where m = mass
a = acceleration
$\ddot{x}$ = second derivative of displacement

The restraining force from the spring is kx. The force moving the mass must equal the restoring force of the spring, or

$$m\ddot{x} = -kx$$

where k = spring constant

The minus sign means that the spring force is opposite the mass force. With no external driving force and zero damping, the equation for a single-degree-of-freedom system under dynamic conditions of motion is at all times

$$m\ddot{x} + kx = 0 \tag{3.1}$$

The solution to Eq. (3.1) is

$$x = x_0 \cos \omega t \tag{3.2}$$

where x = position of the mass at time t
x_0 = maximum displacement
ω = speed in radians per second
t = time

Equation (3.2) is simple and elegant. All the important information about pure oscillatory motion is represented here. The maximum amplitude is x_0; the frequency is ω. The cosine function ensures that the motion repeats itself as time passes. The phase is embodied in the cosine.

Sine Waves

Sine waves and cosine waves are ideal for describing pure oscillations. They are identical except for a 90° shift in position. Refer to Fig. 3.4. Sine waves could be used to describe the motion of vibration if the 90° phase shift were accounted for. Since the sine wave completely describes the motion, some numbers could be assigned to it. The numbers would also describe the motion.

Amplitude, Frequency, and Phase

The first number would be a measure of how much time it takes to complete 1 cycle. This is the period, in seconds. Figure 3.5 illustrates

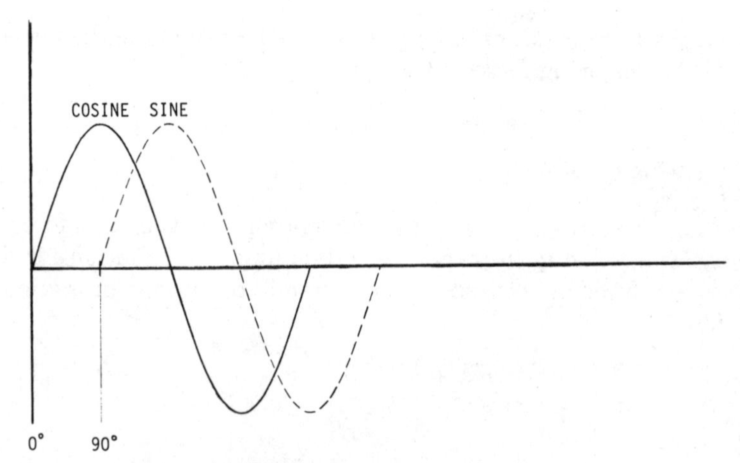

Figure 3.4 Cosine and sine waves.

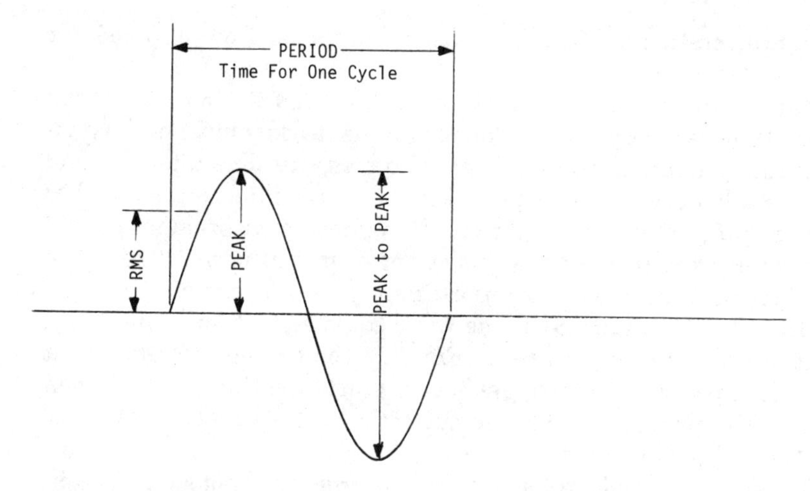

Figure 3.5 Measure of sine waves.

the period. The reciprocal of the period is frequency. Mathematically,

$$\text{Frequency} = \frac{1}{\text{period}} \text{ in cycles per second, or hertz}$$

Other measures of frequency are revolutions per minute and orders. Revolutions per minute = 60 × frequency in hertz. Orders is a unique unit applicable to a spectrum analyzer. The running speed is usually the first order. Multiples of the running speed are the second order, third order, etc.

A measure of how far the mass moves from the neutral position is the amplitude in inches. This is called the *peak amplitude.* A measure of how far the mass moves from one extreme to the other is the *peak-to-peak amplitude.* This is the same as total indicator reading (TIR) on a dial indicator. In fact, a dial indicator can be used to measure peak-to-peak amplitude for slow vibrations, i.e., less than 10 Hz or 600 rpm.

Another measure of amplitude is root mean square, or rms. Root mean square is 0.707 × the peak. The rms is mentioned because that is the unit in which electronic instruments measure amplitude of sine waves. The instruments then can multiply the rms internally by $\sqrt{2}$ and display the peak. The rms is a measure of the energy content of the wave. Europeans report vibration in rms amplitude. In North America, peak amplitude is the standard.

It is important when using electronic instruments to measure vibration to ascertain whether the display is in peak, peak to peak, or rms. It is equally important to use these terms when reporting vibration amplitude.

The third, and last, description of vibration is phase. *Phase* is an angular measurement from some point on the rotor designated as 0°.

Since a full rotation is 360°, the phase is a number from 0 to 360°, or from +180° to −180°.

Vibration instruments cannot measure angles, but they can measure time very accurately. The technique used by most modern vibration instruments to obtain phase is to measure time intervals and perform a calculation. The phase is actually a time measurement that is converted to an angle of rotation. This phase measurement should not be confused with the three phases of electrical power. These are totally separate concepts that unfortunately use the same word.

Figure 3.4 of a cosine and a sine wave could also be two sine waves 90° out of phase to each other. In Fig. 3.6, the timing reference is at the 3 o'clock position. The trigger sensor starts the timing cycle when it detects the passage of the photoelectric tape. The phase angle is then reported in relation to this tape.

In this example, the heavy spot is at 145° from the photoelectric tape. Fifty-five degrees after the tape starts the timing cycle, the heavy spot is at the bottom and causes the maximum negative force on the vibration transducer. And 145° after the tape passage, the heavy spot passes the trigger sensor and is exerting no force on the vibration transducer. Finally, 235° after the start of timing, the heavy spot is at the top and causes the maximum positive force on the vibration transducer.

During balancing, the position of the heavy spot must be ascertained. One available method to do this is to mount a trigger sensor and a vibration transducer and measure the phase angle between them, as Fig. 3.6 illustrates. Another method is to filter the vibration signal at running speed to obtain a clean sine wave, and then use this signal to trigger a strobe light to freeze the motion. This method is further described in the balancing chapter of *Machinery Vibration: Corrective Methods* (1992).

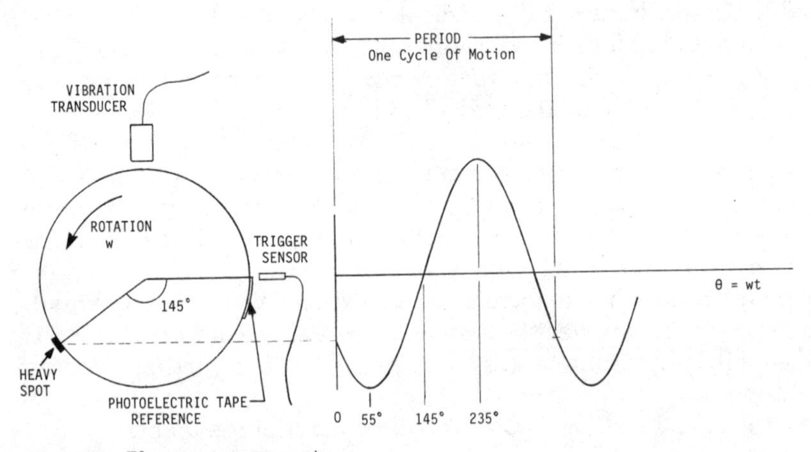

Figure 3.6 Phase measurement.

The three numbers needed to completely describe sinusoidal vibrations are frequency, amplitude, and phase. Sometimes the phase is unimportant. Then, only frequency and amplitude are needed.

Random Vibrations and Shock Pulses

Besides sine waves, which are pure tones, there are two other types of vibrations. These are random vibrations and shock pulses.

Random vibrations look similar to a complex vibration signal except that they do not repeat themselves regularly. Therefore, a frequency is difficult to assign to them. Figure 3.7 is a plot of noise from a random noise generator. A repetitive pattern is difficult to discern from this plot. The motion of fluids (gases and liquids) generates random vibrations when the fluid encounters a stationary object and creates downstream vortices, or otherwise becomes turbulent. Friction also generates random vibrations. Wind, tire noise, and pump cavitation are examples of random vibrations.

Shock pulses are single-event transients such as impacts. The resulting vibrations are repetitive, but they die away due to damping. Figure 3.8 is a shock pulse created by snapping the fingers in front of a microphone. The total duration of this pulse, from initial rise to finally damping out, is about 5 msec (milliseconds).

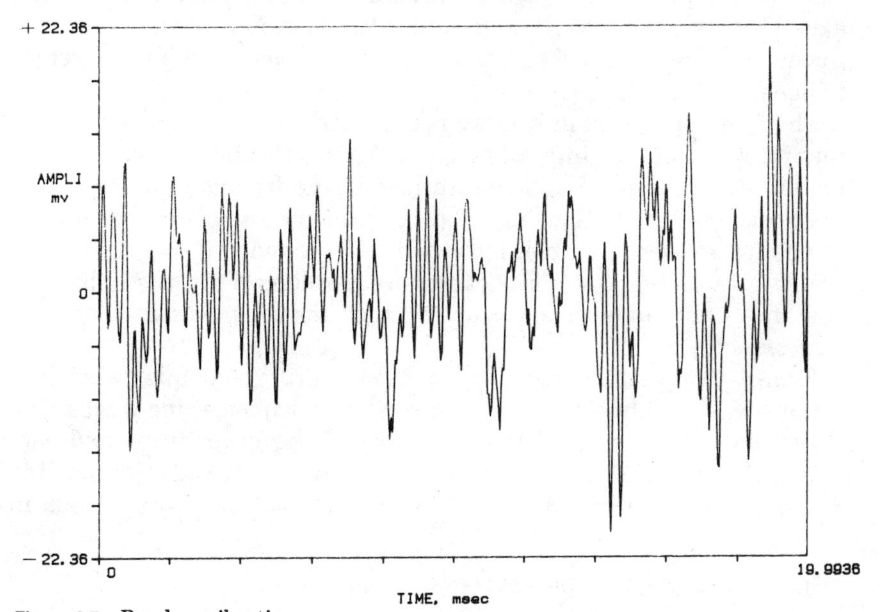

Figure 3.7 Random vibrations.

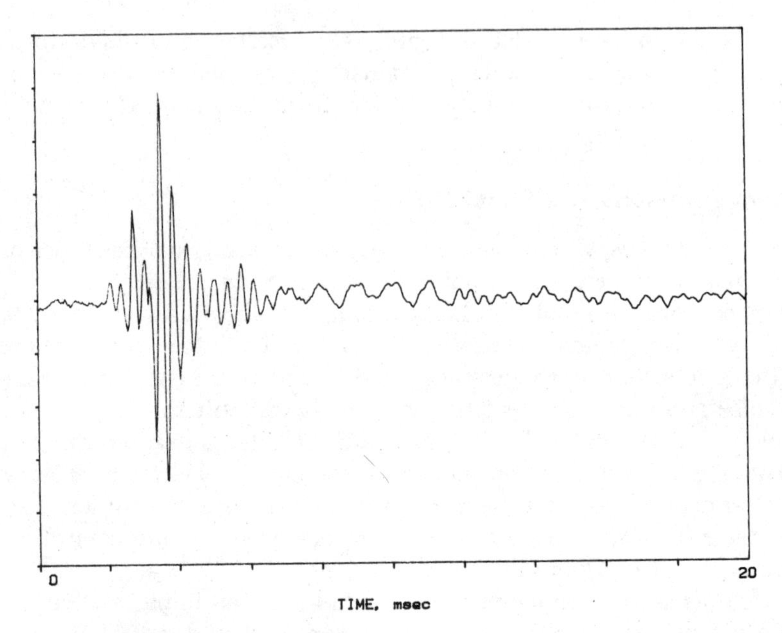

Figure 3.8 Shock pulse.

Time and Frequency Domains

Up to this point, we have been looking at vibrations in the time domain. This is what is displayed on an oscilloscope. The display is amplitude versus time. Amplitude is on the vertical axis, and time is on the horizontal axis. There are other ways to look at vibrations. One of these is in the frequency domain. The frequency domain is a plot of amplitude versus frequency.

The frequency domain display is one of the most powerful windows into machine monitoring. All machine diagnostic and predictive maintenance instruments display vibrations in the frequency domain.

Fourier discovered that all complex harmonic signals can be broken down into a series of simple sine waves.[1] The individual sine waves generally have different amplitudes and frequencies. When the individual sine waves are combined, or added together, the complex signal is reconstructed.

Figure 3.9 illustrates this concept of separating a complex wave into its sine wave components. Figure 3.9*a* shows two pure sine waves. The upper wave is four times the frequency of the lower wave and one-fourth its amplitude. When these two waves are mixed together, the result is shown in Fig. 3.9*b*. Both waves can still be visually seen in

[1]Jean Baptiste Fourier (1768–1830), French mathematician.

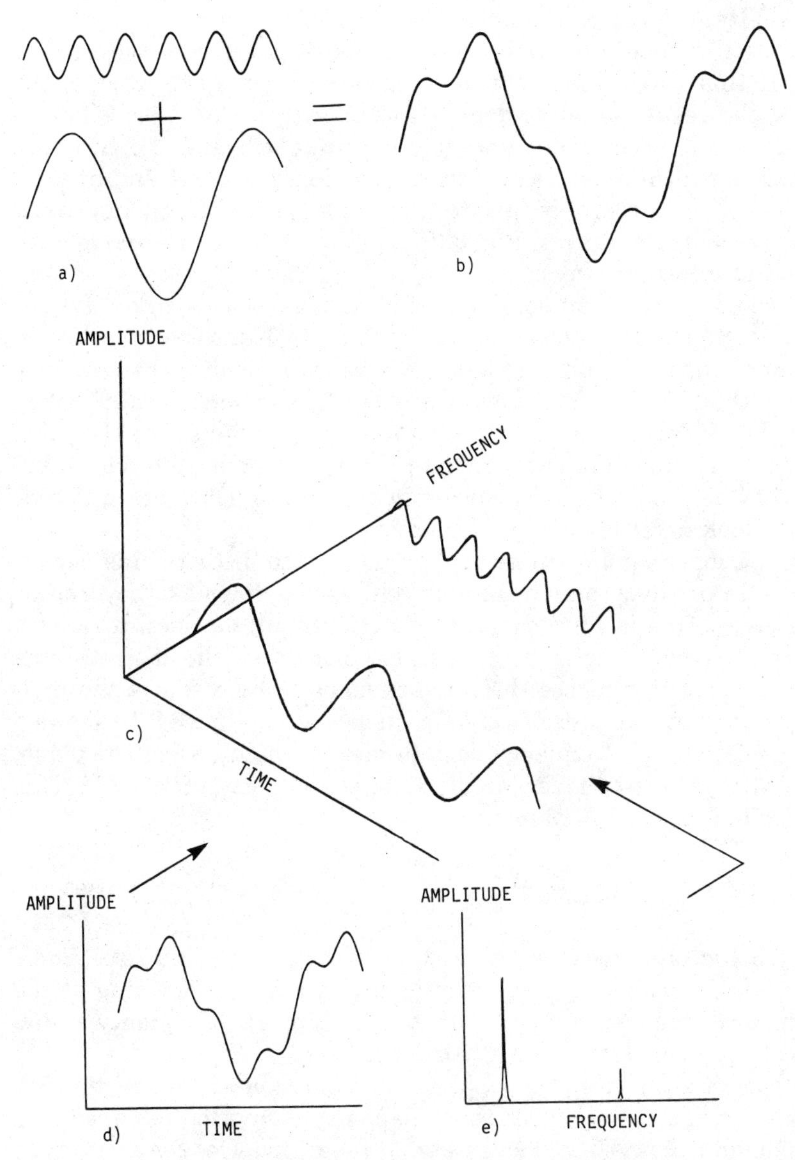

Figure 3.9 (*a*) Two sine waves. (*b*) Two sine waves combined. (*c*) Three-dimensional illustration of different views. (*d*) Time domain view as seen on an oscilloscope. (*e*) Frequency domain view as seen on a spectrum analyzer.

the resultant wave pattern. This is the display as it would appear on an oscilloscope, i.e., amplitude versus time, as depicted in Fig. 3.9*d*. If this wave pattern were separated out into its component waves, the result could be depicted as shown in Fig. 3.9*e*. The display in Fig. 3.9*e* is amplitude versus frequency, and the two sine waves are shown as vertical lines proportional to their amplitudes, positioned in their appropriate place along the frequency axis. These two bottom figures, Fig. 3.9*d* and Fig. 3.9*e*, are the same data viewed in different ways.

If I could draw an analogy, it would be like looking into different windows. Suppose I came upon a farmhouse in Kansas and looked in the front window. I might see a family sitting at the dinner table. Now suppose that I walked around to the side of the farmhouse and looked in a side window to the same dining room. I would also see the same family at the same dinner table, but the view would be different. Likewise for the time and frequency domain views. They are different ways of looking at the same information.

Any periodic signal can be broken down into discrete sine waves, even square waves and triangular pulses—as long as they repeat themselves. This is what the FFT spectrum analyzer does. It takes a complex waveform from a transducer, calculates the discrete sine wave series that compose the complex signal, and displays the individual sine wave amplitudes on a frequency axis. Figure 3.10 shows a time domain and a frequency domain view of middle C on the piano, taken with a microphone. Middle C is a frequency of 256 Hz. This equates to a period of 3.9 msec.

$$\text{Period} = \frac{1}{\text{frequency}} = \frac{1}{256 \text{ cycles/sec}} = 0.0039 \text{ sec, or } 3.9 \text{ msec}$$

In Fig. 3.10*a*, the time interval between major peaks is 3.9 msec. Some other activity at higher frequency is taking place. By looking at the frequency domain view, Fig. 3.10*b*, the fundamental frequency at 256 Hz with harmonics at 512 and 768 Hz is seen.

A true random vibration appears as a horizontal line in the frequency domain. This is because a frequency cannot be assigned to a random vibration. All frequencies are present if viewed for a sufficient length of time, i.e., about 2 min.

We must thank the great French mathematician, Jean Baptiste Fourier, for introducing the theory and the mathematics for this transformation. More recently, an algorithm using modern digital techniques made this transformation "fast."

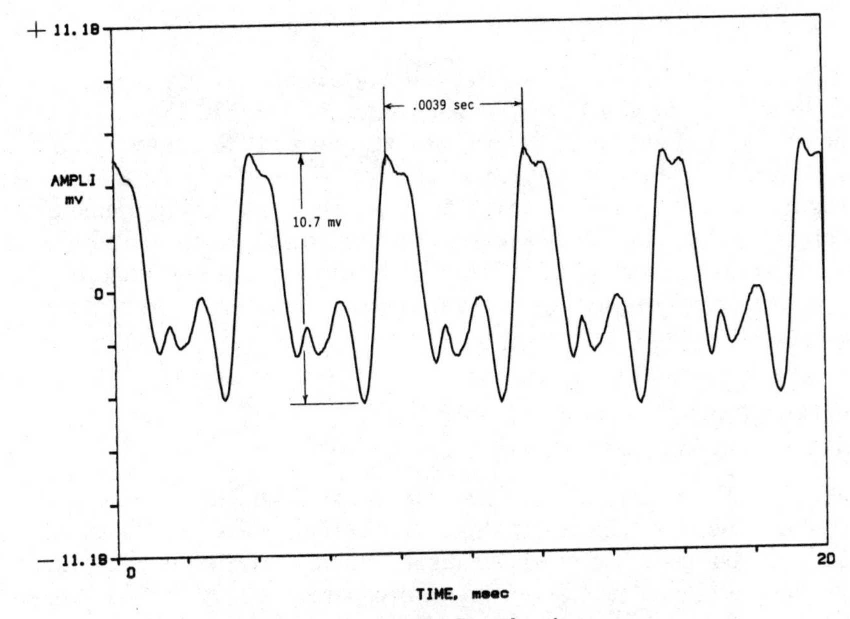

Figure 3.10(*a*) Time domain view of middle C on the piano.

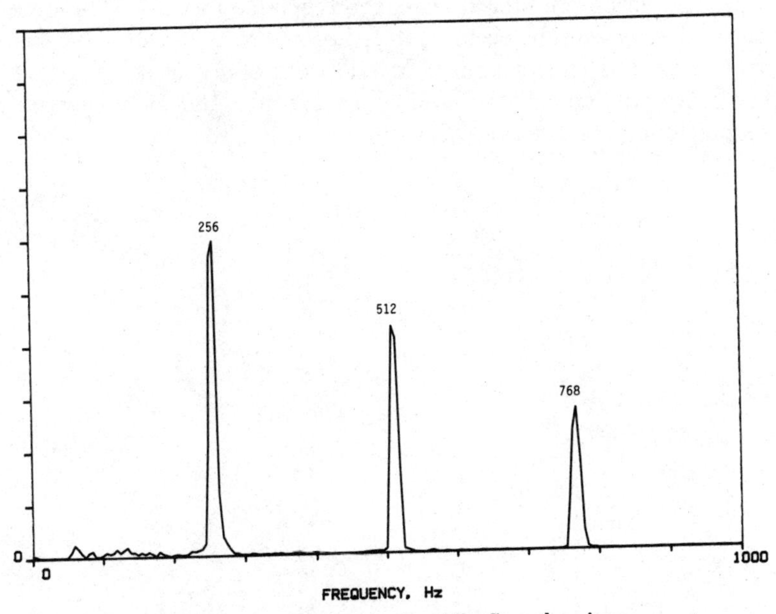

Figure 3.10(*b*) Frequency domain view of middle C on the piano.

Vectors

A vector is a number that has a direction. For example, 30 mi/h is a number; 30 mi/h *north* along San Mateo Avenue is a vector. Figure 3.11*a* shows a vector of length 15 directed at an angle of 30° to the horizontal. The 15 could represent an amplitude, and the 30° could be a phase angle.

Figure 3.11*b* shows another vector of length 15 but now directed at an angle of 120° counterclockwise to the horizontal. Even though the lengths are the same at 15, the two illustrations represent two different vectors because the angles are different. Vectors are used to perform calculations for balancing.

The Relationship of Displacement, Velocity, and Acceleration

Forces are the cause of vibration. The force is the first event that occurs in time. The response to these forces are movements. The movements can be described as either displacement, velocity, or acceleration. Displacement is how we normally perceive motion; however, velocity and acceleration are also valid descriptive quantities for mechanical motion. The displacement, velocity, or acceleration are all responses to some force. The force leads the response in time. The force occurs first, the movement occurs later, i.e., there is a phase lag between the force and the response. In addition, there is a 90° phase difference between displacement and velocity, and a 180° phase lag between displacement and acceleration.

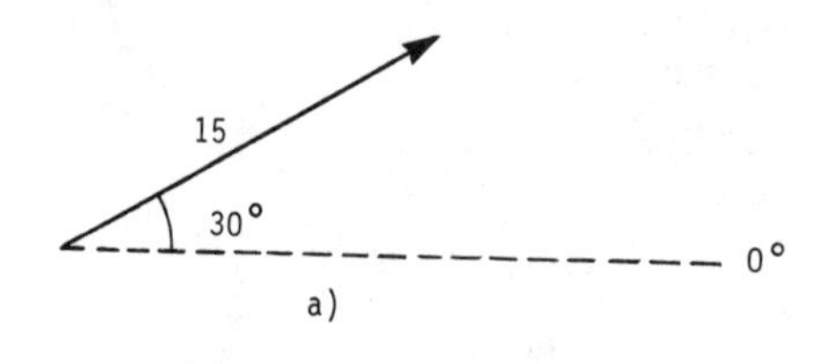

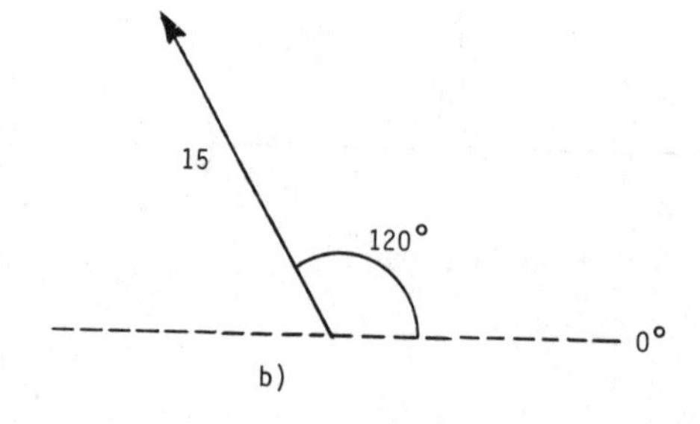

Figure 3.11 (*a*) A vector 15 ∠30°. (*b*) A different vector 15 ∠120°.

A simple way to visualize this is by imagining pushing a child on a swing. This is a common experience that most everyone can relate to. The child on the swing is identical to a weight suspended on a string as in the pendulum of Fig. 3.12*a*. The person pushing the swing is supplying the input force to cause motion. The motion can be described as either:

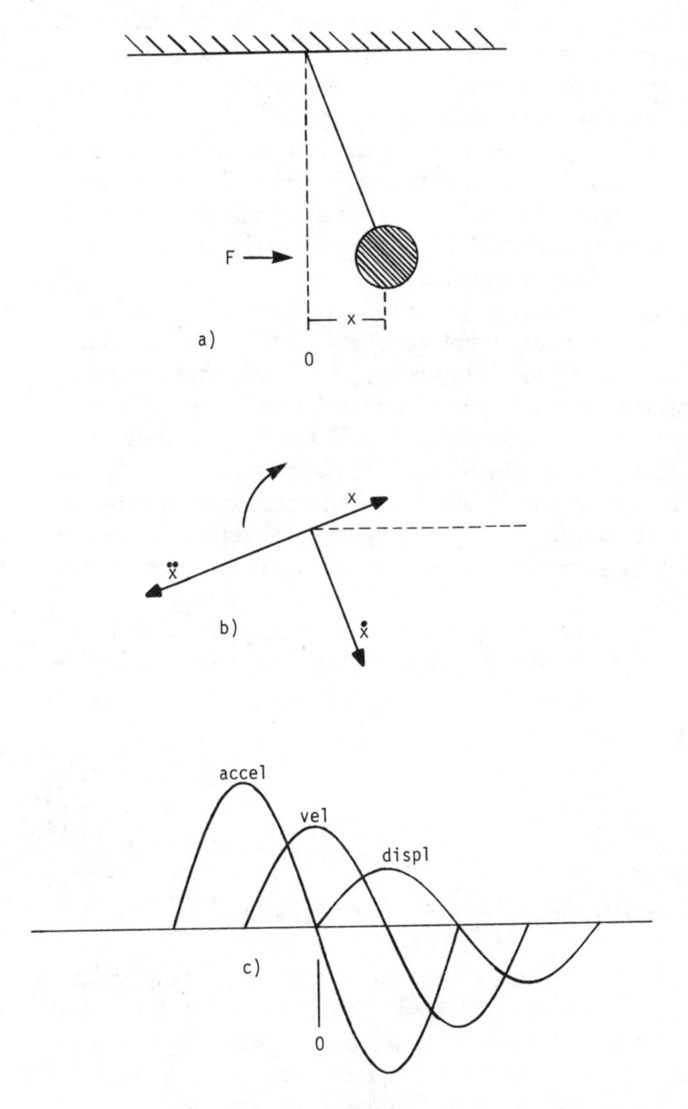

Figure 3.12 (*a*) A force input causes motion. (*b*) A phasor diagram. (*c*) Sine waves to illustrate the relationships of displacement, velocity, and acceleration.

1. A displacement from the neutral position
2. A velocity, the maximum being at the bottom of the swing
3. An acceleration, the maximum being at the instant the swing is changing direction

The swing is usually given a force input during the time it is changing direction at one extreme displacement position. The maximum displacement in the direction of the push comes sometime later, actually 180° later in the cycle. The velocity, however, was at the maximum as the child swung past the bottom position, where the displacement was zero. The velocity, therefore, leads the displacement by 90°.

The maximum acceleration occurs at the same time and in the same direction as the force input. The maximum acceleration occurs at the extreme positions where the displacement is also a maximum. The acceleration, though, is opposite in sign to displacement, so the acceleration is 180° out of phase to displacement, and acceleration leads velocity by 90°.

This can be visualized as rotating vectors, or phasors in Fig. 3.12*b*. The three phasors representing displacement, velocity, and acceleration are 90° apart and scaled to different lengths. They rotate together as a system always maintaining their relative 90° positions. A particular velocity transducer in a fixed position will detect a rise and fall of the amplitude as the velocity phasor rotates past its position. An accelerometer, on the other hand, will sense the acceleration phasor, which is 90° phase-shifted from the output of the velocity transducer. Both transducers, keep in mind, are sensing the same vibration in the same location.

Velocity is the derivative of displacement with respect to time. Acceleration is the first derivative of velocity and the second derivative of displacement.

Mathematically:

$$x = x_0 \cos \omega t \tag{3.3}$$

$$\text{Velocity} = \frac{dx}{dt} = \dot{x} = -x_0\omega \sin \omega t \tag{3.4}$$

$$\text{Acceleration} = \frac{d^2x}{dt^2} = \ddot{x} = -x_0\omega^2 \cos \omega t \tag{3.5}$$

The first derivative of displacement amounts to converting the cosine function into a minus sine function. This amounts to a forward 90° phase shift. That is, the velocity leads displacement by 90°. The velocity also multiplies the maximum displacement x_0 by the speed ω to achieve the maximum velocity $x_0\omega$. The velocity is scaled up in amplitude by the frequency factor.

The second derivative of displacement returns us to the cosine function, but with a minus sign. Therefore, acceleration is 180° phase-shifted from the displacement. The maximum acceleration $x_0\omega^2$ is scaled up by the square of the speed. Figure 3.12*c* attempts to show this. All three quantities of displacement, velocity, and acceleration are sine waves of the same frequency, equal to the frequency of vibration under consideration. They differ, however, in amplitude and phase.

The main point of this discussion is that displacement, velocity, or acceleration can be used to describe vibration. There is a 90 or 180° phase difference between them, and they are scaled to different units. Sometimes the phase difference is unimportant and only the scale difference of amplitude needs to be considered. Acceleration can be converted to either velocity or displacement by using the following formulas:

$$V = 61.7\,\frac{a}{f}\ \text{in/sec peak} \qquad (3.6)$$

$$D = 9.78\,\frac{a}{f^2}\ \text{in peak} \qquad (3.7)$$

where f = frequency, Hz
a = acceleration, g peak

These formulas [Eqs. (3.6) and (3.7)] are valid for sine waves at a single frequency. Keep in mind that the actual velocity or displacement will be larger when other sine waves are added to the one you calculate. Generally, this calculation is valid if there is one prominent vibration peak in the spectrum. For example, I once measured a vibration of 0.9 g peak at 30 Hz:

In velocity:

$$V = \frac{61.7\,(0.9)}{30} = 1.85\ \text{in/sec}$$

In displacement:

$$D = \frac{9.78\,(0.9)}{(30)^2} = 0.0098\ \text{in peak, or 19.6 mils peak to peak}$$

By any measure, this was a serious vibration problem. In fact, the support system for this 100-hp fan was already cracked.

Velocity is the displacement multiplied by the frequency. So the velocity display is scaled up by the frequency. This emphasizes higher-frequency components. Similarly, the acceleration scales up the displacement by the square of the frequency. The acceleration emphasizes the higher-frequency components even more so.

Figure 3.13 shows a constant 0.1 in/sec velocity across the frequency spectrum, with the corresponding displacement and acceleration. The displacement view of vibration amplitude emphasizes the lower-frequency components and deemphasizes the higher frequencies. Acceleration, on the other hand, emphasizes the higher frequencies and deemphasizes the lower ones. This is all for a constant amplitude of 0.1 in/sec velocity across the entire useful frequency range.

Notice that the three lines representing displacement, velocity, and acceleration cross in the range of 10 to 1000 Hz. This is the range of the most interesting machine vibrations, and any amplitude display could be used. For very low frequency measurements, the displacement display is favored because it emphasizes the low frequencies. For high-frequency measurements, i.e., above 100 Hz, the acceleration display is a better choice. For general machinery vibration measurements between 10 and several hundred hertz, velocity is a good choice.

A good analogy is a sound system with bass and treble controls. The bass and treble controls at the neutral position are equivalent to the velocity display. If the bass were turned up and the treble turned down, then the low frequencies would be emphasized. This is equivalent to the displacement display. If now the bass were turned down and the

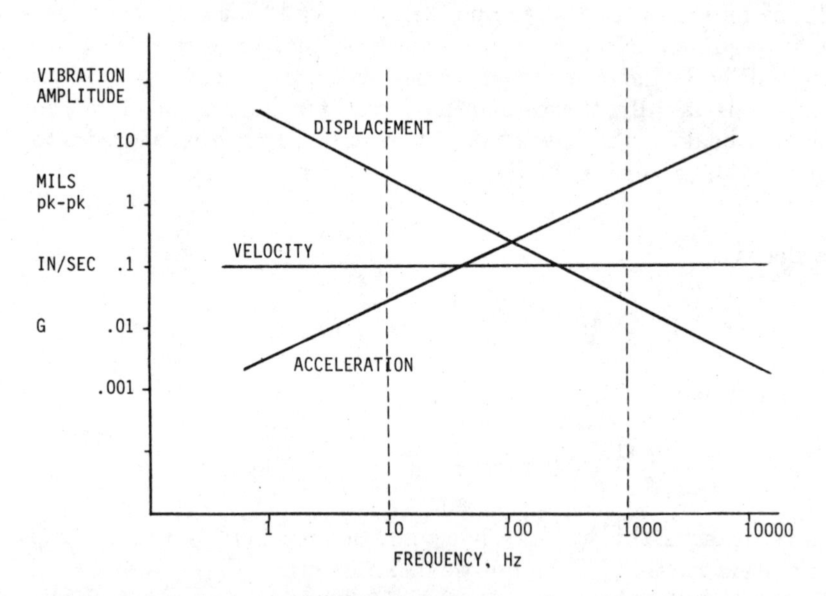

Figure 3.13 Vibration amplitude in displacement and acceleration corresponding to a constant 0.1 in/sec velocity.

treble control were turned up to emphasize the higher frequencies, then this would be analogous to the acceleration display.

To make this point clear, Fig. 3.14 shows three views of the same vibrating machine. The machine is a ½-hp tabletop demonstration machine that runs at 1792 rpm (approximately 30 Hz). It has an imbalance at 30 Hz and a strong 120-Hz motor hum. Figure 3.14*a* is the displacement display. Figure 3.14*b* is the same machine under the same conditions displaying velocity. Notice the change in the relative heights of the 30- and 120-Hz vibrations. Figure 3.14*c* is the acceleration display of the same machine. Since acceleration emphasizes the higher frequencies, some vibrations can be seen above 120 Hz that were not visible in the displacement display. The peak at 60 Hz is approximately the same height in all three displays. Sixty hertz can be considered as the pivot point for vibration measurements in the frequency spectrum. Below 60 Hz, displacement looms larger. Above 60 Hz, acceleration shows the peaks better. All these displays were acquired with an accelerometer as a transducer. Modern instruments have the capability to switch from one display to another without changing transducers. This will be covered more in depth in the next chapter.

Displacement is normally measured in mils, millimeters, or micrometers (microns). A mil is 0.001 in. A millimeter is 0.001 m, and a

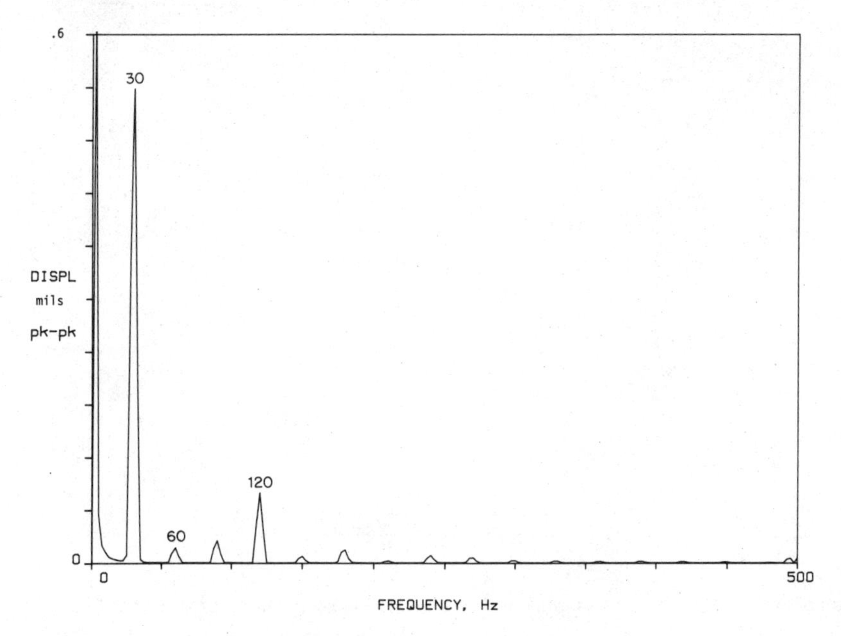

Figure 3.14(*a*) Displacement display of a vibrating machine.

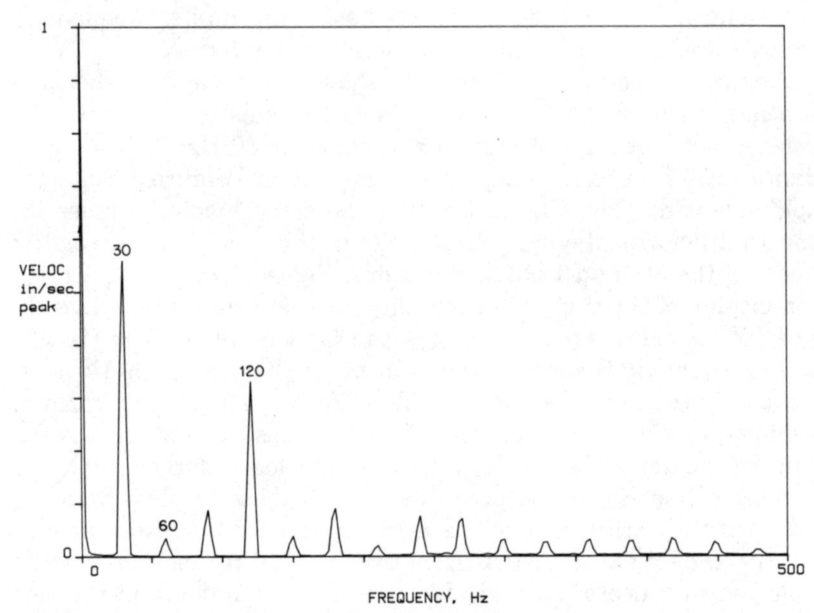

Figure 3.14(*b*) Velocity display of the same vibrating machine as in Fig. 3.14*a*.

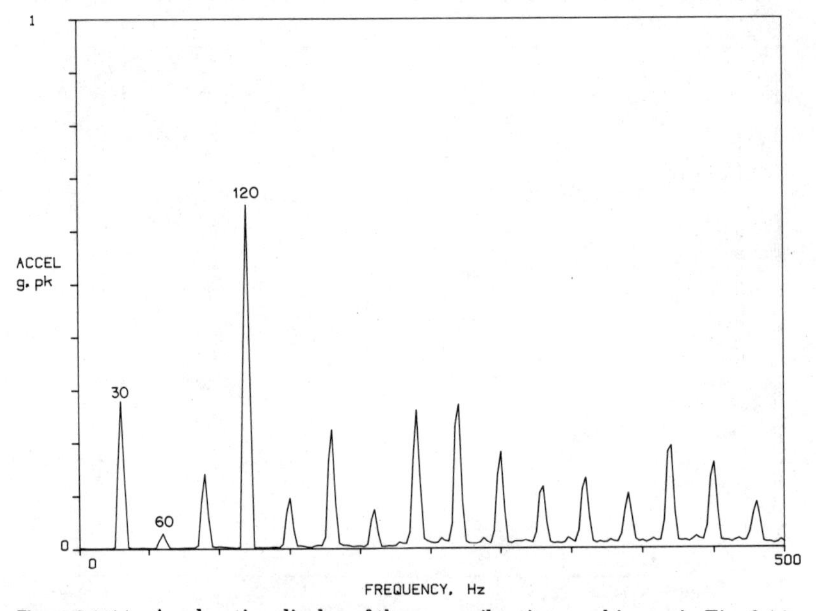

Figure 3.14(*c*) Acceleration display of the same vibrating machine as in Fig. 3.14*a*.

micrometer is 0.000 001 m. Velocity is normally measured in inches per second in English-speaking countries, or in millimeters per second elsewhere. Acceleration is almost always measured in g's everywhere. A g is the local acceleration of gravity. This unit is the same in English-speaking countries as well as the rest of the world. One very good reason for using g's to measure and report vibration is that it is a universal unit throughout the world and it is compatible with the International System of Units (SI). Incidentally, the confusion over measurements taken and reported with different amplitude units that are frequency dependent is the most significant problem in vibration analysis today. Standardizing on a common unit and display would do much to simplify a technology that is complicated enough without this use of different "languages" to measure amplitudes. Frequency measurements, fortunately, are not as complicated, with only hertz, revolutions per minute, and orders to deal with.

We will return to this concept of the relationship of displacement, velocity, and acceleration when we discuss transducers in the next chapter.

Natural Frequencies and Resonance

All physical objects "ring" at certain frequencies when they are tapped with a hammer. Metals are good ringers. A tuning fork is an example of a good ringer. These unique tones that objects ring at depend on the material stiffness, its shape, and its mass. These unique tones are called *natural frequencies*.

If you "ring" a tuning fork or a bell, you will notice that the tones hold at the same pitch (or frequency), but they eventually die away. If the same object is forced to vibrate at those natural frequencies, by some external driving force, then its amplitude of vibration can build up to very large values. This is called *resonance*.

For a simple spring mass system, the natural frequency is defined by ω_n:

$$\omega_n = \sqrt{\frac{k}{m}} \tag{3.8}$$

where k = spring stiffness
m = mass

This formula states that the natural frequency depends on the ratio of the spring stiffness to the mass. Using a stiffer spring will raise the natural frequency. Adding more mass will lower the natural frequency. This applies to machine parts as well. Stiffening a pipe section will make it ring at a higher frequency, and probably with lower amplitude. Welding weights to the same pipe section will add mass and lower its tone.

The vibration response of any mechanical system is highly dependent on the frequency, or speed of excitation, of the input force. At a frequency below resonance, the force input is used up mostly to compress the spring in the system. Masses move readily at low frequencies, so below resonance the system is stiffness dominated. Above resonance, the mass inertia of the system becomes dominant. Since masses don't like to move at higher frequencies, most of the force input is consumed in overcoming inertia of the system.

The force due to stiffness kx is not dependent on frequency. Therefore, the stiffness force is constant with frequency. This is shown in Fig. 3.15 as a horizontal dashed line labeled "stiffness force."

The force due to inertia is proportional to the acceleration which has a ω^2 term in it. The inertia force, therefore, increases with frequency in a quadratic manner. The inertia force and the stiffness force are always 180° out of phase. This is due to the nature of springs, which always provide a return force for the mass. The magnitude of the inertia force increases with frequency and at some point it equals the stiffness force magnitude. When this occurs, there are two forces of equal magnitude and 180° apart. They essentially cancel each other,

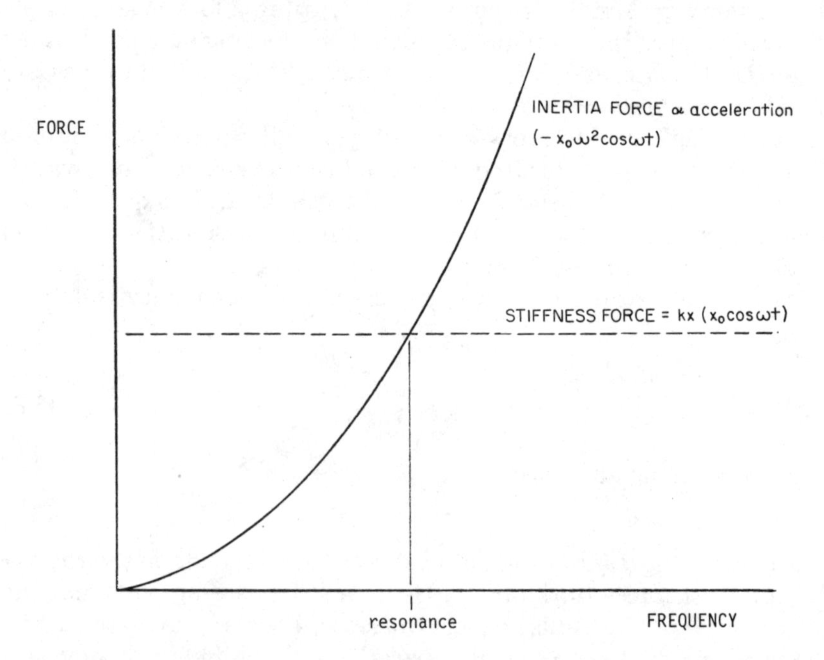

Figure 3.15 Resonance occurs at the crossover frequency where the stiffness force equals the inertia force.

and we have a situation with a driving force input and no restraints. This is a condition of resonance.

Resonance is a condition when the input force frequency F equals the natural frequency of the system ω_n. The ratio of the two frequencies then equals 1. This is shown in the resonance curve of Fig. 3.16. The amplitude of vibration increases dramatically at resonance.

At driving force inputs of lower or higher speed, the amplitude response is much lower, but at resonance, the amplitude is magnified. Figure 3.16 shows a magnification of 4, but on metal parts it is possible to easily achieve magnifications of 10 to 100X. Theoretically, the amplitude can build up to infinite values. The only thing that prevents this buildup to infinity is damping.

Pushing a child on a swing is a resonant condition. The small input force causes the "pendulum" to build up to a large amplitude. The key criterion is that the small input force is time to be input at exactly the natural frequency of the pendulum.

This condition of resonance should not be confused with the "sounding-board" phenomena that every musician is familiar with. The sounding board is a panel that is well coupled to the vibrating part. Due to its flexibility, the board picks up the vibrations from a vibrating string and then vibrates itself at the same frequency as the string. Because it is a panel, it sets more air into motion than the string is capable of doing by itself. Human hearing is extremely sensitive to the smallest pressure fluctuations. The sounding board does not vibrate at resonance, it simply acts as a diaphragm that picks up a small vibration and couples it well to the air. The human ear does

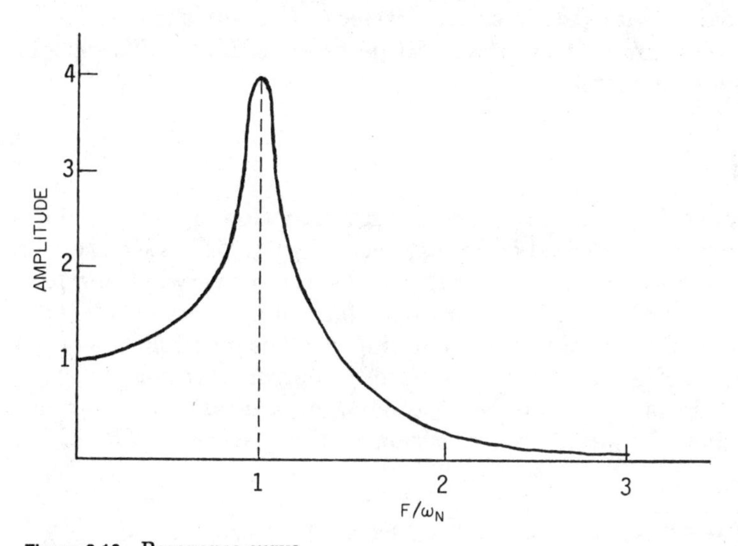

Figure 3.16 Resonance curve.

the rest. The panel actually moves a very small amplitude, but it moves over a large area.

Many machine parts are made up of metal panels that can act as sounding boards. Fans are a good case in point. They can pick up a small bearing or blade-pass frequency and acoustically couple it to the air such that it sounds terrible. This may be an acoustical problem, but it is not necessarily a vibration problem. If, however, the panel were vibrating at its own natural frequency, then this would be a different situation, namely, resonance.

There are formulas for calculating the natural frequencies for common simple shapes like beams, plates, and other simple systems. A good source for such formulas is the *Shock and Vibration Handbook.*[2] These formulas are used mainly in a design situation before anything is built, to obtain an approximation of the natural frequencies. It must be emphasized that these formulas are only approximations. They only consider very simple geometries which rarely exist on actual machinery. The person making the calculations must make assumptions about the end restraint conditions and damping. For these reasons the computations of natural frequencies do not correlate well with the actual hardware after it is built. A difference of 20 percent is considered good. Machine manufacturers have empirical formulas of their own that are more accurate. However it is done, the calculation of natural frequencies is a nontrivial task, best left to the engineering department. This knowledge is of critical importance, for machines that are built and operated at or near their natural frequencies will not run properly at best, or could have disastrous consequences at worst. For machine parts that exist, it is much easier to measure the natural frequencies with simple tests that will be presented later. These tests yield precise numbers.

Damping

A damper is a device that converts energy of motion into heat. All materials have some internal damping, even metals. This is what causes their vibrations to die away. Metals, however, have very low damping, and their natural vibrations can linger for a long time, evidence the bell. Other materials have high damping, such as plastics, rubber, paper, and soil. These materials are used to control vibrations.

Figure 3.17 is a plot, in the time domain, of a vibration that dies away in about 30 msec. This vibration is highly damped. The rate of

---

[2]Cyril M. Harris, *Shock and Vibration Handbook,* 3rd ed., McGraw-Hill, New York, 1988.

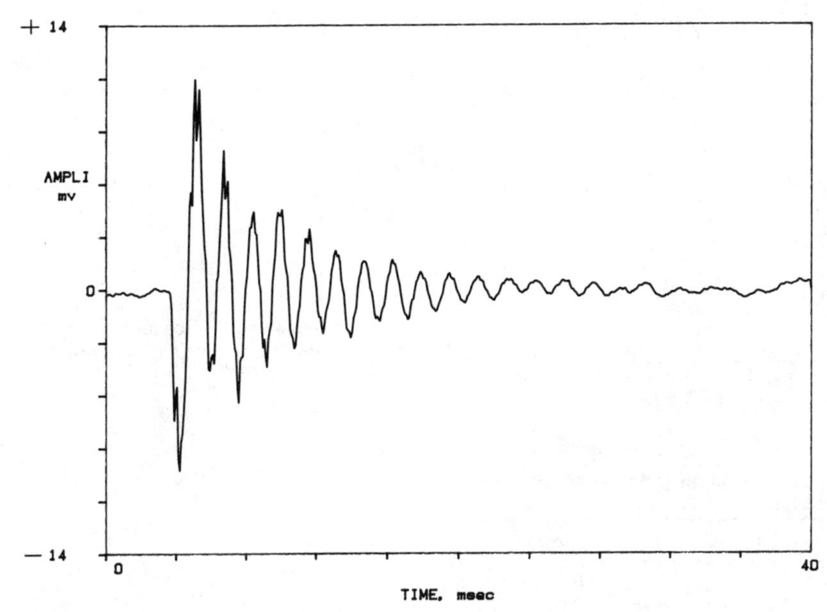

Figure 3.17 Damped vibration.

decrease is a measure of damping. Figure 3.17 is a time capture of a hand clap. Notice in this plot that the time interval from peak to peak is the same as the peaks diminish in amplitude. This means that the period, and therefore, the frequency, remains constant. In the frequency domain, a damped vibration appears as a single vertical line when initiated. This vertical line drops in amplitude as the vibration dampens out.

Without damping, the world would continue to reverberate with every vibration ever produced. All real systems have some degree of damping. When operating at or near resonance, damping is the only way to control the amplitude of vibration.

Figure 3.18 shows three resonance curves with three levels of damping ratios: 0, 0.2, and 0.4. These numbers refer to the fraction of critical damping (C_c). At critical damping, there is no oscillation and the system returns to its neutral position in less than 1 cycle. Damping above 0.5 is considered heavy damping. Damping below 0.5 is considered light. Table 3.1 contains some common materials and their typical damping ratios.

In Fig. 3.18, the peak of the curve shifts to the left with higher damping. Damping does introduce some frequency shifting, but for typical machines made of iron and steel, this damping is so low that the frequency shift is negligible. This means that the vibration from a

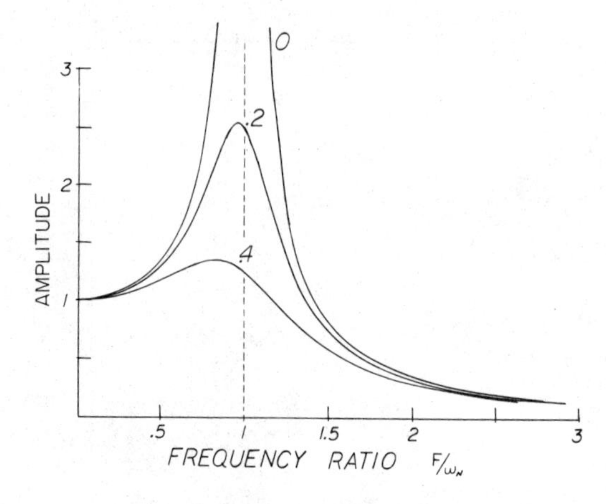

Figure 3.18 Resonance curve with three levels of damping.

TABLE 3.1 Damping Ratios of Common Materials

Material	C/C_c
Steel	0.001
Concrete	0.01
Lead	0.02
Natural rubber	0.05
Butyl rubber	0.05–0.5

bearing deep within a machine will transmit to the outside casing with virtually no change in frequency.

In electrical circuits, the resistor performs the same function as a damper does in mechanical systems. The resistor resists the flow of electric current. The damper resists the movement of velocity.

With damping, the equation of motion for a single-degree-of-freedom system becomes:

$$m\ddot{x} + c\dot{x} + kx = 0 \tag{3.9}$$

where c = damping

The solution is significantly more complicated with damping, and will not be considered further here.

Mechanical Impedance

The forces of vibration are generated inside machinery on the rotating parts and find their way to the outside casing. The forces travel through the structural parts of the machine—along the rotor and shaft, through the bearings, and along the machine casing. On their way out, the forces pass through different materials and metal interfaces. All along the way they get reduced, or attenuated. This is the concept of mechanical impedance.

Figure 3.19 shows a force traveling through three structural parts and two interfaces. Materials introduce some energy loss due to their internal hysteresis and damping. For steel this energy loss is low. The majority of losses on machines and structures occur at the interfaces, or joints. There the forces diminish in amplitude.

The frequency does not change. Not very much anyway. Damping does introduce minor frequency changes, but this shift is insignificant in machinery composed mostly of metal. The vibration signal that gets to the outside of the machine is a combination of the source vibration and everything it must pass through on the way out. The farther away from the source of the vibration, the more it is attenuated. Or, the corollary, the closer to the source, the stronger is the amplitude. The frequency is characteristic of the source. The amplitude is characteristic of the path.

Mechanical impedance can be defined as the property of machines and structures that resists the transmission of oscillating forces. The impedance varies with frequency. Figure 3.20 contains three plots that portray how impedance changes with frequency for the ideal mechanical components. A damper has constant impedance with frequency. The impedance for a mass increases with frequency. A mass does not pass high frequencies very well. This is intuitive, because a mass cannot move very fast in oscillation. The impedance of a spring decreases with frequency. The surprising result is that pure springs pass high-frequency vibrations very well. Mechanical springs are poor isolators for high-frequency vibrations.

A stiff link in the force transmission path will pass on the vibration energy to the next link without going into significant motion itself.

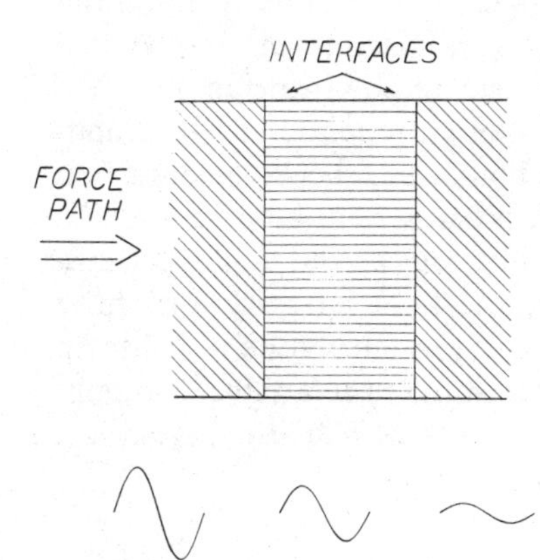

Figure 3.19 Mechanical impedance: The force amplitude is attenuated at each interface.

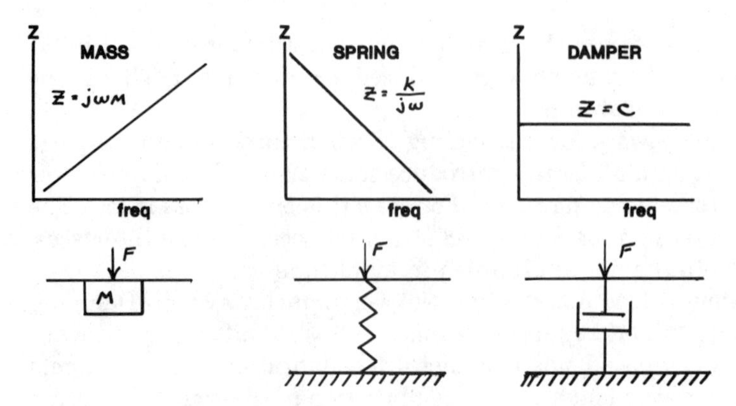

Figure 3.20 Mechanical impedance Z versus frequency for ideal components.

This was impressed upon me one day when measuring the structural vibrations coming from fans. The fans rotated at 20 Hz, and the top panels of the fans had a resonance at 8 Hz. The panels vibrated with amplitudes of 7 mils at 8 Hz. The fans rested directly on 10-in steel beams that formed the structure of this three-story factory. The 8-Hz vibration was measurable on these beams, but about 1000 times smaller. All around the fans were walkways with handrails. The handrails vibrated significantly at 8 Hz. In this situation, the steel beams formed a very stiff link between the fan top panels and the handrails. The beam did not move very much itself, but it served to transport this energy to a more flexible part. In this case, the steel beams presented a low mechanical impedance path to the 8-Hz vibration. The entire force transmission path was steel, from flexible fan top panel to flexible handrail, with steel joints. Vibration energy takes the path of least resistance through a structure.

One method to mitigate dynamic effects is to intentionally assemble materials of different densities. This creates mechanical impedance mismatching which introduces further losses at the interfaces.

Using the concept of the path of vibration and attenuation of the amplitude through materials and interfaces, it is possible to check the joints in a structure. Certain types of mechanical wear occur at joint interfaces where there is micromovement. Most structures fail first at the joints, because joints are usually weak links. It is possible to check the integrity of joints by exciting the structure on one side of a joint and measuring the response on the other side. The measured response is the sum of the excitation amplitude and the attenuation through the joint interface. The attenuation is a characteristic of its mechanical condition.

Isolation

With the concept of impedance, we can insert materials into the force transmission path to reduce the amplitude of vibrations. The frequency remains the same. Figure 3.21 illustrates the concept of an isolator. Generally, any material with a lower stiffness than the adjacent materials will function effectively to attenuate the force. Isolators work equally well in either direction. That is, the same isolator providing 95 percent isolation efficiency from machine to concrete floor will provide the same 95 percent efficiency in blocking floor vibrations from reaching the machine.

Typical isolator materials are mechanical springs, air springs, neoprene rubber, cork, felt, lead, and fiberglass. The performance of isolators is frequency dependent. Therefore, the selection of isolators must consider the frequencies to be attenuated, in addition to weight and deflection. Vibration isolation is discussed in depth in a separate chapter of *Machinery Vibration: Corrective Methods* (1992).

Critical Speeds

Rotors also have natural frequencies of vibration. When the rotor is spun at this frequency, it begins to flex and go into a condition of resonance. These resonances of rotors are called *critical speeds.* The critical speeds of rotors are very nearly the same as the resonant frequency when not spinning as detected by a bump test. The critical speed is not exactly equal to the natural frequency of the rotor because the rotor and its bearings generally have speed-dependent dynamic characteristics. The second critical speed is not an exact multiple of the first critical speed. The second critical speed is typically 2.5 to 4 times the first critical speed, depending on

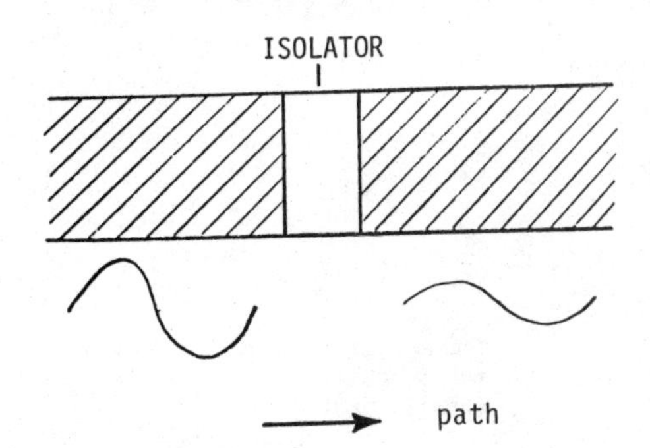

Figure 3.21 An isolator in the force transmission path.

its cross-section shape and the end-restraint conditions. Figure 3.22 illustrates the characteristic bending shapes of rotors at the first, second, and third critical speeds.

At critical speeds the amplitude of vibration builds up to very large values, then decreases dramatically once past the critical speed. There is also a 180° phase change when passing through the critical speed. Figure 3.23 is a plot of phase angle and amplitude versus speed. At each critical speed, there is a peaking of the amplitude and a 90° change of phase. There is an additional 90° phase change beyond critical speed, for a total of 180° from below to above critical speed. The vibration response at the critical speed is usually limited only by the damping in the bearings. Hence, bearing damping is a significant characteristic on rotors that must pass through critical speeds.

Calculation of the critical speed is a nontrivial matter that depends on the material properties, the rotor geometry, and end-restraint conditions. Even after this calculation, the result is only approximate. As an example, I calculated the critical speed of a large, horizontal shaft motor. The shaft diameter had an average 3-in outside diameter. The rotor weighed 400 lb, and the distance between bearings was 4 ft. The first critical speed was calculated to be 2133 rpm. The motor operated at 1800 rpm and showed no signs of being anywhere near its critical speed. It operated just fine. The calculation, to remind you, was only approximate.

These considerations of critical speeds are the responsibility of the equipment manufacturer and design engineers. You should understand the concept and recognize a critical-speed condition when it is encountered. This is not as easy as it sounds. Structural resonances on stationary parts show the same symptoms as critical speeds, i.e., vibration amplification and phase reversals during speed changes. Resonance tests are sometimes required to identify the true cause.

The best operating guidelines are to be 20 percent away from any critical speeds. It is also best to accelerate and pass through critical speeds as quickly as possible. Some equipment does get sold and set up

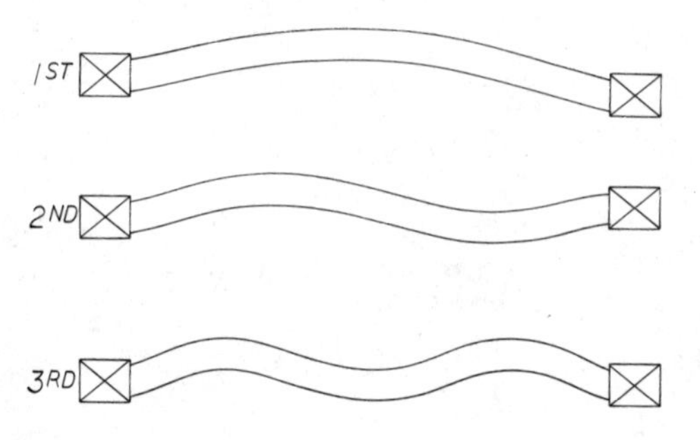

Figure 3.22 Bending shapes of rotors at the first, second, and third critical speeds.

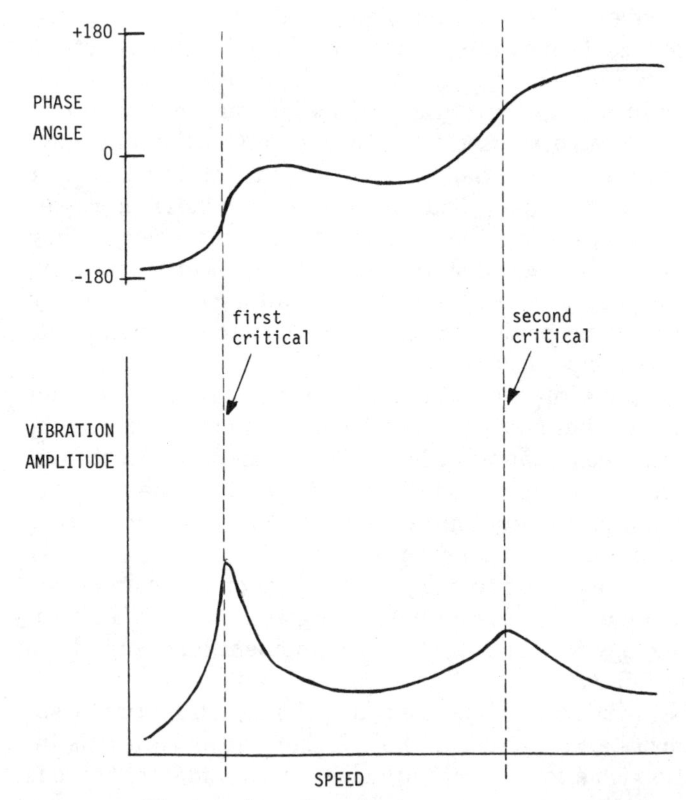

Figure 3.23 Phase angle and amplitude versus speed for first and second critical speeds.

operating dangerously close to critical speed. This is a damaging and dangerous condition to operate at. Also bear in mind that equipment modifications that alter the mass or shaft stiffness will change its critical speeds.

Life Cycle of Machinery

To complete this discussion of basic concepts, the life cycle of machinery should be traced. Environmental effects will also be introduced.

Every machine part starts as an ore in the ground. The metal atoms are locked up as an oxide of iron, copper, zinc, or any of the other oxides of metals. The ore must be mechanically and chemically refined to separate out the atoms that are to be used. At some point, the process heats up the refined particles to above the metal's melting temperature. This further separates and refines the metal and allows it to be poured into molds. During the cooling and solidifying, crystals of the metal form and grow toward each other, finally freezing into each other, forming grains. Immediately, the metal begins a process of oxidation to return to its previous condition in the ground as an ore.

The next steps in manufacturing of a machine part introduce mechanical stresses. These processes could include any of the following: remelting and casting, rolling (hot or cold), forging, machining, heat treatment, grinding, or other straining processes. These processes leave residual stresses in the metal. Metals have a memory. They want to return to an unstressed state. Immediately after fabrication, each machine part begins a process to relax its internal stresses by changing shape. This is sometimes called "creep," and is usually accelerated at higher temperatures.

The next step is machine building. The parts are assembled together to form a working machine. Surface defects are introduced at this stage and can become latent failures. This is the last opportunity to inspect the individual parts and their surfaces since the machine is usually sealed up. Contaminants of dirt, moisture, and other unwanted material are sealed up within it.

The machine is transported to the job site. This presents an opportunity for shock damage due to rough handling and impacts. Unfortunately, our technology for handling heavy equipment lags behind the technology of building it.

The machine may be stored for some time prior to installation. Usually this is outdoors and exposed to the weather, but even if it is indoors, construction job sites are not very clean environments. There is a potential here for introducing contaminants again into the machine. Moisture and dirt can get in past seals and openings. During storage, the machine sits idle and is under the influence of gravity. This constant load can introduce a permanent set into the shaft. And finally, storage environments are not benign. The machine may be exposed to nearby vibrations that will accumulate damage to the bearings.

The next stage in the life cycle of a machine is to set up and start it. This means more handling and initially energizing it. If a machine survives this initial start-up, then it will probably run okay. The majority of the defects in the machine have already been built in. This start-up exposes the machine to the largest stresses it will see during its lifetime. If it runs for 30 min without problems and thermally stabilizes, then you can relax with a 50 percent confidence that it will survive to maturity.

After start-up, most machines exhibit a failure rate over time that looks like the familiar "bathtub curve" of Fig. 3.24. This is a plot of failure rate versus time. If 100 machines were started at the same time, then this curve would predict their rate of failure. Initially, a large number would stop working because of built-in defects. After

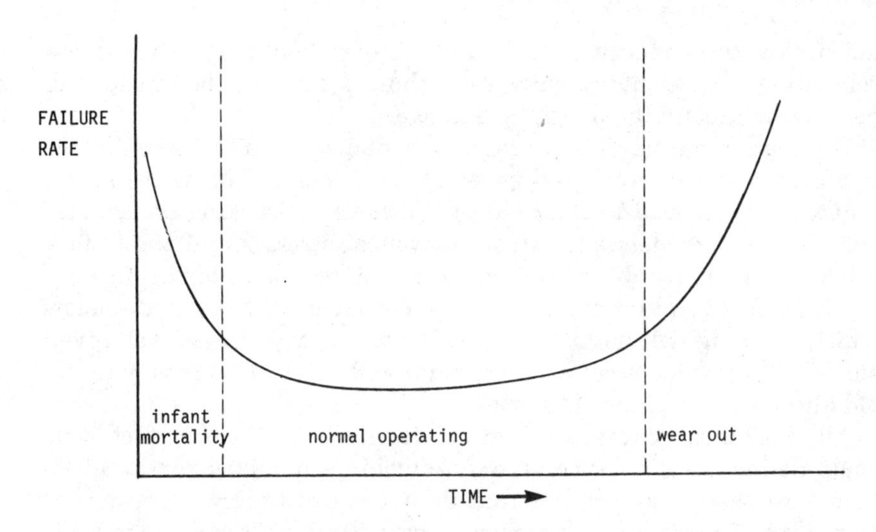

Figure 3.24 Bathtub curve of failure rate.

some operating time, these infant mortalities would be corrected and the machines placed back on-line. The time line for these repaired machines must then be reset back to zero since they will have new parts and they may also have experienced some new built-in defects due to additional handling.

The machines would then enter the normal operating range where the failure rate is the lowest if people would just keep their hands off the good running machines and let them run. The normal operating part of the curve may have a slight upward slope when approaching the wear-out period.

During the wear-out period, the failure rate increases rapidly. This is called a *bathtub curve* because its shape resembles a nineteenth-century bathtub. This curve is based on a statistical sample of machines, but it can also be viewed as the probability of failure for a single machine. If a machine survives the initial running in, or infant mortality period, then it can be expected to have a low probability of failure until the wear-out period begins. This is predicated on the assumption that the machine is not handled unnecessarily.

Whenever a machine fails, decisions are made to rebuild it or replace it. This could be the end of the life cycle for some machines, i.e., the first failure. At each subsequent failure, these same decisions concerning rehabilitation versus retirement are made anew. Or the end of life for a machine may come when the process or building is retired from service, and a perfectly good machine goes with it.

The metal of which the machine is composed could be recycled. In this case, it is remelted and the manufacturing process is reset. If the machine is to be rebuilt using some of the original parts, then stress relaxation and geometric changes of those parts must be considered, along with accumulated fatigue damage.

Fatigue is a mechanism of progressive damage to metal parts under cyclical stresses (vibration) that results in cracking. This is one mechanism that can lead to wear-out of a machine. Cracking on a metal part is strong evidence of excessive cyclical stresses and points to a certain vibration problem. It must be mentioned that nonferrous metals (specifically aluminum) do not have an endurance limit below which cracking will not occur, as steels do. At *any* stress level, given sufficient numbers of cycles, aluminum will crack. This is why flying old aluminum airplanes is so risky.

All machines generate vibration during operation. In fact, this could define the difference between machines and structures. Structures are passive assemblies that do not generate any vibrations of their own. These vibrations originating within machines make their way to the outside of the machine and can be measured there. These vibrations are characteristic of the machine's health and can be measured at any time during the machine's life. The vibration, however, is only known at the transducer location. It must be inferred elsewhere. This is the process of analysis. It begins with a measurement, so measurement instruments are the subjects of the next chapter.

Chapter

4

Instruments

Ears, Hands, and Watch

Humans are sensitive to vibrations. Prior to any instruments becoming available, all vibration analysis was done by listening and feeling. This method is still used by those who do not have access to instruments. We have some built-in sensors in our skin and our ears. These biological transducers that we come equipped with served a survival function and still do. The human ear is amazingly sensitive to the smallest pressure change. The pressure sensors in our skin can detect a constant pressure, and also an oscillating pressure, or vibration.

When mechanical equipment came along, humans did not have any built-in sensors to judge their quality, but rather adapted the ones that they already had for another purpose. We do not come equipped with the ability to measure horsepower or flow rate, or speed, but we can directly detect loudness and smoothness. These, then, became the criteria for mechanical quality. They are good ones too. By listening and feeling, we make decisions on whether to buy or not. Listening and feeling is the first step in my analysis procedure when approaching a new vibration problem before any electronic instruments are connected. This is a valid sensing method for vibration analysis if certain precautions are taken, namely, calibration and frequency analysis. I will explain these.

As a crew member at a missile launch complex, our daily work schedule was to walk through all levels of the silo observing the mechanical equipment. We were instructed to place our hand on the equipment and feel for unusual temperatures and vibration. The assumption was that we could detect a change in vibration and schedule repair work before serious damage occurred.

In reality, we usually came upon a vibrating machine after the damage occurred. It was leaking, belts were broken, or bearings were already seriously worn, and maintenance was usually repair rather than preventative. The reason this system did not work as well as it should is that humans maintain poor calibration. There were no vibration measuring instruments in the maintenance organization of this air force wing. Severity judgments were based on human perceptions.

It was difficult to compare a machine with vibration "readings" separated by several days, especially if many similar machines had been seen during the interval. The "measurements" were also highly subjective. One person's judgment of "rough" could be another person's "acceptable." This system of human perception was an overall vibration reading. There was no attempt to distinguish frequencies. The best that could be obtained with this hand sensing of vibration was a crude, overall, uncalibrated, subjective vibration measurement that sounded an alarm when equipment failure was eminent. This method of hand sensing does work satisfactorily for some organizations. I occasionally receive trouble calls to analyze and diagnose a piece of equipment that feels "rough." After examining the equipment, I find that in 70 percent of the cases, there is indeed trouble developing. A diagnosis is given, corrective action is taken, and a catastrophic failure is avoided. In the other 30 percent of the cases, there is no real problem, usually only a worsening sheet metal vibration due to load changes, or heightened anxiety on the part of the owner.

The reason this method of hand sensing works satisfactorily to some organizations is that some human calibration techniques are employed. First, one person is given responsibility for specific equipment. They are his or hers and no one else's. So the human variable is removed. Second, this one person checks the equipment on a daily basis, so there is not a long interval between "measurements." And third, there are usually identical pieces of equipment nearby to compare against. Using these methods of calibration, a company can be successful in avoiding most catastrophes with only human hand feeling measurements. They will not, however, be very successful in detecting early predictors of failure. To do this, frequency analysis is necessary. Humans are equipped with a frequency analyzer.

The combination of the human ear and brain is actually a pretty good spectrum analyzer and is extremely sensitive. Human hearing has a bandwidth of about 40 Hz, or less for people with a good musical ear. The human ear is sensitive to air vibrations from about 20 to 20,000 Hz. This is also the frequency range of most annoying vibrations of mechanical equipment. Figure 4.1 shows some typical vibrations and their positions on the frequency spectrum. This frequency spectrum extends to only 10,000 Hz. There are few, if any, mechanical vibrations of machinery beyond 10,000 Hz. The vast majority of it is

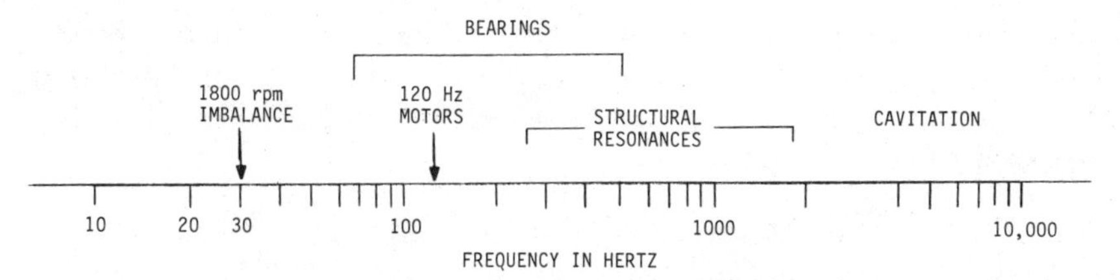

Figure 4.1 The frequency spectrum and typical machinery vibrations.

below 5000 Hz. Pumps at 1800 rpm generate vibrations at 30 Hz. This 30 Hz is barely audible for most people, but it certainly can be felt with pressure sensors in the fingertips. A metal object, such as a coin, can be held between the fingertips while probing for vibrations. Evidently, there is some amplification from a metal object pressed against a vibrating surface. Hand sensing works for low-frequency vibrations less than 100 Hz.

For higher frequencies, listen to the tones. All motors and transformers give off an electrical buzz at 120 Hz in North America (100 Hz in parts of the world where the power line frequency is 50 Hz). An imbalance in a 3600-rpm machine will be a vibration at 60 Hz. Rolling element bearings give off four discrete tones. The three higher tones are typically between 50 and 500 Hz for normal machines running at 1800 or 3600 rpm. Metal-to-metal impacts typically generate shock pulses at 1000 to 10,000 Hz. These are mostly detected by their very short duration, high-amplitude "metal-crash" type of sounds. Most metal structural resonances occur between 100 and 5000 Hz. Fluid cavitation shows up as broadband vibration, typically around 3000 to 5000 Hz. As you can see, the majority of mechanical vibrations are within the frequency range of the human ear to detect.

Using the human ear for frequency analysis of machinery is more effective when the ear can be coupled directly to the machine. This means using a stethoscope, a wooden stick, or a screwdriver. The skin is more sensitive to lower-frequency vibrations. Therefore, imbalance at 30 or 60 Hz can best be detected with hand feeling. For vibrations above 100 Hz, the human ear is a better frequency analyzer. Above 100 Hz listen for the tones. Any vibration analysis should always be done first by listening and feeling to give the analyst a subjective frame of reference. If no other instruments are available, then significant vibration analysis can be done with only listening and feeling.

If you cannot feel any vibration, or hear anything objectionable, then you can be confident that there probably is not a vibration problem in terms of machine wear. Humans are well calibrated, in this respect, to machine damage criteria, but this judgment depends on no interfering background noise.

Figure 4.2 is a chart based on statistical data of the human perception of vibration. It is somewhat frequency dependent, but it shows that human judgment of vibration severity can be quantified by the level of acceleration:

0.001 ~ 0.01 *g* is the threshold of perception.

0.1 *g* is considered unpleasant.

0.5 *g* is considered intolerable by most subjects.

Notice that this chart only goes to 100 Hz. Above this frequency the vibration is clearly audible, and there is a noise problem. Some of my work is in response to complaints about excessive noise. The source of all noise problems is a mechanical vibration that couples to the air and continues through the air as noise. The airborne noise contains the same frequencies as the mechanical vibration.

The point of this discussion is that vibration problems and human sensitivity to them are well correlated in the absence of background vibration. That is, if vibration is uncomfortable to people, then it is probably causing serious damage to the machine.

As vibration analysts, we are looking for vibration. We want to amplify it and take a closer look at it. This is the whole purpose for vibration instruments—to amplify vibration and display it in a way that we can better understand it. Instruments increase the signal-to-noise ratio. In addition to the vibration of interest, there are other background vibrations that we want to ignore.

Imagine yourself in a mechanical room where multiple machines are running. The machines may be a boiler, a chiller, three motors and pumps, two fans, and a large transformer. Suppose you wanted to examine the vibration from only one of the three pumps. Standing there in the center of this cacophony, you are exposed to all the noise in summation. You could approach the pump of interest and place

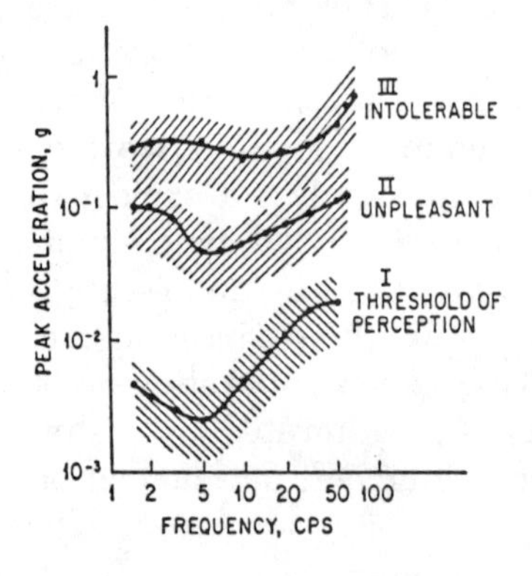

Figure 4.2 Human perception of vibration. Exposure of 5 to 20 min. Shaded areas are about 1 standard deviation on either side of mean. (*D. E. Goldman,* USNMRI Report 1, *NM 004 001,* March 1948. *Reproduced with permission of McGraw-Hill, Inc.*)

your hand on it, or couple your ear to its housing with a wooden stick. This will increase the signal of interest above the background noise. Alternatively, you could turn everything else off and run only the pump of interest. This will reduce the background noise and again increase the signal-to-noise ratio. By placing a transducer directly on the machine of interest, we couple directly to that source of vibration and separate out the background vibrations. The signal-to-noise ratio is increased.

To perform any significant vibration analysis, the important parameters to measure are frequency and amplitude. The hands and ears can make both of these measurements subjectively. In addition, a watch is useful to determine the period of slow vibrations. The speed of rotation is the most fundamental number required to do any vibration analysis, and I suggest that a stroboscope with an accurate speed readout be the first instrument purchased. The stroboscope is not reliable for stopping motion below 300 rpm. For these slow speeds, a watch is the best.

I once measured the speed of a cooling-tower fan by counting the revolutions over a 60-sec period. It was done this way because the speed was slow enough and there was no other practical way to make the measurement. It was too bright in the sun for a strobe, and there was no safe way to attach a mechanical or photo tachometer.

Many vibrations have a characteristic rhythm that comes and goes. The period of this rhythm, and hence its frequency, can easily be measured with a watch. Using this method, it is possible to identify a cause to an effect some distance away if they have the same rhythm. An example would be a steel platform hundreds of feet away from a marine diesel engine, that thumps at the same frequency as the engine.

The methods described above, using the hands, ears, and a watch (and perhaps a screwdriver), are not yet outdated. This is all that is available in many maintenance departments, and they can be used effectively. I even know several consultants who own no instruments and provide advice to clients on the vibration condition based on these human transducers. Keep these tools available in your toolbox and use them frequently to correlate with your instruments, as a check. Let's now proceed to electronic instruments.

Transducers

The ears and hands are very subjective when sensing vibrations. Most of us want a hard number that we can compare to. Transducers along with a readout instrument can provide this number. Vibration transducers are more sensitive than the human skin, have a response to lower frequencies, and a narrower bandwidth. Bandwidth will be discussed shortly in the subject of filters.

A *transducer* is a device for converting the mechanical motion of vibration into an electric signal, commonly called a *pickup*. Vibration transducers measure motion. There are generally three kinds of transducers: displacement, velocity, and acceleration. Figure 4.3 shows a representative of each type that is commonly used for machine monitoring.

Other types of transducers are used for earth motions and other applications outside of routine machine monitoring. These other transducers will be briefly introduced after the three basic types are discussed. The most important advantages and disadvantages of each type will be covered.

The most common type of displacement transducer is the proximity probe (Fig. 4.4), which operates on the eddy current principle. It sets up a high-frequency electric field in the gap between the end of the probe and the metal surface that is moving. In reality, the proximity probe includes an oscillator-demodulator and a cable. The proximity probe senses the change in the gap and therefore measures the relative distance, or dis-

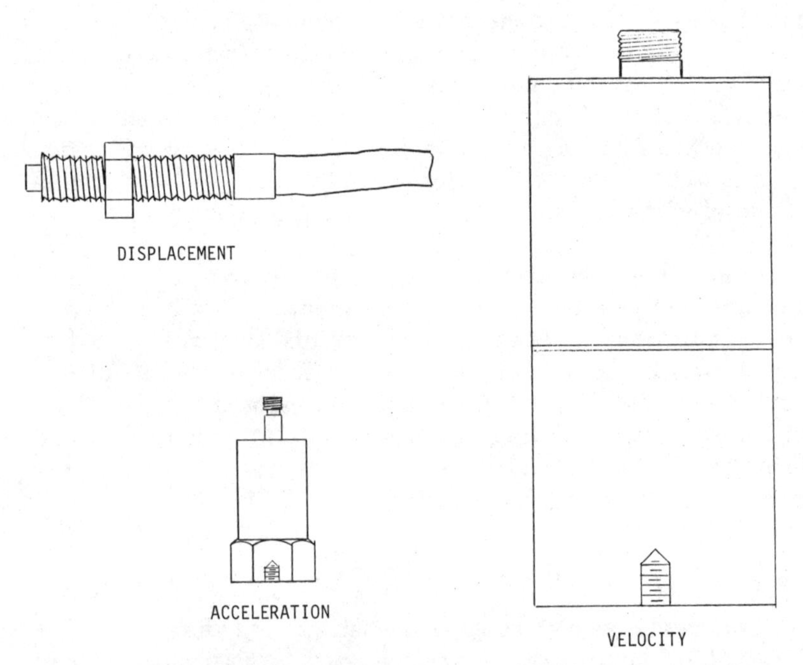

Figure 4.3 Three types of transducers commonly used for machinery vibrations.

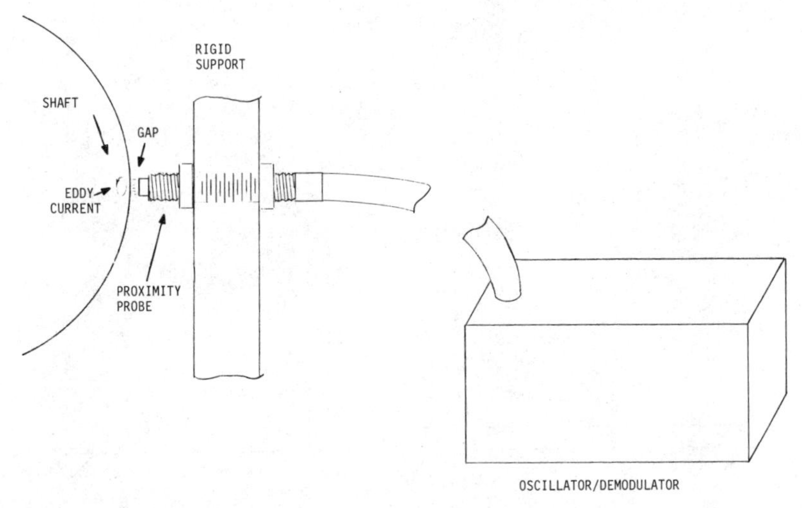

Figure 4.4 The proximity probe in operation.

placement, between the probe tip and the surface. It is important to recognize that the proximity probe measures relative displacement, not absolute displacement. It is usually permanently mounted in a bearing housing looking at the surface of a rotating shaft. The probe is useful for gaps from about 10 to 90 mils. The probe must also be calibrated for the specific shaft material.

In addition to shaft relative displacement, the proximity probe is also sensitive to shaft surface defects such as scratches, dents, and variations in conductivity and permeability. The output from the proximity probe is the sum of shaft relative displacement and surface variations.

The proximity probe also senses shaft runout, and it has a very difficult time distinguishing vibration from runout. A slow roll sometimes has to be performed during which an electronic circuit memorizes all the shaft imperfections, including runout, and subtracts these out from the signal the proximity probe reports at running speed.

The practical maximum frequency of proximity probes is about 1500 Hz. The minimum frequency is zero. It can measure static displacement.

A useful application of proximity probes is to measure very slow relative movements, like thermal expansion. Figure 4.5 shows a setup for measuring the movement of a machine as it warms up to operating temperature. In addition to thermal growth, the proximity probe can

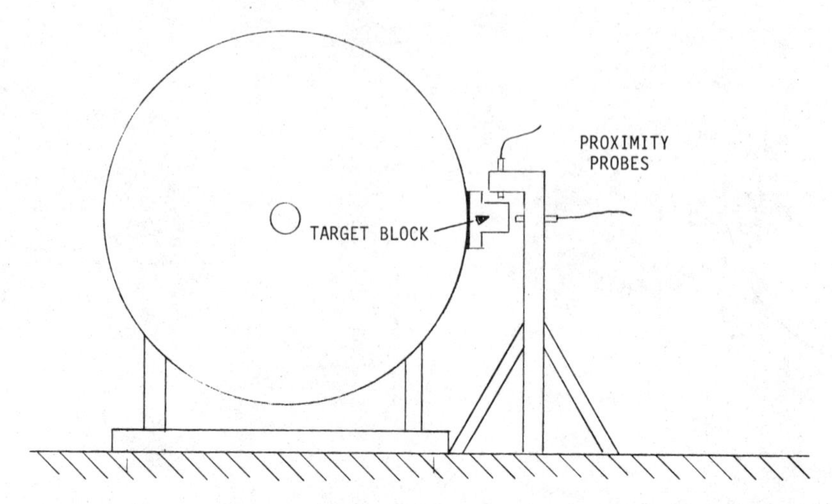

Figure 4.5 Typical setup for using noncontact proximity probes to measure thermal growth.

be used wherever relative motion information is needed, not only for very slow motion but also for high-speed motion up to the maximum frequency response of the transducer.

Since the proximity probe does not actually touch the vibrating part, it is useful in situations where the vibrating part cannot tolerate the mass of the pickup. Some applications are where a thin, light part is in motion, and this motion needs to be measured without touching it, such as printed circuit boards, computer disks, and sheet metal panels. The object must be electrically conductive for the proximity probe to properly set up an electric field and sense the gap.

There are some obstacles to overcome in using a proximity probe. Holes need to be drilled and tapped, or some other type of fixturing needs to be put in place. The gap and voltage outputs need to be adjusted and the probe calibrated for the material. The shaft must be smooth and polished with a minimum of surface defects. Finally, a slow roll should be done to compensate for runout.

The velocity transducer is one of the oldest electronic vibration sensors. For many years it was the only practical sensor for machinery monitoring; therefore, many standards are composed around the information that this device provides. The velocity transducer is an adaptation from a voice coil in a speaker. It consists of an internal mass (in the form of a permanent magnet) suspended on springs. Surrounding the mass is a damping fluid, usually oil. A coil of wire is attached to the outer case. Figure 4.6 illustrates the internal construction of a velocity transducer.

In operation, the case is held against the vibrating object. The outer

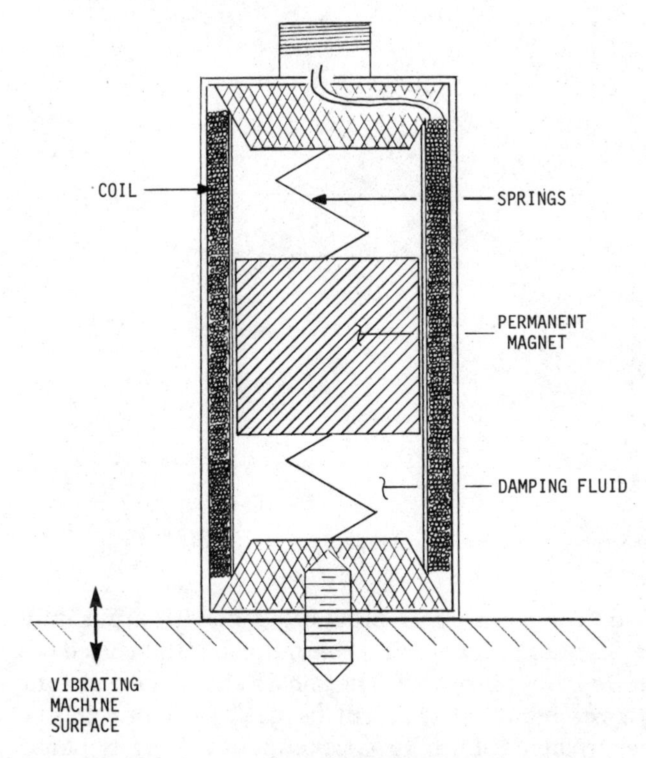

Figure 4.6 Velocity transducer internal construction.

case moves with the vibrating object while the internal mass remains stationary suspended on the springs. The relative motion between the permanent magnet and the coil generates a voltage that is proportional to the velocity of motion, hence velocity pickup.

The velocity pickup operates just like an old electrodynamic microphone. The voice coil in speakers can be used as a velocity pickup. In fact, the first velocity pickups used by Westinghouse in the 1930s were modified speaker coils.

The velocity pickup is self-generating and produces an output that can be used by analysis instruments without any further signal conditioning. It can be connected directly into an oscilloscope or other readout instrument. Also, the velocity output is an absolute velocity referenced to inertial space. For this reason it is called a *seismic pickup*.

The velocity transducer has an internal natural frequency of about 8 Hz (those sizes that are used for machine monitoring). This natural frequency is simply the resonance of the single-degree-of-freedom internal mass suspended on springs. Figure 4.7 illustrates this frequency response. The response at resonance is highly damped because

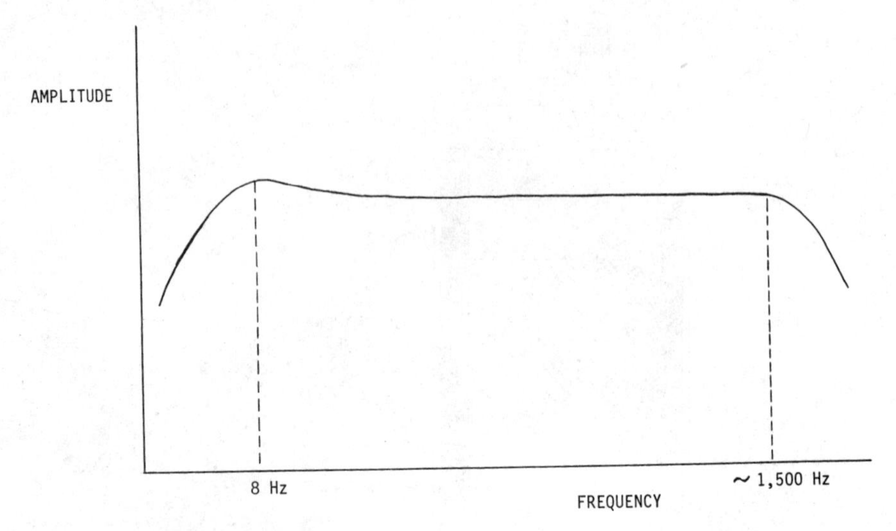

Figure 4.7 Frequency response of a velocity transducer.

of the internal fluid. This transducer produces a linear output only above this resonant frequency (above 600 rpm). It can still be used below that speed, but its output drops off. The sensitivity is not constant below this internal resonance. It can still be used for balancing or other relative measurements down to approximately 1 Hz if phase measurements are not used. Its phase measurement capability is no longer accurate in this resonance region because of the phase shift that occurs. For absolute measurements, of amplitude below 10 Hz, correction charts are available. Its upper frequency limit is about 1500 Hz, the same as the proximity probe.

The velocity transducer is rather large. On small devices this added mass can significantly affect the vibration output. It also has internal moving parts that are prone to failure and sticking. Lastly, the coil in the velocity pickup is sensitive to external electromagnetic fields. It makes a good antenna. This coil will detect electromagnetic waves, add them to the vibration signal, and send the sum on to the analysis instrument. Care must be taken when working around large motors or other EM fields. This care involves checking for 60 Hz or other interference. A simple method of checking is to hold the velocity pickup close to, but not touching, the surface you want to measure. The presence of a 60-Hz signal with no mechanical motion input to the transducer (other than hand motion) confirms interference. In fact, a good vibration analyst will always be suspicious of a strong 60-Hz signal, with any type of transducer.

The most common acceleration transducer is the piezoelectric accelerometer. It consists of a quartz crystal (or barium titanate, a man-made quartz) with a mass bolted on top and a spring compressing the quartz. Figure 4.8 illustrates its internal condition.

A property of a piezoelectric material is that it generates an electric charge output when it is compressed. In operation, the accelerometer case is held against the vibrating object and the mass wants to stay stationary in space. With the mass stationary and the case moving with the vibration, the crystal stack gets compressed and relaxed. The piezoelectric crystals generate a charge output going positive and negative as the crystals are alternately compressed tighter and relaxed about the preload. The charge output is a faithful reproduction of the motion of the surface in the direction of the accelerometer's sensitive axis. The charge output is proportional to force, and by Newton's second law

$$F = ma$$

it is also proportional to acceleration; hence the name, *accelerometer*. It is useful, however, to consider the output of this transducer as proportional to the internal forces in the machine. If the acceleration level is high, then the internal forces are high. Forces are the cause of oscillation, and forces are the cause of excessive wear and premature failure.

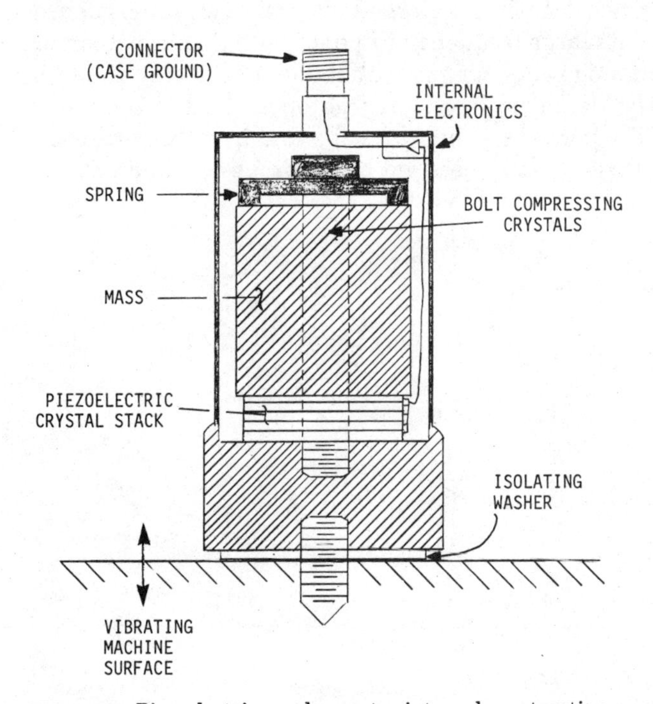

Figure 4.8 Piezoelectric accelerometer internal construction.

The accelerometer's sensitivity response versus frequency is illustrated in Fig. 4.9. Because of the stiffness of the crystals and steel spring, the typical accelerometer used for machinery monitoring has a very high natural frequency, typically 25,000 Hz. Its response is linear for about one-third of this range. Unlike the velocity pickup, the accelerometer operates below its first resonant frequency. It has a useful frequency range of from about 5 to approximately 10,000 Hz depending on its size. Therefore, the accelerometer is sensitive to vibration frequencies much higher than the proximity probe or the velocity pickup.

The sensitivity and frequency response of accelerometers are directly related to their size (Fig. 4.10). Small accelerometers (with small crystals and small masses) have low sensitivities but higher operating frequencies (because their internal natural frequencies are high). Large accelerometers (with large crystals and large masses) have high sensitivities but lower high-frequency limits (sometimes only 800 to 1000 Hz). Hence, big accelerometers are used to measure ground motions, which are low in frequency and low in amplitude. Small accelerometers are used to measure high-frequency vibrations. Some are useful above 50,000 Hz.

The accelerometer is the transducer of choice for machine health monitoring. The benefits of the accelerometer are its extreme ruggedness, its small size, its large frequency response, and its large dynamic range. A large dynamic range means that it can detect very small vibrations and not be damaged by very large ones. This is best illustrated by Fig. 4.11, where the dynamic range and frequency response is plotted for the three transducers we have discussed thus far. The

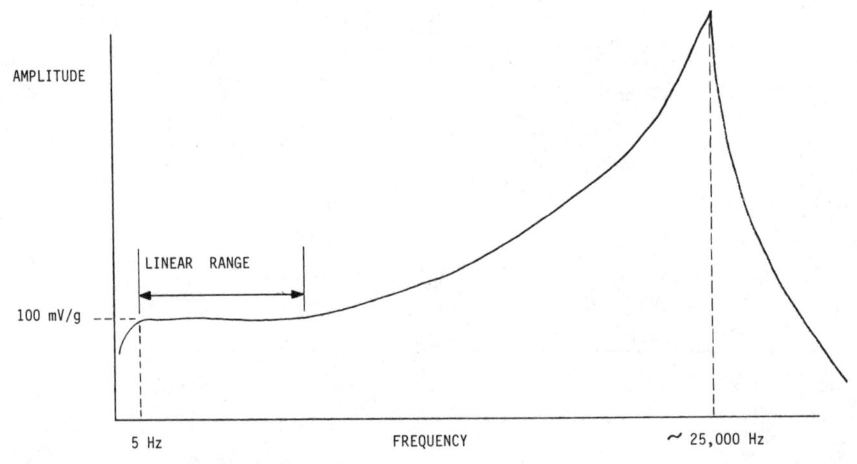

Figure 4.9 Sensitivity versus frequency for typical, general-purpose accelerometers used for machinery monitoring.

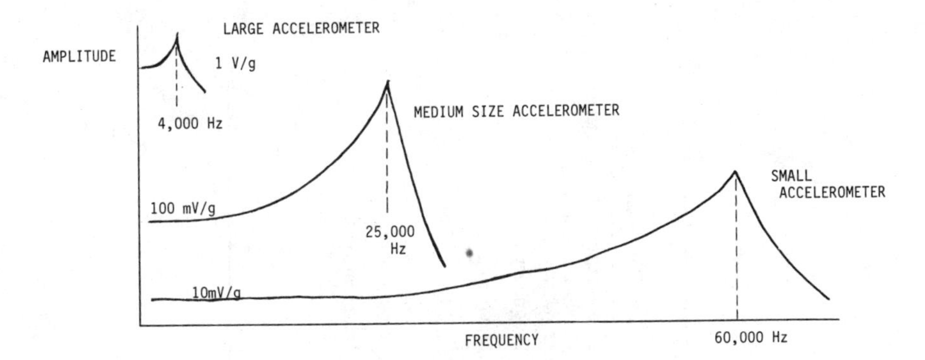

Figure 4.10 Relationship of sensitivity to natural frequency for three sizes of accelerometers.

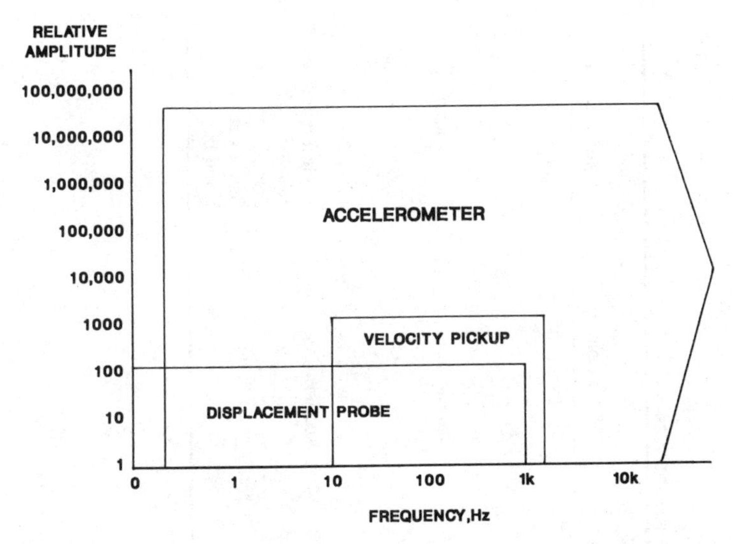

Figure 4.11 Dynamic range and frequency response of common vibration transducers.

displacement and velocity transducers have frequency and amplitude limits for machine monitoring. The accelerometer has no such limits within the range of machine vibrations. The accelerometer can detect very small vibrations and not be damaged by large shocks. It can also measure low-frequency vibrations from a fraction of a hertz to beyond 10,000 Hz, not necessarily with the same accelerometer. This data is further summarized in Table 4.1.

The accelerometer is the preferred transducer for machine monitoring for several other reasons. First, it is the only transducer that can accurately detect pressure wave vibrations. These high-frequency vi-

TABLE 4.1 Summary of Common Vibration Transducers

Transducer type	Useful frequency range	Measurement	Advantages	Disadvantages
Proximity probe	0–1,500 Hz	Displacement	Noncontact	Senses surface imperfections Conductive parts only Mounting difficulty Frequency limits
Velocity pickup	8–1,500 Hz	Velocity	Self-generating Seismic Good indicator of machine condition Hand-held	Moving parts Large Massive Senses EM fields Frequency limits
Accelerometer	1–20,000 Hz	Acceleration	High frequencies Rugged Seismic Small size Hand-held	Temperature limits

brations are best seen in the time domain, but if a frequency domain view is desired, then the acceleration display is the display of choice. Recall from Fig. 3.13 that the velocity and displacement displays suppress the high frequencies. The acceleration display emphasizes high frequencies. Many defects manifest themselves first as high-frequency vibrations. The accelerometer's output is proportional to acceleration, so why not look at its voltage output directly without electronic manipulations that introduce errors.

Modern spectrum analyzers have the capability to produce an acceleration display from a velocity transducer, or even a displacement transducer. This electronic differentiation does not come without errors, and it cannot overcome the frequency limits of the transducers. More will be said concerning this in the section of FFT spectrum analyzers.

Second, the accelerometer output is proportional to forces which are the cause of internal damage. Third, the acceleration unit *g* is a universal unit applicable in the English engineering or metric system of measurements. No further conversion needs to take place for technicians of different countries and different languages to interpret the data. Fourth, the acceleration value is closely related to human uncomfortableness. Fifth, the accelerometer has high-frequency sensitivity for detecting bearing faults, and the acceleration display is the best way to view them. For these reasons, I will dwell longer with the accelerometer than with the other transducers. The objective is to give the reader enough information to select an accelerometer for his or her purposes and to make valid measurements with confidence.

The accelerometer is self-generating, but the signal output is of such high impedance that it is unusable by most analysis equipment. This is like trying to inject a red dye into a pressurized water line. If a port in the water line were opened, then water would come gushing out. It would be hopeless to try to pour some dye into that port. A pressure chamber should be constructed with two valves, similar to an airlock. The dye can be placed in the chamber, the outside valve closed, and then the inside valve opened, admitting the dye into the pressurized fluid stream. This is a crude analogy to impedance matching. The output impedance of the accelerometer must be matched to the impedance of the readout instrument for it to accept the signal from the accelerometer. This is like matching pressures with an airlock. The output signal from the accelerometer must be converted to a low-impedance signal by special electronics. This electronic circuitry can be outside or inside the accelerometer. Figure 4.12 illustrates the two general types of accelerometers. The charge-mode accelerometer must have a charge amplifier nearby, which provides the proper impedance matching. The short piece of cable between the

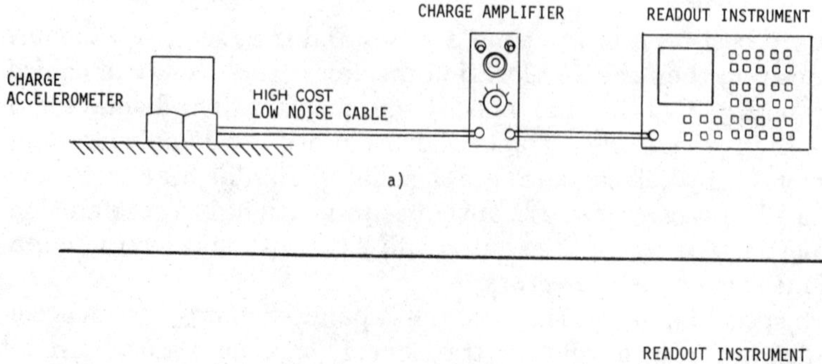

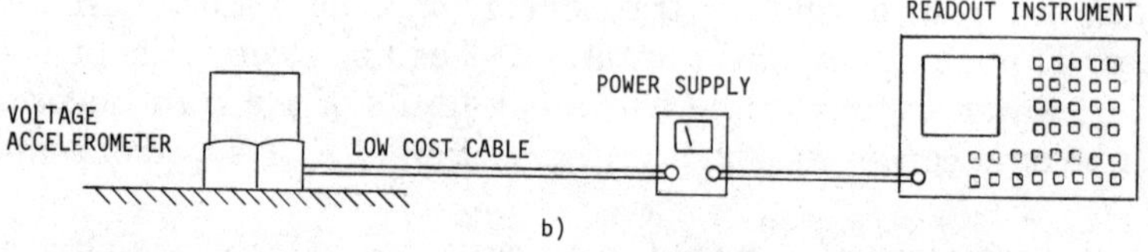

Figure 4.12 The two general types of piezoelectric accelerometer systems. (*a*) Charge system using a charge amplifier. (*b*) Voltage system with electronics internal to the accelerometer (called *ICP* type by some).

accelerometer and the charge amplifier is critical. It must be a low-noise cable, and its length cannot be changed. The accelerometer, cable, and charge amplifier must be calibrated as a unit and not changed.

The voltage-mode accelerometer has the impedance matching electronics built inside. It needs no charge amplifier, and the cable length is not critical. All it needs is a low-cost power supply (typically batteries) to power the internal electronics. For machinery applications, it is superior to have the electronics inside the accelerometer. Then regular low-cost cables can be used, and the cable length can be changed at will—a 6-ft cable or a 600-ft cable will cause negligible change in signal.

It is important to use charge amplifiers with charge accelerometers and power supplies with voltage accelerometers. The two types are not compatible and will lead to erroneous data. When the electronics are built into the accelerometer, this simplifies their use. The voltage-mode accelerometers are labeled *ICP (integrating circuit piezoelectric)* by some manufacturers, and some readout instrument manufacturers provide built-in power supplies to power their electronics. Their power requirements are typically a nominal 27 VDC at 2 mA. There is, however, a temperature limitation of about 250°F for accelerometers with internal electronics. Higher temperatures will overcook the internal electronics. For higher-temperature applications, consider the charge-mode system.

The accelerometer is also a seismic device and measures absolute motion relative to inertial space similar to the velocity pickup. Unlike the velocity pickup, it is practically unaffected by external electrical or magnetic fields. It is, however, sensitive to ground loops as the other pickups are also. This can be easily eliminated by providing ground isolating washers at the accelerometer.

The primary considerations in selecting an accelerometer are sensitivity and frequency response. If high-amplitude motions are to be measured, i.e., greater than 10 *g*, such as in shock measurements, then a low-sensitivity accelerometer is appropriate—10 mV/*g* or less. If very low level motion is to be measured, such as building or structural motions at low frequencies, then a high-sensitivity accelerometer should be chosen—1 V/*g*. For most machinery monitoring, 100 mV/*g* sensitivity accelerometers provide the right balance of sensitivity and frequency response.

Other considerations in accelerometer selection, or any transducer, are temperature exposure, linearity, transverse sensitivity, damping, strain sensitivity, size and weight, and mounting configurations. Temperature limitations for the piezoelectric accelerometer have already been discussed. Linearity is expressed as the percent deviation from a constant value of the sensitivity. Transverse sensitivity is the ability of the transducer to detect motion in directions perpendicular to its sensitive axis. Damping is very low in piezoelectric accelerometers but can be significant in other types, such as piezoresistive accelerometers. This may even be a desirable property if high-frequency signals wish to be suppressed. Strain sensitivity is the ability of the transducer to generate a signal when the base is distorted, such as when it is clamped against a nonflat surface. Piezoelectric accelerometers can also respond to airborne acoustic energy, and if it is to be used in a strong acoustic environment, then due consideration should be given to this parameter. For normal machinery monitoring, the amount of acoustic energy sensed compared to the surface vibration is negligible. For floor vibration measurements in a recording studio, it would not be negligible.

There are situations where one type of transducer has advantages over the others. Displacement transducers are the most sensitive to very low amplitude and low-frequency vibrations. For this reason, they are used in very sensitive earth motion sensors. A large mass is suspended on springs as an inertial mass, and capacitive or eddy current probes detect its motion. The displacement transducer is also the best sensor to detect shaft motion directly where the rotor is light and the bearing supports are massive. This is typical for high-speed turbines and compressors. In this situation, very little motion is transmitted to the outside casing or bearing supports because the bearings are journal bearings where the shaft rides on a fluid film and has some clearance to move about in. The proximity probe can be mounted to the bearing, or its support, and it can sense the relative motion of the shaft to the bearing.

On small- to medium-size machines with rolling element bearings, the casing and bearings move with the rotor. In this case, a seismic transducer on the bearings will be a good indicator of shaft motion. Accelerometers and velocity transducers are convenient to use here, since they can be attached directly to the outside of the bearing with a magnet. They can also provide useful information on the bearing vibrations. These seismic transducers can also be used on journal bearing machines on the outside casing. Useful information will be detected. It must be realized that the motion will be smaller on the outside casing because of the mass of the casing and the reduced transmissibility through the fluid film. Useful frequency information for analysis can be obtained.

At times, the high-frequency data from an accelerometer is undesirable. I can remember balancing a very large fan at a cement plant. The bearings and supports were massive, so there was little motion on the bearings. In addition, there was very high amplitude vibrations at high frequencies, 1000 to 5000 Hz, due to turbulence in this high-pressure fan. The low amplitude at running speed of 20 Hz was not a problem to detect for the accelerometer. What was a problem was the high-frequency turbulence; it was causing overloading of my spectrum analyzer. I could have done without it. It would have been better to use a velocity transducer, which is naturally insensitive to these high frequencies. Unfortunately, one was not available. The solution in the field was to use a tuneable bandpass filter to eliminate everything but the running speed. This worked well for the amplitude measurements, but the filter had to be very carefully tuned for accurate phase measurements. There are two other solutions to this problem of unwanted high-frequency data from an accelerometer.

The first is a mechanical filter. This device is simply a rubber pad mounted under the accelerometer. Figure 4.13 illustrates its use, and Fig. 4.14 is a drawing of the one I made. This mechanical filter has a resonant frequency of approximately 1000 Hz with a 30-g accelerometer mounted on it. It effectively filters out vibrations above 1400 Hz. Lower-frequency cutoffs can be obtained by using a softer rubber or a larger mass accelerometer. I now carry a couple of these in my toolbox to eliminate unwanted high frequencies.

The second method would be to use a piezoelectric velocity transducer. This is a piezoelectric accelerometer with integrating electronics built in. Its output is proportional to velocity. It has all the benefits of an accelerometer, such as small size, ruggedness, and dynamic range, but it produces a velocity output which deemphasizes the higher frequencies.

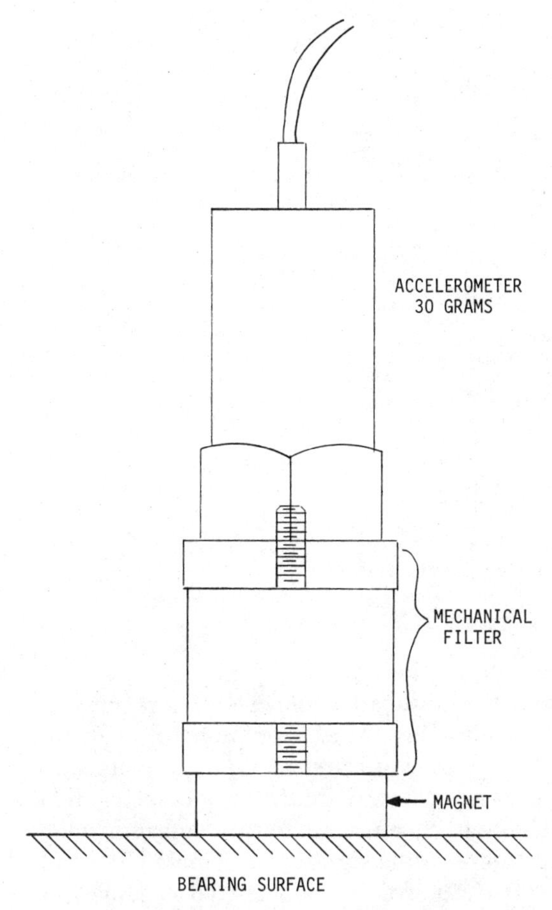

Figure 4.13 Mechanical filter used under an accelerometer to eliminate unwanted high frequencies.

In addition to transducer selection, there is another related selection that must be made in relation to how to display the data. It is always best to choose a display corresponding to the transducer output. An acceleration display should be chosen for an accelerometer and a velocity display for a velocity pickup. This practice prevents errors associated with electronic integration and differentiation.

With modern digital analyzers, any transducer input can be converted to any other value. For example, you can put an accelerometer input in and integrate that value once to get the velocity and twice to get displacement. The displays will have the same frequency content, but the amplitudes will be scaled according to the frequency. Remember the relationship between displacement, velocity, and acceleration from Figs. 3.13 and 3.14.

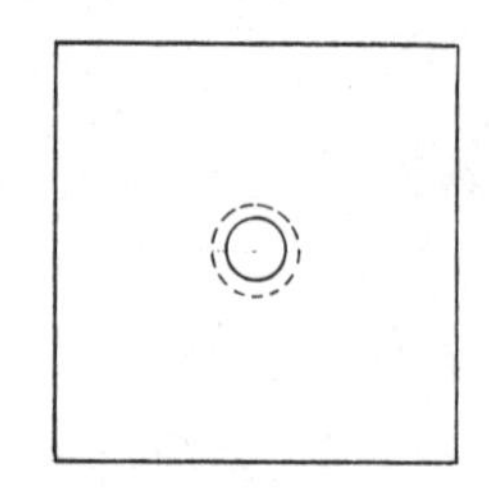

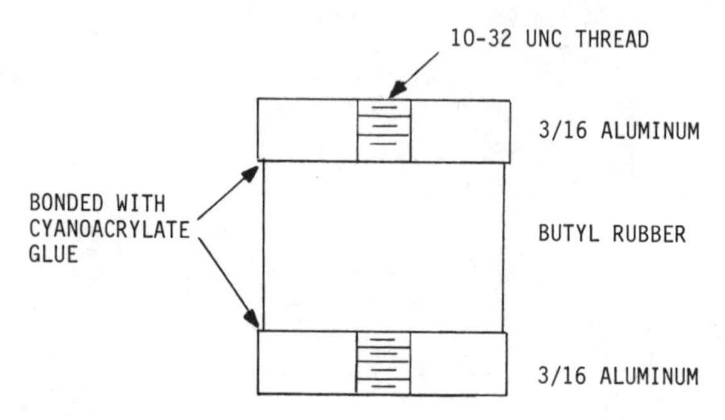

Figure 4.14 Design of a mechanical filter.

Accelerometers emphasize the higher frequencies. There may be large peaks at higher frequencies that an accelerometer senses, but they may not be serious. Defects show up first at higher frequencies, so the acceleration spectrum is a good early indicator of defects, and a good choice for trending. The analyst must make this determination, i.e., whether high-frequency vibration is significant or not. The acceleration display deemphasizes the lower frequencies, so it somewhat filters out low-frequency traffic and building motion.

The velocity spectrum is a good indicator of seriousness for machinery based on absolute amplitudes. It is presently used more than any other display for general machine condition, and many standards promote the velocity unit. Displacement is favored by some for balancing low-speed equipment less than 3000 rpm, because it produces a tall peak in the frequency spectrum. This is really a matter of perspectives. Velocity or acceleration could be used just as well, and equivalent results will be achieved because it is actually the same data, just viewed differently. An inordinate amount of effort has been expended in the past fixing low-frequency vibration problems like imbalance because the available instrumentation emphasized this low-frequency data. The high-frequency information was always there, but deemphasized. The velocity and displacement displays put blinders on us and did not allow us to see this high-frequency data. Imbalance is the

Number 1 cause of excessive vibration and can be corrected by measuring in displacement, velocity, or acceleration. To see other defects at higher frequencies, acceleration is the better choice.

There is another transducer that is still used today for vibration. This is the strain gauge. Its use for machinery monitoring is very limited because it is not rugged enough. The most usual application for a strain gauge is to measure the actual strain at a location and correlate this to dynamic stress. With this information, along with material properties, the fatigue damage can be evaluated. This type of measurement and analysis is usually done in an engineering design and testing environment. You may, however, encounter a strain gauge previously installed on machinery, and it is useful as a vibration transducer. It has a frequency response that limits its use to low-frequency measurements, from dc to approximately several hundred hertz.

There is another type of accelerometer available that has a linear response down to dc, or zero frequency. That is, it can measure a constant unchanging acceleration. It is made by depositing a strain gauge on a beam of silicon. It is called a *piezoresistive accelerometer*. The entire package is very small and rugged. These accelerometers can be made smaller than piezoelectric accelerometers, but they do not have the high-frequency response because of internal damping.

Another new development, just within the past few years, is piezo film. Figure 4.15 illustrates this film packaged as a transducer with leads. This is a plastic film that has piezoelectric properties, that is, it generates a charge when squeezed, stretched, bent, or deformed in any way. It has a frequency response similar to conventional piezoelectric accelerometers. There is some difficulty in determining the direction of the applied force, but it has some intriguing applications in communications, security, and medicine. It is extremely light, and it can be formed to stick to complex contours, like a spherical surface. It is very sensitive and low in cost compared to other motion sensors. For these

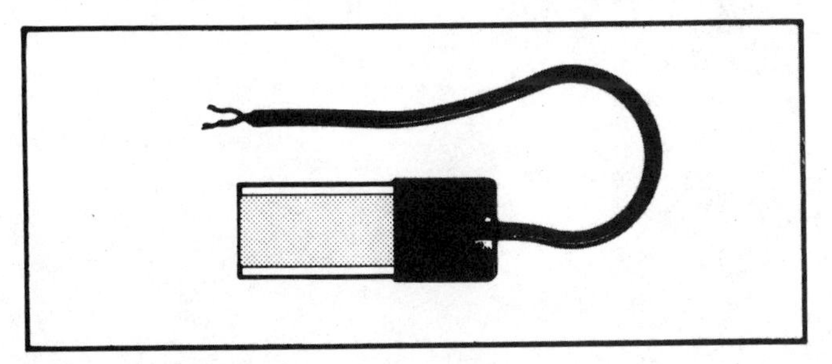

Figure 4.15 KYNAR piezo film; shielded and laminated with coaxial cable. (*Courtesy Atochem Sensors, Inc.*)

reasons, it has strong potential for applications as imbedded sensors in machinery.

Shock and vibration recorders have been around for some time. They have been used for monitoring the transportation system for damaging exposure. A good example of this use was in the case of a $200,000 machine tool shipped from Switzerland that was rejected upon arrival in Colorado. At the airport the prospective owner examined the vibration-shock recorder that was attached to the machine casting at the factory. It revealed that the machine had been exposed to a severe impact shock, as would occur if the crate were dropped. The instrument also recorded the exact time of this event. The purchaser refused to accept this damaged machine, and he had data to justify his case.

Modern shock and vibration recorders packaged in smaller sizes are more sensitive and can operate on batteries for several months. Some are programmable to activate at a preselected level. They will record the level and time, or digitize and store the waveform. Being programmable, their data can be offloaded to a computer and reused after their batteries are recharged. Figure 4.16 shows one model.

The next type of transducer I want to discuss is the vibration switch (Fig. 4.17). This is not really a vibration sensor, but a shock sensor. It is a relay that will latch over when it senses a large enough bump (or

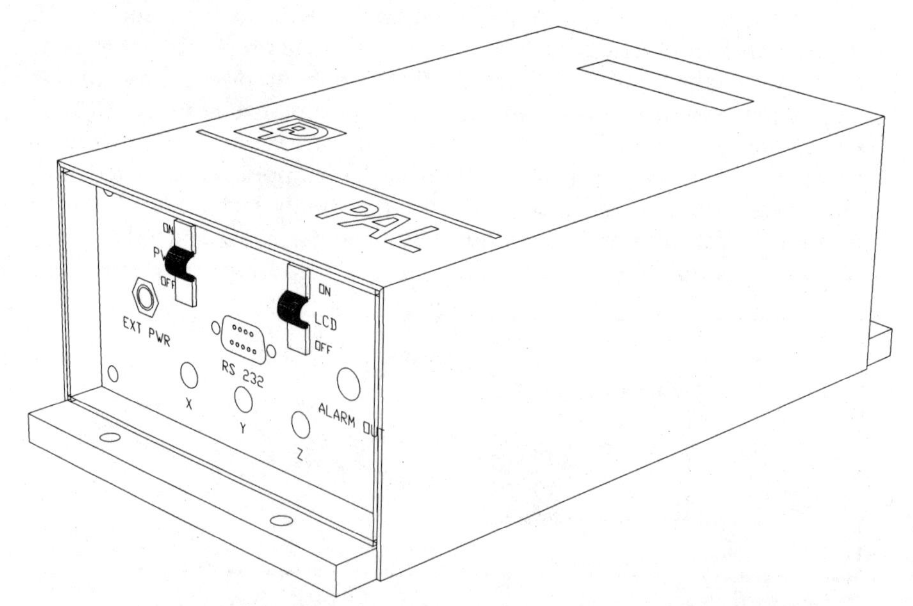

Figure 4.16 Portable shock-vibration recorder. (*Courtesy Dallas Instruments, Inc.*)

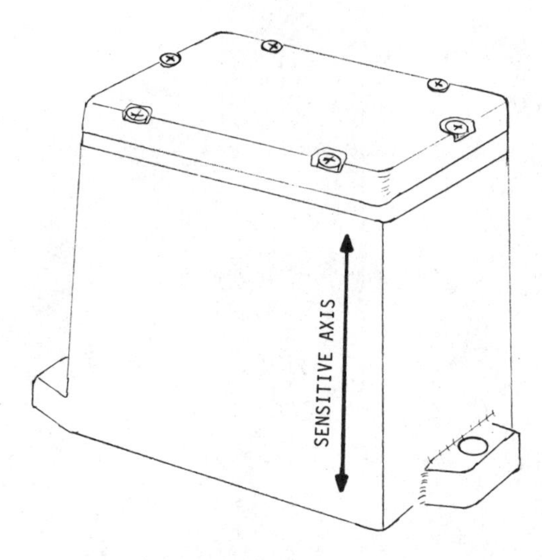

Figure 4.17 Vibration switch.

shock) in a specific direction. The output of the relay is used to trigger an alarm or a shutdown sequence.

The original idea was to shut the machine down before it damaged itself. This is useful for machines that tend to abruptly lose or gain weight, like a propeller fan in cooling towers. If a blade came loose, or if it picked up a rag, the fan would be so far out of balance that a severe shock pulse would be sensed by the switch. The relay would trip, disconnecting power to the motor. A vibration switch would save the motor and gearbox and prevent further damage to the fan.

In practice, vibration switches are subject to nuisance trips, and the technicians inevitably adjust the switches to be less sensitive. They provide no useful information for analysis or trending. Their cost is very similar to an accelerometer, so why not use an accelerometer? The accelerometer is smaller and lighter. The accelerometer's output can be used to trigger an alarm or shutdown if necessary with appropriate monitoring circuitry, and during daily operation its output can provide vibration data for analysis and trending.

This brings up the subject of sensitivities for transducers. Transducers produce an electric signal to represent a mechanical measure of vibration amplitude. The electric signal output is usually in millivolts (thousandths of a volt). This is a very small value, but amplifiers in modern instruments bring this up to a useable value. Vibration is measured in mils (0.001 in) for displacement, in/sec for velocity, or *g* for acceleration. The sensitivity of a transducer is expressed as a ratio of output in millivolts to the input quantity. The sensitivity of charge-mode accelerometers is measured in picocoulombs per *g*. Figure 4.18

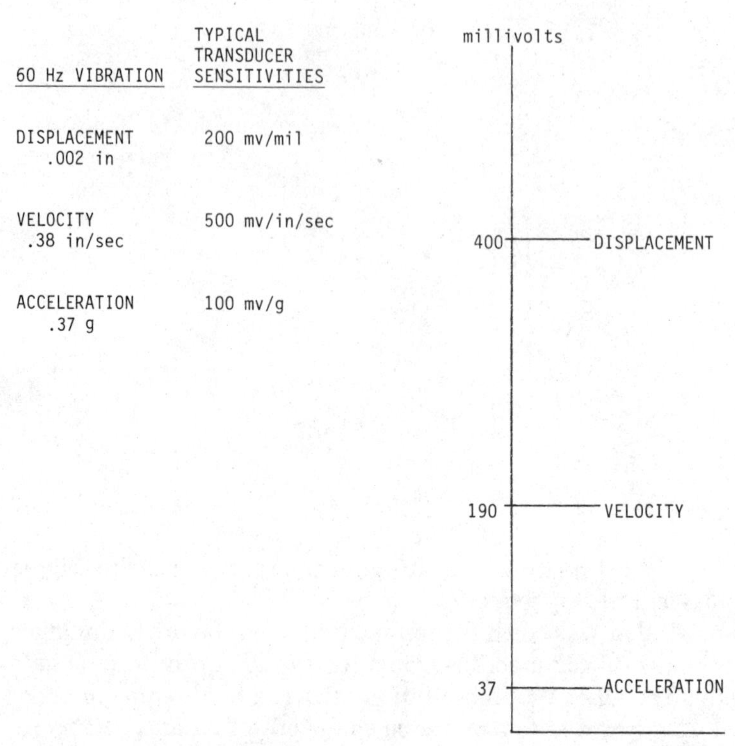

Figure 4.18 Typical transducer sensitivities and resulting voltage output for a 60-Hz vibration.

contains a table and a chart which illustrate this for the three common transducers. A proximity probe has a typical sensitivity of 200 mV/mil. If this proximity probe is exposed to 2 mils peak-to-peak displacement at 60 Hz, then it will produce an output of 400 mV peak to peak at 60 Hz. By using Eqs. (3.6) and (3.7), this displacement can be converted to acceleration and velocity.

$$a = \frac{D(\text{peak})f^2}{9.78} = \frac{(0.001)(60)^2}{9.78} = 0.37\,g$$

$$v = 61.7\,\frac{a}{f} = \frac{6.17\,(0.37)}{60} = 0.38 \text{ in/sec}$$

The same 0.002-in vibration at 60 Hz is equivalent to a velocity of 0.38 in/sec at 60 Hz. A velocity pickup with a typical sensitivity of 500 mV/in/sec will generate 190 mV. An accelerometer will only generate 37 mV for the same vibration. So for the same vibration, the proximity probe generates the most voltage and the accelerometer the least. This is not a problem because modern instruments have excellent input amplifiers that can boost these small signals to very useable values. The noise floor of modern instruments is typically 0.3 μV, or 0.000 000 3 V. Thirty-seven millivolts is approximately 100,000 times larger than 0.3 μV.

The accelerometer is the transducer of choice for machine monitoring, not based on sensitivity or voltage output but based on convenience of use, small size, ruggedness, frequency response, and dynamic range.

This sensitivity of transducers is a value that can change over time or if the transducer is damaged. Therefore, it is important to check this sensitivity periodically. This transducer sensitivity is the same as the calibration constant. Transducers should be calibrated at least every 6 months. The sensitivity should be checked across the entire useable frequency range of the transducer.

The sensitivity of velocity probes can change due to temperature because of oil viscosity changes. The sensitivity of accelerometers can change if the crystal is cracked, such as might happen if it were dropped on a concrete floor.

Cables

The next link in the measurement chain are cables to connect the transducers to the readout instrument. Included in this link are also signal conditioning devices such as:

- Oscillators-demodulators for proximity probes
- Charge amplifiers for accelerometers
- Transducer power supplies

Cables are not to be taken lightly. Cables incompatible to your system can seriously degrade the measurement quality. For example, a charge amplifier senses the charge on the cable in addition to the charge output from the piezoelectric accelerometer. This has introduced measurement errors in the past and is the primary reason for moving the electronics inside the accelerometer as in voltage-mode, or ICP, types. Cable length and cable capacitance considerations are negligible in voltage-mode accelerometers. The best approach is to buy cables from the same supplier where you purchased the transducers and signal conditioning electronics.

Cables are delicate cords, they tend to be laid on the hard floor during measurements, and maintenance men usually wear big boots. This is a perfect setup to damage cables by crushing. Be alert for instrument cables on the floor during vibration measurements, not only to protect the cable, but to avoid tripping over them.

Cables are the least expensive link in the measurement chain, but also the most fragile. If problems develop with intermittents, the first place to investigate should be the cables and their connectors. Every cable mating and demating flakes off metal particles. These metal particles accumulate on the connectors and will eventually cause intermittents and shorts. It is a good practice to flush out these particles with a spray cleaner-degreaser for electronic parts.

Temporary cable connections, as with alligator clips, will act as an antenna and pick up any stray electromagnetic fields. You should always be suspicious of 60-Hz signals. You can test for stray electromagnetic field and your cable noise by a simple disconnect test. If the cable is disconnected at the transducer, then any remaining signal is electromagnetic noise and not any mechanical motion of interest.

Meters and Filters

The simplest type of readout instrument is a meter (Fig. 4.19). This displays a single number that represents the vibration amplitude. The display can be a needle deflection or an LCD digital readout. The

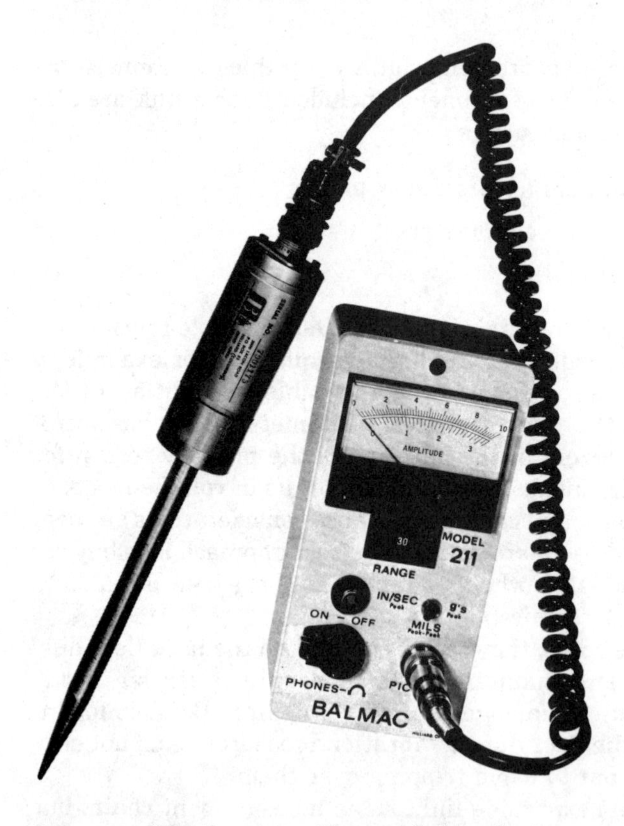

Figure 4.19 Vibration meter. (*Courtesy Balmac, Inc.*)

number displayed is an overall value that represents the energy content of all the vibration frequencies. For a single-frequency sine wave, the meter will display the rms value. But when multiple-frequency components are present (as they usually are), the meter displays a single value that represents all the vibrations. Figure 4.20 illustrates a typical vibration spectrum. There is more than one frequency present. In fact, there is a range of broadband vibration from 500 to 800 Hz. The overall value of the sum of all these vibrations is 0.29 *g*. This is the number that a simple hand-held meter will display. The meter is simple to use and useful for comparing machinery and for trending in maintenance programs.

To perform any kind of useful vibration analysis, the frequency components of the vibration must be separated out. The first way that this was done was with a tuneable bandpass filter. This is just like tuning your car radio. The dial for changing stations controls a tuneable bandpass filter that accepts, and passes through, the frequency that you want to hear, and rejects all the rest. A rotating machine generates many vibrations from many independent sources—imbalance, bearings, electrical, gears, couplings, resonances, and others. Each defect is generating a vibration at its own particular

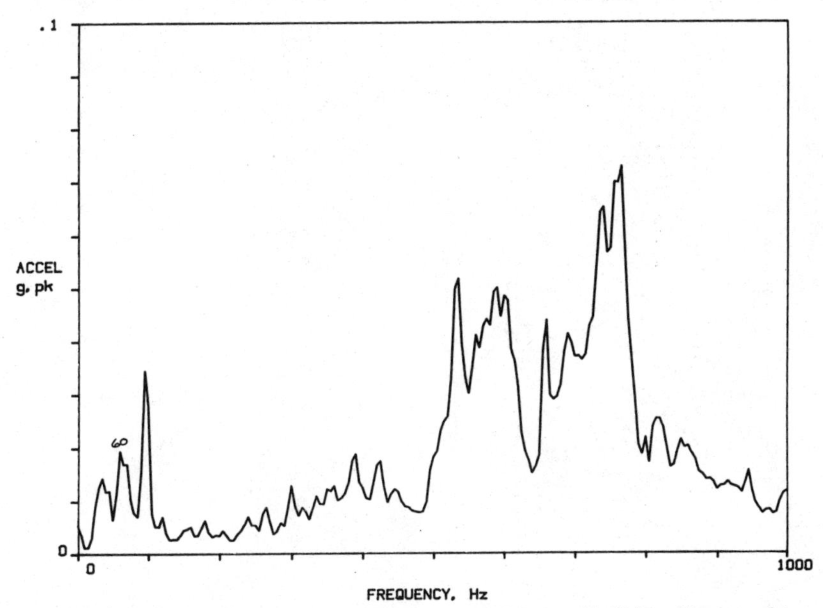

Figure 4.20 A typical vibration spectrum from a 3600-rpm fluid-handling machine. Overall is 0.29 *g*.

frequency of force repetition. These vibrations all mix together and arrive at the transducer as a cacophony.

A filter is a device that passes some frequencies and rejects all the rest. Visualize the filter as a window that has a certain width. You can see everything that appears through the window, and you cannot see anything outside the limits of the window. Figure 4.21*a* illustrates an ideal bandpass filter. This illustration is ideal because the cutoff occurs sharply at fixed frequencies, and is depicted as straight vertical lines. In reality, filters are not perfect. The cutoff occurs over a range of frequencies, and the window has sloping sides as shown in Fig. 4.21*b*.

A tuneable bandpass filter is one which can be tuned along the frequency axis and bring different parts of the spectrum into view. In this way simple frequency analysis can be done with a tuneable bandpass filter and a meter. The instrument connections are shown in Fig. 4.22. The meter display will change as different vibration peaks come into view. In place of the meter, an oscilloscope can be used as the readout device. Appendix A contains a schematic of a simple tuneable bandpass filter suitable for vibration work. Figure 4.23 shows a commercially available tuneable filter instrument. The important features that allow you to recognize such an instrument are a tuning dial to change the center frequency of the filter, a means to read the frequency, and a meter to display amplitude. These types of

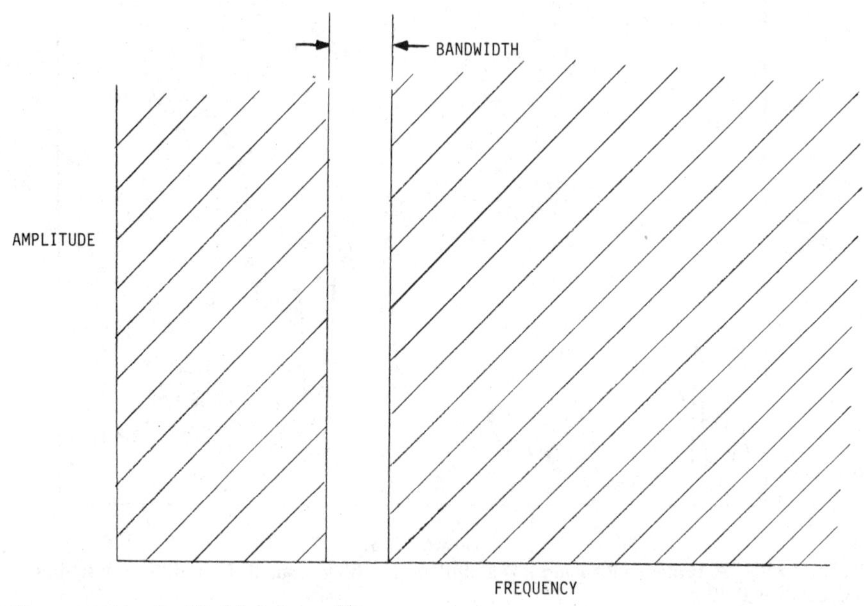

Figure 4.21(*a*) An ideal bandpass filter.

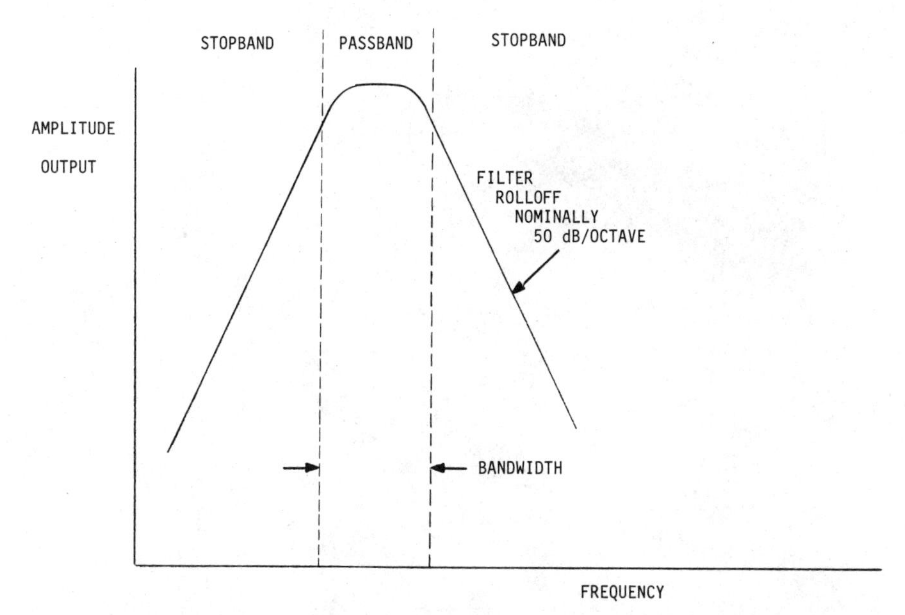

Figure 4.21(*b*) A real bandpass filter.

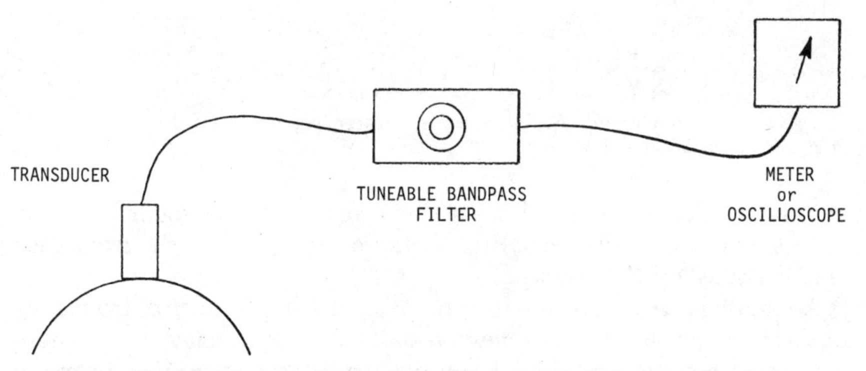

Figure 4.22 Diagram of a tuneable bandpass filter used to do frequency analysis.

instruments can do frequency analysis, and with the addition of phase measurement capability, they become excellent for balancing also.

It is important to know what frequencies are present in a complex vibration signal because each specific frequency is associated with some internal defect. The amplitudes diminish from the source as a vibration signal emerges to the outside casing because of mechanical impedance, but the frequencies come out unchanged. There is some minor frequency shifting due to damping. Frequency measurements from a constant frequency source could be shifted a few hertz at re-

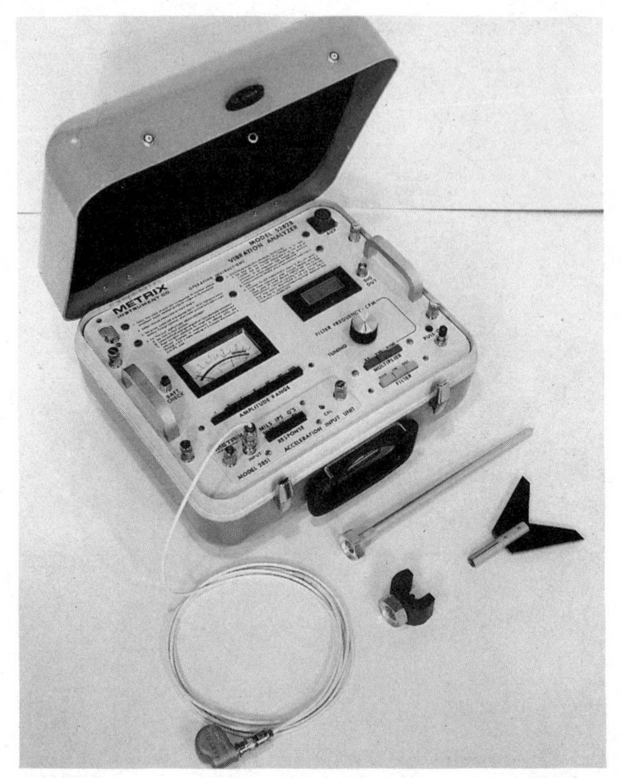

Figure 4.23 A commercially available tuneable filter instrument. (*Courtesy Metrix Instrument Company.*)

mote measurement points, but for the purposes of machine monitoring in localized areas with very little damping in the path, the frequency is essentially unchanged.

An analogy can be drawn to emphasize the importance of frequency analysis. Suppose I walked down a hallway past a noisy room. There were people crowded in the room, each talking or listening in many individual conversations. From the overall noise that I perceived in the hallway, I could not discriminate the purpose of the gathering. Is it a birthday party, a wedding shower, a political campaign, or a funeral gathering? To determine the purpose, I would need to go inside and listen to a conversation, while ignoring the overall noise. I cannot talk, only listen. The overall noise level cannot determine the purpose; an individual conversation might. Likewise, the overall vibration level cannot determine the cause of vibration. An individual conversation with the machine, focusing on frequencies, has a much higher potential for success. This is the purpose of the tuneable bandpass filter—to ignore the overall and focus on specific frequencies, one at a time.

Frequency analysis with an automatically tuneable bandpass filter is sold commercially as swept filter analyzers. The filter is tuned automatically across the frequency spectrum, and the result is displayed on a CRT screen that holds it, or the data is printed on a strip of paper. This type of analyzer has been available for about 30 years.

Remember that there were two ways to look at any complex signal (including vibration):

1. In the time domain
2. In the frequency domain

The time and frequency domain views are like looking into a room through different windows.

The oscilloscope displays signals in the time domain. We are now beginning to discuss the instruments that display in the frequency domain. The swept filter analyzer displays vibrations in the frequency domain, i.e., with amplitude along the vertical axis and frequency along the horizontal axis.

To speed up the process, rather than have a single bandpass filter and tune it, there are instruments that have many filters. These parallel filter analyzers process the signal simultaneously through many filters. The result is an output from each filter determined by the amplitudes of vibration residing between its lower and upper cutoff frequencies. The display is a bar graph across the spectrum as shown in Fig. 4.24. The height of each bar represents the output from each filter. These parallel filter analyzers can process data and display re-

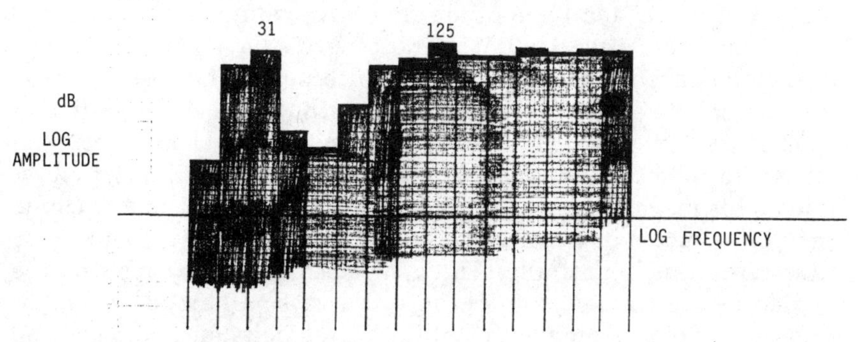

Figure 4.24 Display of a parallel filter analyzer, one-third octave. This data corresponds to the spectrum of Fig. 3.14*c*.

sults much faster than a swept filter analyzer. They are typically used in audio or sound work. They can be used for vibration analysis. Figure 4.24 is the parallel filter data for the same machine vibration that is shown as a spectrum in Fig. 3.14*c*.

FFT Analyzers

We are now going to discuss the most powerful type of vibration analyzer that has been available only during the last decade—the FFT spectrum analyzer. Figure 4.25 shows two models.

All the previous instruments that we discussed were analog instruments, i.e., they processed the signal itself. By contrast, the FFT analyzer first digitizes the input signal, then performs all subsequent operations digitally on numbers. This introduces some errors, but it makes the display in the frequency domain much faster.

An FFT analyzer acts like a parallel filter analyzer but with hundreds of filters. There are at least three good reasons for looking into the frequency domain. First, specific defects show up as specific peaks. I prefer to use the term *defects* because a perfect machine would generate no vibration, and both the time and frequency domain views would be straight lines. This correlation of peaks to defects is crucial to successful analysis and is not easily accomplished in the time domain.

Second, surveillance measurements are best done in the frequency domain. An overall number can be assigned and trended. Envelope analysis can be done. (Envelope analysis will be covered in a later section of this chapter.) This feature of envelope analysis, with a computer doing the bulk of the sifting, is what has made these instruments so popular in a maintenance environment.

Third, small signals are not hidden in the frequency domain. It is difficult to see a small amplitude sine wave riding on top of a large amplitude sine wave on an oscilloscope display, but in the frequency domain it shows up as a definite peak. The result is that mechanical defects can be seen when they are still small and tracked as they progress. Small vibrations can be clearly seen and measured in the frequency domain in the presence of large vibrations. This means that a small gear defect can be seen on a machine with large motor imbalance and bearing noise present. Figure 4.26 illustrates this beautifully. The gear-mesh frequency may not be visible in the time domain because it is a small-amplitude–high-frequency signal riding on top of the motor imbalance and bearing noise. Nevertheless, it is present and measurable if the proper transducer and data display are chosen.

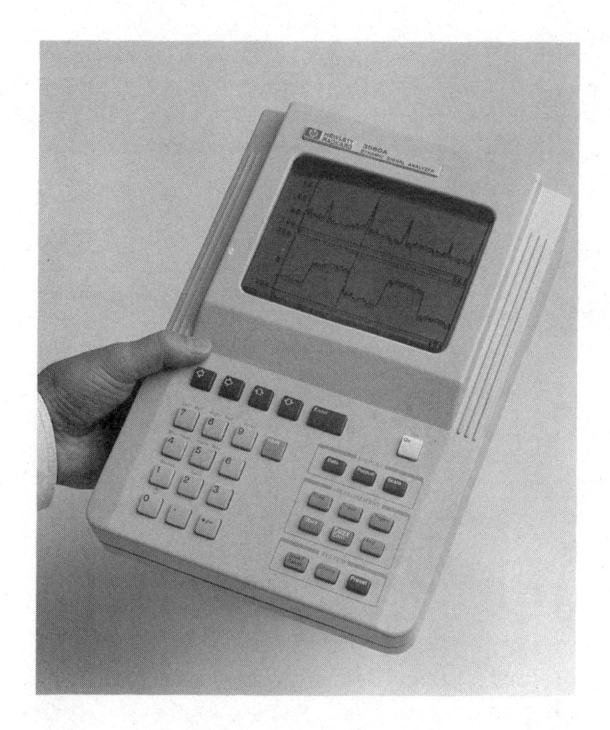

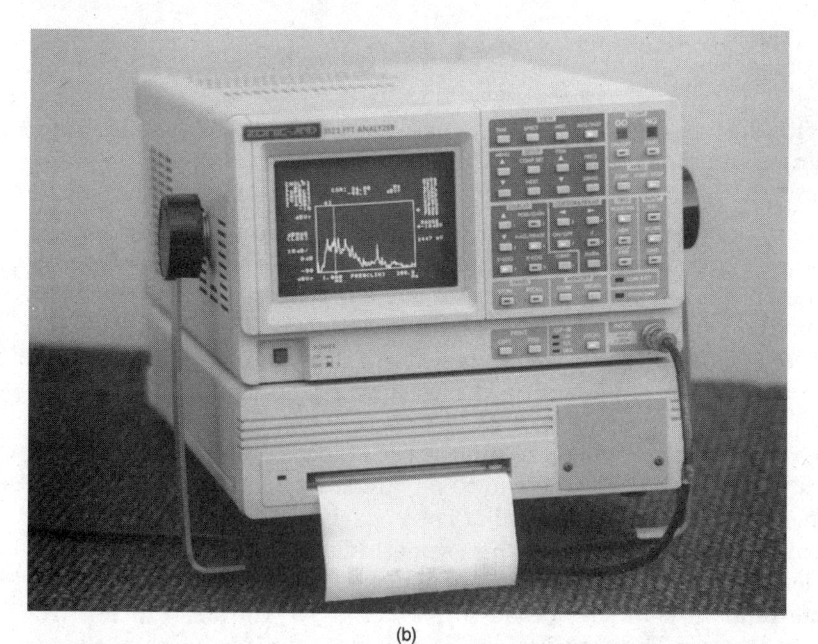

(b)

Figure 4.25 FFT spectrum analyzers. (*Courtesy Hewlett-Packard and Zonic A&D Company.*)

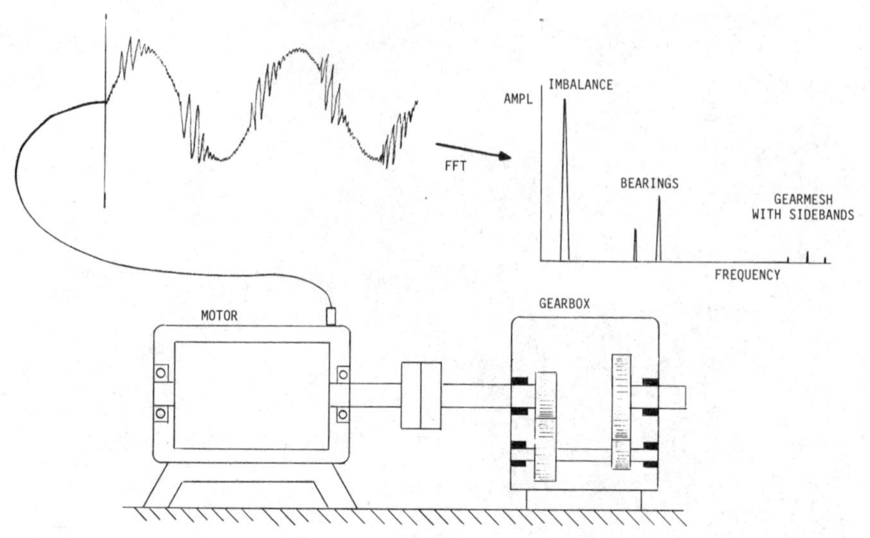

Figure 4.26 Amplitude and frequency views of a motor and gearbox.

There are equally important reasons for digitizing the data. The benefits of digital processing are:

1. Accurate measurements can be made.
2. Transient events can be captured.
3. With data in digital form, data can be stored, recalled, and manipulated in other ways.

The remainder of this chapter will cover some fundamental aspects of the FFT spectrum analyzer and some features. This knowledge will enable an analyst to set up and use one effectively.

The frequency domain is just another way to look at the same data. All the original information is preserved in the transformation. Some examples of different signals and how they appear in the time and frequency domains are in Fig. 4.27.

The sine wave is typical of a pure imbalance. The frequency is the running speed. A pure sine wave is a single line in the frequency domain. A square wave is typical of nonlinearities as when the machine bumps up against its stops when something is loose. Notice the harmonics of running speed in the frequency domain. Transients, such as metal-to-metal contacts, produce a widening of the base in the frequency domain. The amplitude rises quickly, then decreases as the vibration dampens out.

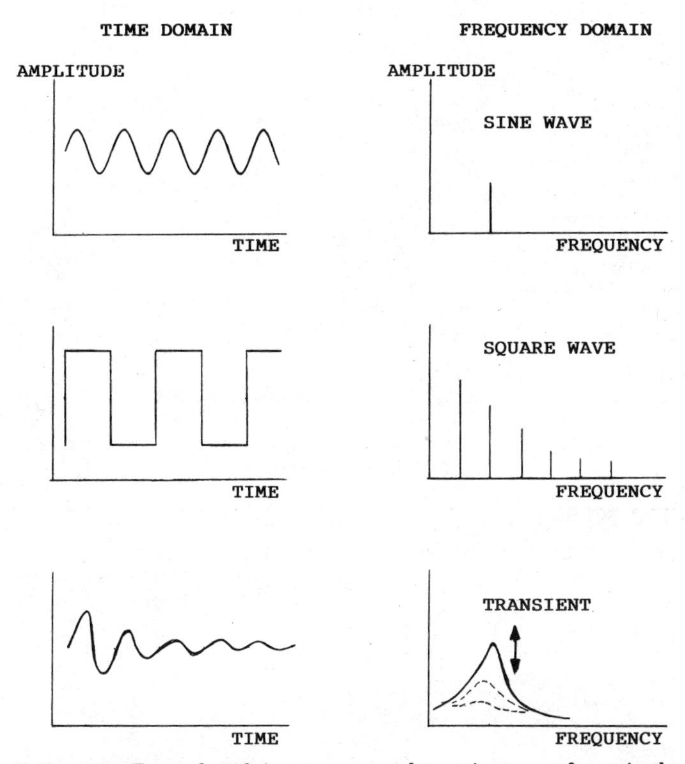

Figure 4.27 Examples of sine, square, and transient waveforms in the time and frequency domains.

Sampling, Aliasing, and Windows

Figure 4.28 is a simplified block diagram of an FFT spectrum analyzer. This diagram shows that the input waveform, a simple sine wave in this case, is first amplified, then put through an antialias filter, which is a low-pass filter. The purpose for this will be evident shortly. A time block of data is captured and held in the sampling process while an analog-to-digital (A/D) converter digitizes the data. In digital form, it can be displayed as a time waveform or further processed into a frequency spectrum. To accomplish the latter, a window function is applied to remove errors, the FFT transformation is done, and the resulting data can be displayed in the frequency domain. In this section, we will discuss these processes in more detail. Let's first look at the sampling and digitizing functions.

The analyzer cannot perform the Fourier transformation continuously. It has a dedicated internal computer that is very fast to perform these calculations, but it must look at a time block of data. This is the purpose of the sampling section. The sampling section holds the am-

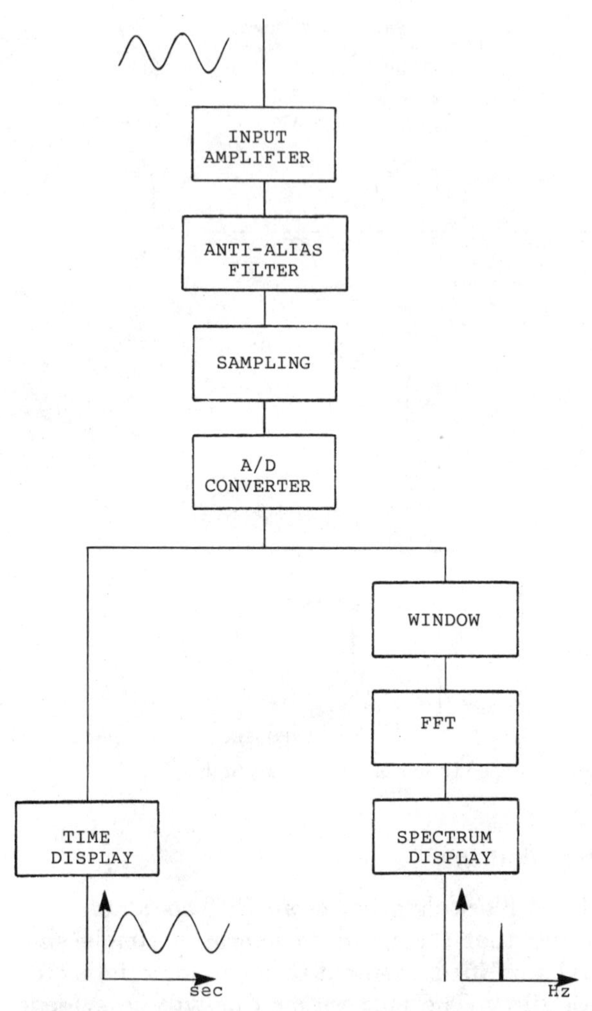

Figure 4.28 Simplified block diagram of an FFT spectrum analyzer.

plified and filtered waveform in short-term memory while voltage readings are taken. The voltage readings are taken to convert the time waveform into a table of numbers. Figure 4.29 illustrates this sampling and digitizing process for a simple sine wave. When complete, the time block of waveform data resides in computer memory as a table of numbers that contains both amplitude and phase information. This is necessary because the fast Fourier transform is a digital mathematical process that operates with numbers.

The typical spectrum analyzer takes 50,000 voltage readings per second, or more, on the input waveform. The analyzer measures this

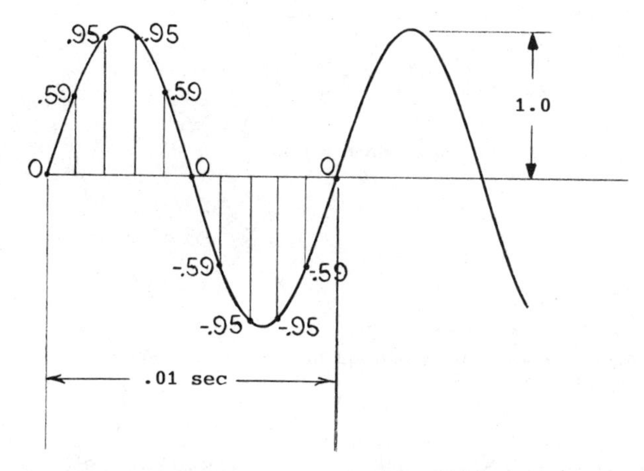

VALUES STORED IN ANALYZER AFTER A/D CONVERSION

0
.59
.95
.95
.59
0
-.59
-.95
-.95
-.59
0

Figure 4.29 Digitizing a 100-Hz sine wave. A voltage reading is taken every one-thousandths of a second, or 10 measurements per cycle.

fast to overcome a problem of being able to track high-frequency signals. An analyzer that has a frequency range of 0 to 20,000 Hz must measure the voltage of the input signal at least twice as fast (or 40,000 Hz) to be able to see changes at 20,000 Hz. This is the *Nyquist criterion,* that is, more than two measurements per cycle.

High-frequency signals can form false peaks in the frequency domain. This is called *aliasing.* This is a byproduct of the digitizing process. Look at Fig. 4.30*a,* a plot of temperature variation. Suppose that this was the variation of temperature in your room over a 7-day period. You can see daily peaks and troughs as the sun rose high, or your heating and/or cooling systems operated according to their programmed instruction. If you took a measurement once a day at about the same time, you would read about the same temperature, and your conclusion might be that the temperature in your room is constant and never changes. The measuring rate of the data is exactly equal to the frequency of variation, and the result is a constant value as plotted in Fig. 4.30*b.*

Now suppose that you took measurements of the data at a faster

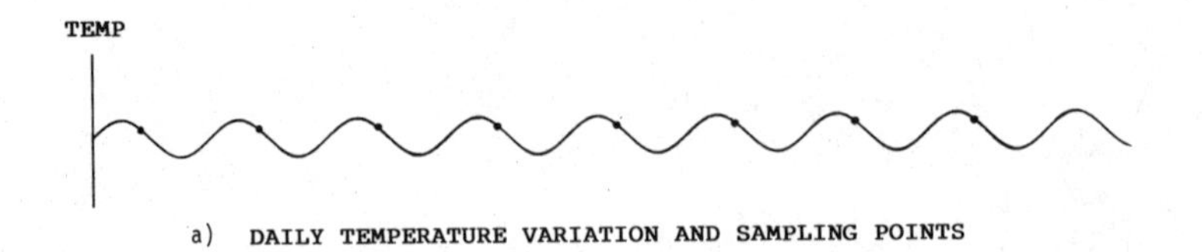

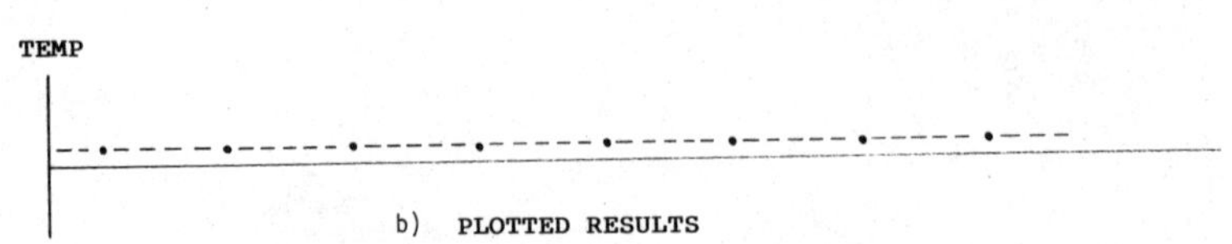

Figure 4.30 (*a*) Measurements taken at the same frequency as the variation. (*b*) The result is a constant value.

rate but still slower than the Nyquist criterion of more than 2 measurements per cycle. Assume that you took 12 measurements in 9 cycles, as in Fig. 4.31*a*. The result is plotted in Fig. 4.31*b*. It would appear to you that the signal is much slower, or lower in frequency, than it actually is.

This is what happens when the measuring rate is too slow. High-frequency signals get mirrored into the lower-frequency range. They appear as vertical lines in the frequency domain of an FFT spectrum analyzer. In reality, they are phantoms and not real data. They are the ghost images of high-frequency data that is input to the analyzer, when the measuring speed is too slow. Figure 4.32 illustrates this effect.

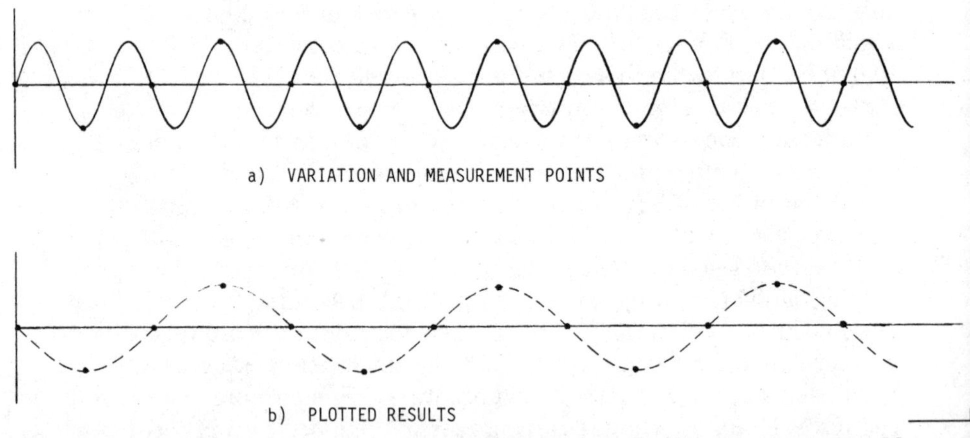

Figure 4.31 Measurements taken at a rate which is less than the Nyquist criterion. (*a*) Measurements taken at 4⁄3 the rate of variation. (*b*) The result is an apparent signal that has a frequency less than the real rate of variation, 1⁄3 the true frequency in this case.

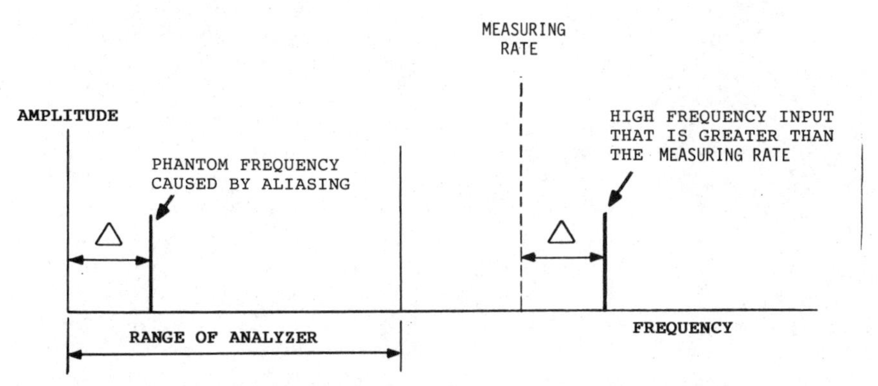

Figure 4.32 The effect of aliasing: Phantom peaks are created in the frequency display.

A good example of aliasing is using a strobe light to stop high-speed motion. The strobe flashing rate is the measuring speed. When the flash rate equals the rotational speed, you can stop the motion. This is analogous to a constant value. By varying the flash rate slightly, you can make the motion move forward or backward in slow motion. This is the phantom image of low-frequency data due to high-speed motion when the measuring speed is too slow. Another good example is old time western movies where the spokes of wagon wheels appeared to roll backward at slow speeds. In this case, the camera frame speed was too slow, and again, false low-frequency data appeared.

Aliasing must be overcome by two methods. The first is to measure the input data at more than twice the highest frequency of interest. That is, if the analyzer frequency range is set at 20,000 Hz, then the measuring speed should be more than 40,000 measurements per second, or 40,000 Hz, to avoid false low-frequency data. The second method, used in conjunction with the first, is to limit the frequencies that can enter the analyzer. Low-pass filters are applied to the input waveform to remove high-frequency data that could cause phantoms. These are called *antialias filters.* The person pushing the buttons on an FFT spectrum analyzer has the capability to turn the antialias filters on or off. There are situations where it is desirable to have them off. For example, the antialias filters may be off in the time domain so that high-frequency data is not missed. But in the frequency domain, the antialias filters are almost always on. Figure 4.33 illustrates the consequences of not using the antialias filters in the frequency domain.

Figure 4.33*a* is a spectrum of the vibration taken on a precision gyroscope. This spectrum is from 0 to 2000 Hz, and some prominent peaks are displayed at 880, 960, 1000, and 1630 Hz. Note the small

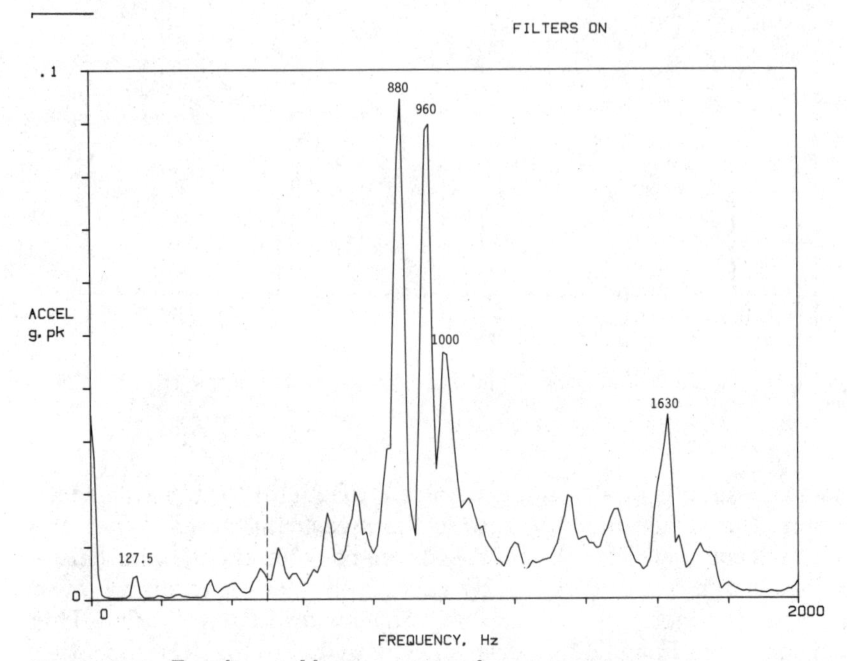

Figure 4.33(*a*) Two thousand-hertz spectrum of a gyroscope.

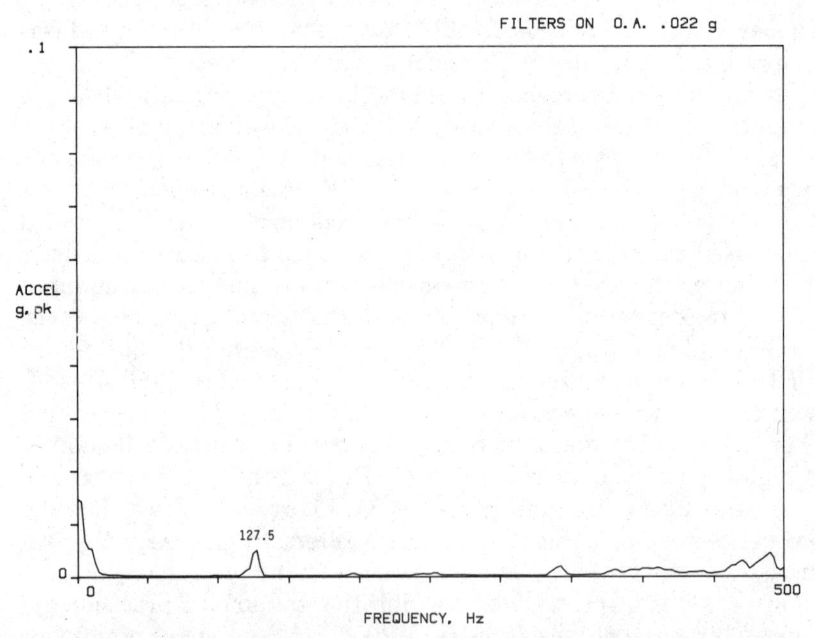

Figure 4.33(*b*) Five hundred-hertz spectrum of a gyroscope with antialias filter on. Overall is 0.022 *g*.

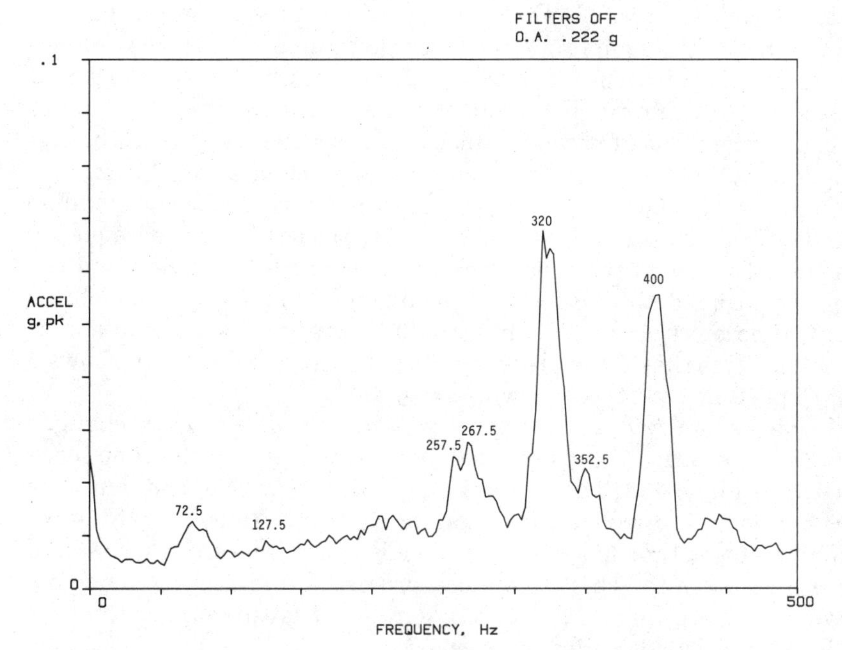

Figure 4.33(*c*) Five hundred-hertz spectrum of a gyroscope with antialias filter off. Overall is 0.222 *g*.

peak at 127.5 Hz and the dashed line at 500 Hz. The next two spectrums, Figs. 4.33*b* and 4.33*c*, were taken on the same precision gyroscope with the antialias filters on and off, respectively. In Fig. 4.33*b*, notice that the peak at 127.5 Hz is present, as expected, and the overall value of 0.022 *g*. In Fig. 4.33*c*, the antialias filters have been turned off. The peak at 127.5 Hz is hardly visible, the overall is now 0.222 *g*, and some significant new peaks have appeared. These new peaks are ghost images of the vibration above 500 Hz when the antialias filter is disabled. They are not real data. The important point to be made with this example is that the overall shows a dramatic change of 1000 percent with the simple error of taking two measurements with the antialias filters configured differently. There was no change in the machine vibration. The implications for machine monitoring are obvious. The antialias filters should always be in the same configuration—preferably on—when trending machines using an FFT spectrum analyzer. Otherwise, changes could be due to measurement errors and not any real change in machine condition.

There is another property of FFT analyzers that the operator needs to be aware of. This is windowing. Recall that the digital sampling takes place during discrete blocks of time. The FFT analyzer assumes that the signal present in that block of time is present for all time before and after the sampling. This block processing is necessary in real-world data acquisition. We cannot wait for infinity to see if the signal is periodic to infinity. The Fourier transformation, however, makes this assumption, and it can produce an error called *leakage* if the signal does not match up at the beginning and end of a time block.

Figure 4.34*a* shows a sine wave that is matched at the ends, and no window function is required to process a sharp peak into the frequency domain. The end of the time block shows a sine wave at zero and rising. If continued, this wave could wrap around to the beginning of this same time block and the signal would be continuous with no abnormalities. This time block processed into the frequency domain shows a sharp peak at 130 Hz as shown in Fig. 4.34*b*.

If the frequency from the signal generator is now slightly changed to 120 Hz, the signal does not match up at the ends of the time block shown in Fig. 4.35*a*. The signal is close to zero at each end, but they are not going in the right direction for a smooth transition. If the beginning of this time block was placed adjacent to the end, there would be a discontinuity. This time block processed into the frequency domain shows a peak at 120 Hz, but significant broadening at the base (Fig. 4.35*b*). This is called *leakage*.

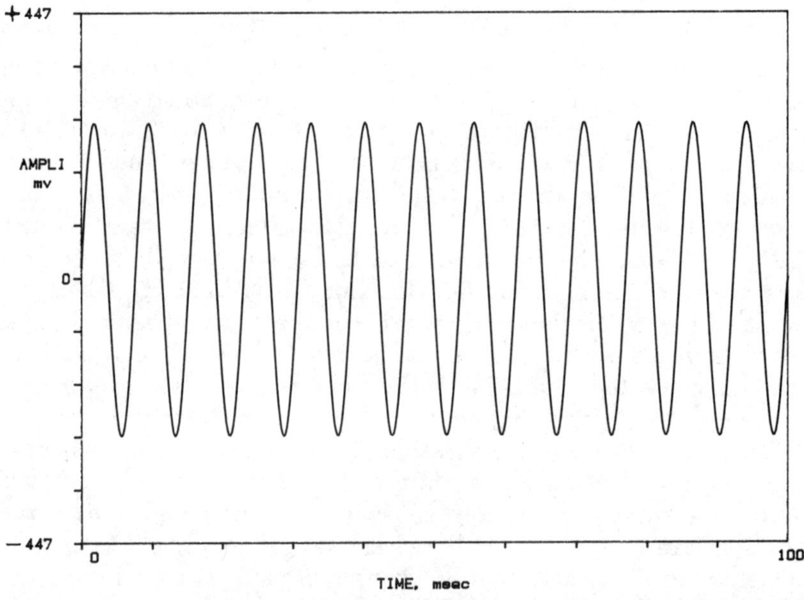

Figure 4.34(*a*) A time block capture of a 130-Hz sine wave where the signal is matched at the beginning and end.

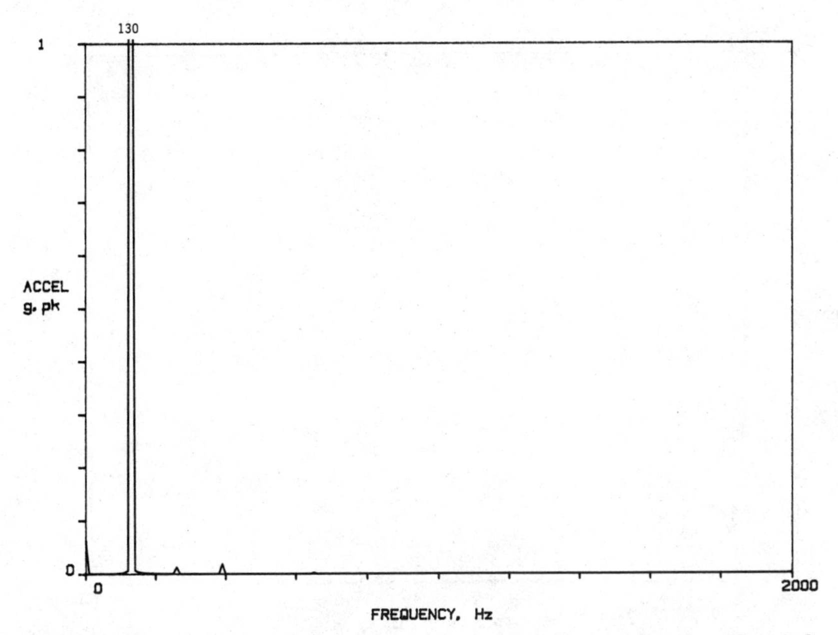

Figure 4.34(*b*) A sharp peak is processed into the frequency domain. (No window function applied.)

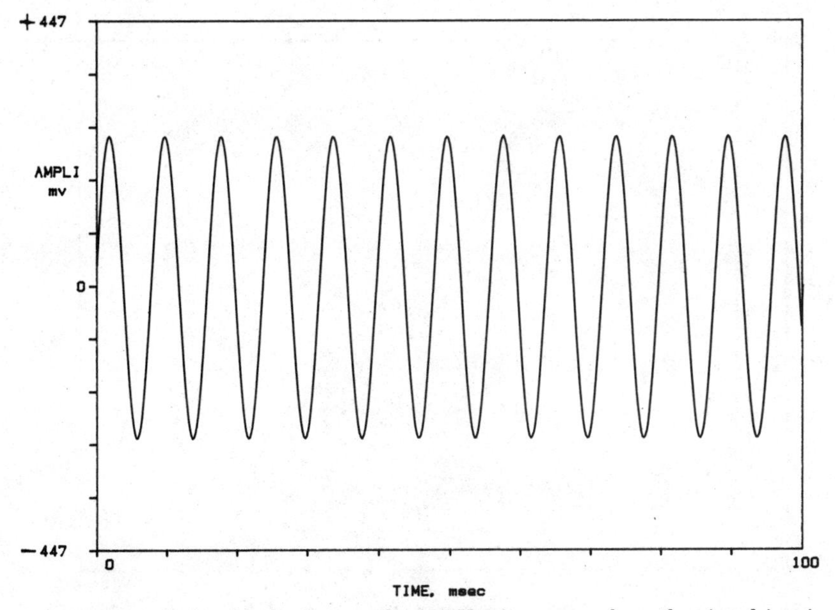

Figure 4.35(*a*) A time block capture of a 120-Hz sine wave where the signal is mismatched at the beginning and end.

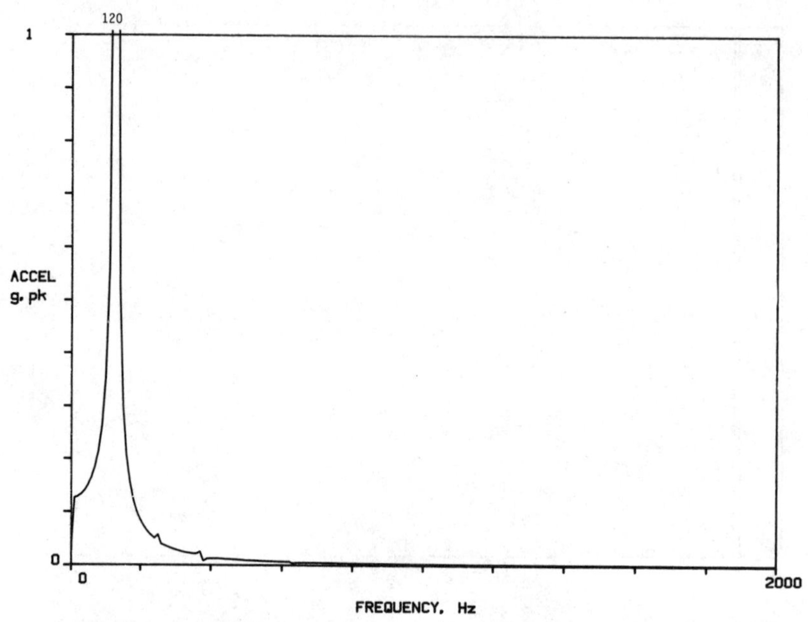

Figure 4.35(*b*) The 120-Hz sine wave processed into the frequency domain with no window function.

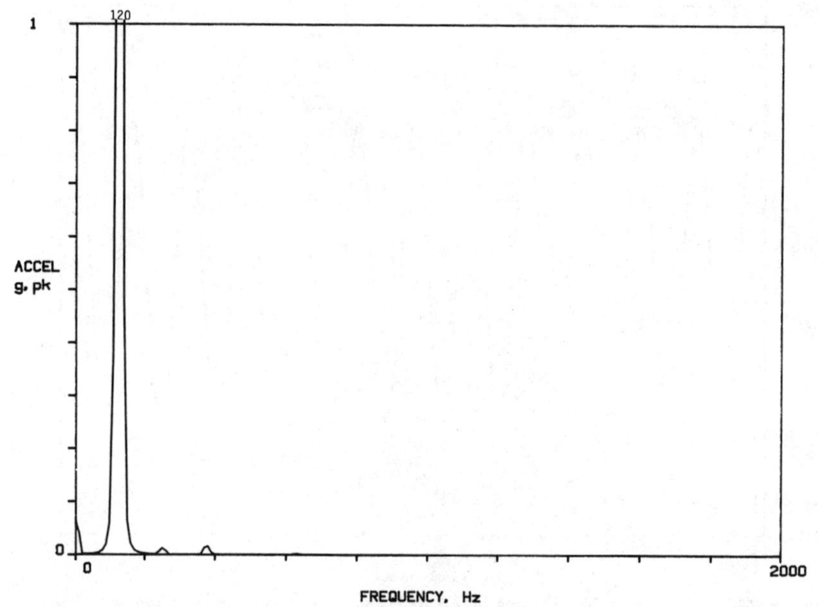

Figure 4.35(*c*) The 120-Hz sine wave with a Hanning window.

This leakage is undesirable. It hides low-level signals making the spectrum analyzer less useful. To avoid this leakage, the input data is weighted with a mathematical function that favors the data in the center of the time block and reduces the data at both ends to zero. Since the data is now zero at both the beginning and end, it will match up properly and cause no leakage. For vibrations that are truly repetitive, i.e., the same vibrations appear in the middle as well as the ends of the time block, windowing will not cause significant errors. There will be a slight decrease in amplitudes and a slight loss in frequency resolution, but these errors are less objectionable than leakage. Figure 4.35*c* shows the same 120-Hz sine wave that is mismatched at the ends, but now with a Hanning window applied. Notice that the leakage is substantially reduced.

There are times when the operator may want the window function turned off, or a different window function applied. The operator has control. There are three useful window functions:

1. *Hanning window:* The most common window for typical measurements during machine monitoring
2. *Uniform window:* For transient events and maximum frequency resolution (also called *rectangular window*—the rectangular window is actually no window function at all, and causes maximum leakage)
3. *Flattop window:* For maximum amplitude accuracy

These problems of aliasing and leakage are inherent in all FFT spectrum analyzers and are due to the digital processing techniques. These problems have been overcome with antialias filters and window functions. There are times when it may be desirable to have the antialias filters and windows turned off to get a different view of the data. The instrument operator has the control to turn them off. If the antialiasing filters are inadvertently turned off, or a different window is used, then the operator should be aware that he or she has manipulated the data and the results will be different.

Averaging

Now that we have covered some of the idiosyncracies of FFT spectrum analyzers, let's move on to discuss some of their more useful features. One very useful feature of these instruments is averaging. Averaging is the ability to combine time records (blocks of data) with previous data to smooth out the display. This is very useful, since vibration data is not usually stable—it is always changing. Averaging allows you to smooth out the spectrum with data over a long period of time—thus statistically averaging it.

Figure 4.36 is an instantaneous spectrum of a gyroscope—a snapshot. The overall is 0.25 *g*. The very next snapshot, a second later, would be different. The overall can vary plus or minus 25 percent from one snapshot to the next. It would be similar with the same major peaks present, but the low-level random vibration would be very different. Averaging eliminates this low-level random noise, and the amplitudes are representative of stable operation.

Figure 4.37 is the same point on the same gyroscope. Here the spectrum display is actually 32 spectrums averaged. This averaging took place during a 20-sec period while the transducer was held on the gyroscope. In this averaged spectrum, the random noise is eliminated, and the remaining peaks in the spectrum are truly representative of the vibration in steady-state operation. The overall is 0.21 *g*, and this value is also repetitive. If nothing changes on the gyroscope, then 1 h later, or 1 month later, I should be able to acquire a very similar averaged spectrum.

The summation averaging mode is the one normally used for machine monitoring. Eight to sixteen averages are all that is necessary to get a stable spectrum, and this normally takes less than 10 sec to acquire. For periodic monitoring, as in predictive maintenance programs, it is strongly recommended to do summation averaging.

A special form of averaging is peak hold. This keeps the maximum signal amplitude at every frequency. This feature of peak-hold averaging is very

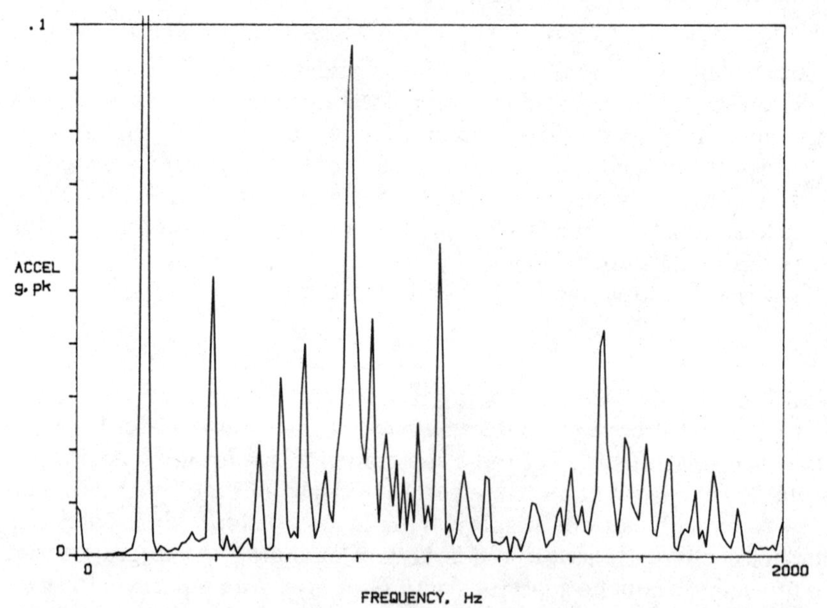

Figure 4.36 Instantaneous spectrum of a gyroscope (a different gyroscope than Fig. 4.33). Overall 0.25 *g*.

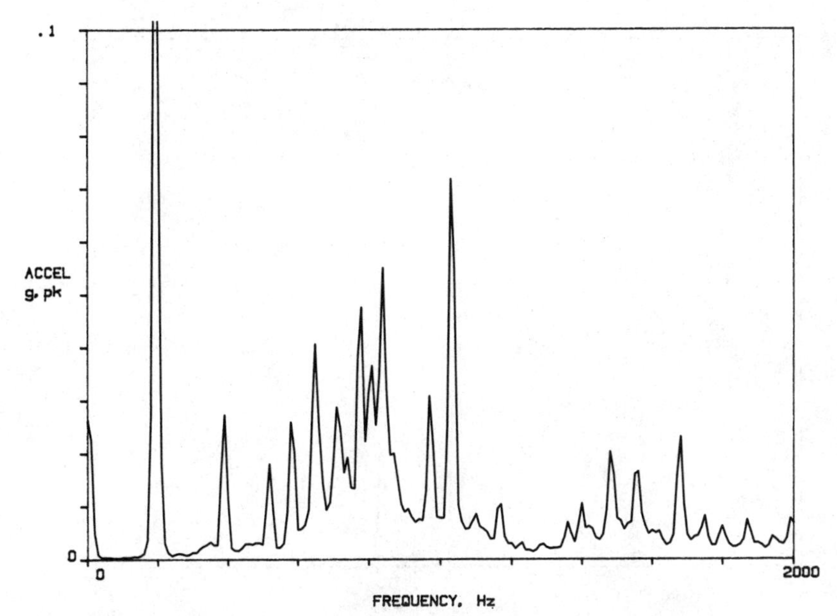

Figure 4.37 Averaged spectrum of the gyroscope of Fig. 4.36. Thirty-two spectrums, summation averaged, during approximately 20 sec. Overall 0.21 *g*.

useful when looking for the maximum vibration over a long period of time. The analyzer can be set up and left in this peak-hold averaging mode at a specific location. You can come back 24 h later and see what maximum vibration occurred during that time period and read out its frequency components. You will not know from this display when the peak occurred, but you will know its amplitude and frequency. In this way, the FFT analyzer can be used as a vibration recorder.

A third common form of averaging is exponential averaging. This weights the most recent data most heavily. This is useful during changing conditions, like speed changes, when it is desired to know when conditions stabilize, and the values at those conditions.

Zoom

Zooming is the capability to expand a portion of the spectrum, which is useful in separating out closely spaced frequency components. The two spectrums in Fig. 4.38 illustrate the usefulness of zoom. Figure 4.38*a* shows a peak at 1470 rpm and another one to its left smaller in amplitude. By zooming into a narrower frequency span, the closely grouped peaks are further separated. The zoomed spectrum in Fig. 4.38*b* shows the smaller peak to be at 1398 rpm. The peak at 1785 rpm is also seen split into two separate frequencies. Zooming is also useful as a tachometer to measure speed accurately.

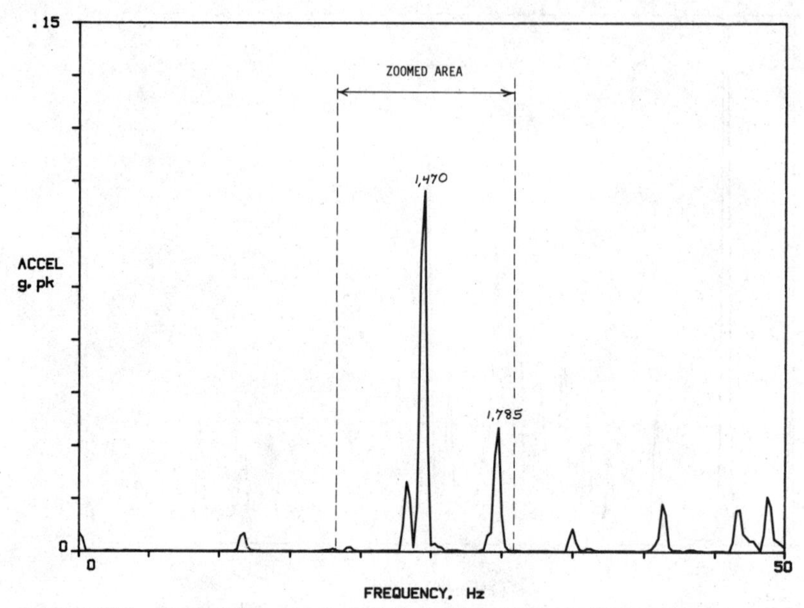

Figure 4.38(a) A normal spectrum from 0 to 50 Hz of a belt-driven rotor. The motor turned at 1785 rpm (29.75 Hz), and the rotor turned at 1470 rpm (24.5 Hz).

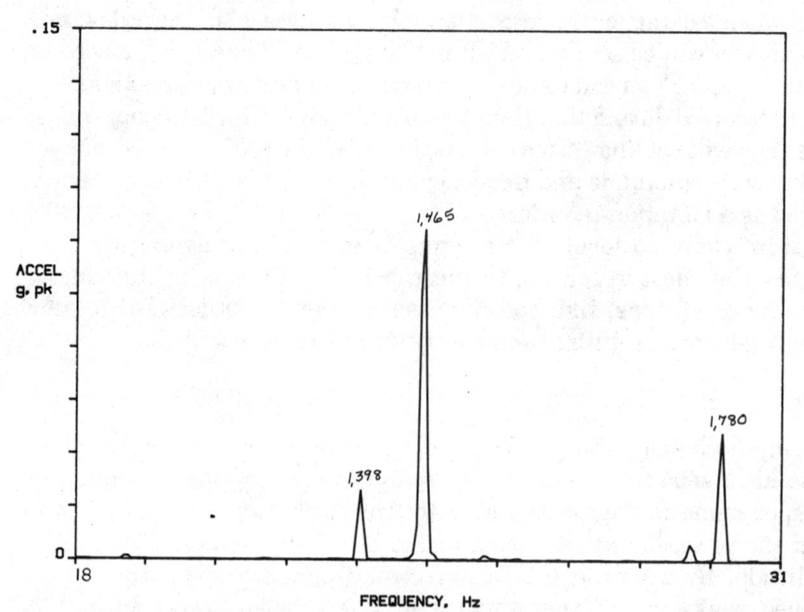

Figure 4.38(b) Zoomed spectrum 18–31 Hz.

Linear and Log Scaling

There is another data manipulation that the FFT analyzer can do because the data is in digital form. This is the capability to display the data in logarithmic scales for both the amplitude and frequency axis.

The FFT analyzer has the capability to measure large and very small signals simultaneously. With a linear amplitude scale, the large and small signals cannot be displayed simultaneously. The solution is to change the scale. The large signals are compressed and the small ones are expanded upward to bring both into view at the same time. The scale used is a logarithmic amplitude.

Figure 4.39*a* is a linear amplitude plot of a strong 90-Hz vibration with harmonics. The 90 Hz is above the scale, and there are some low-level vibrations at the bottom at 30 and 120 Hz. Figure 4.39*b* is the same vibration, now in a logarithmic amplitude display. The 90 Hz is well within the display. The 30- and 120-Hz vibrations are more clearly visible, and there are now many other low-level vibrations that have been raised into the display which were not visible in the linear scale plot.

This is a good place to sidetrack temporarily to discuss the logarithmic scale. The unit bel is named after Alexander Graham Bell who

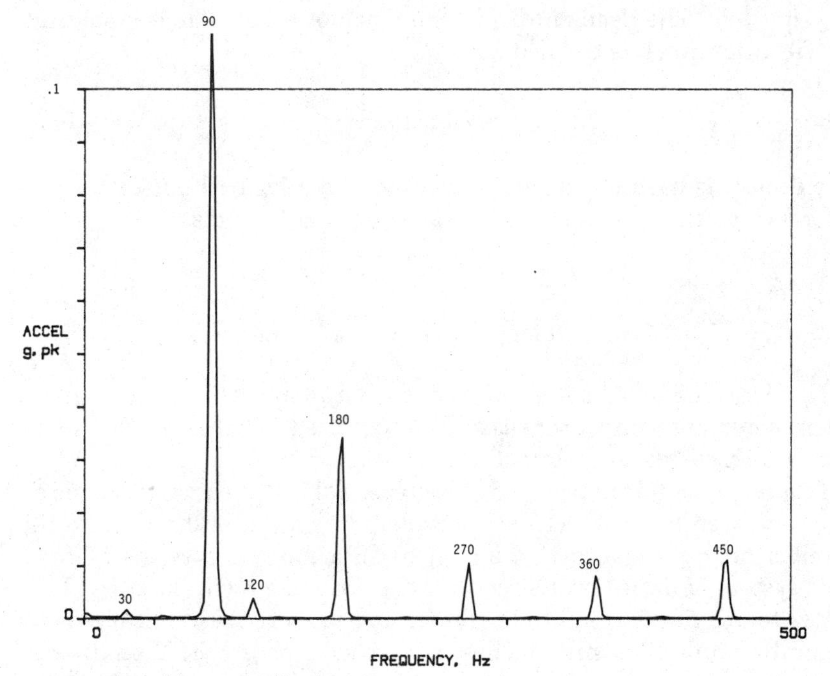

Figure 4.39(*a*) Linear amplitude.

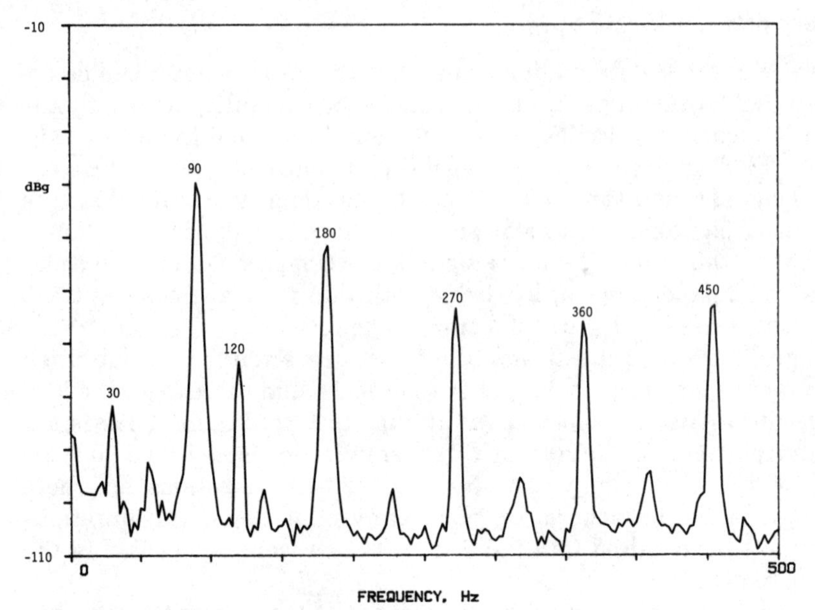

Figure 4.39(*b*) Logarithmic amplitude (0 dB_{REF} = 10 *g*).

discovered that the human ear responds logarithmically to sound power levels. The decibel (dB) is one-tenth of a bel. The decibel unit for vibration work is defined as:

$$\text{dB} = 20 \log_{10} \frac{V}{V_{\text{ref}}} \tag{4.1}$$

The decibel is basically a ratio of voltages on a logarithmic scale. The reference voltage on most modern analyzers is 1 V rms.

A voltage of 0.01 V rms is −40 dB

$$\text{dB} = 20 \log_{10} \frac{0.01}{1} = 20(-2) = -40$$

The data in Fig. 4.39 is displayed in *g* units and dB*g* units. The accelerometer used has a sensitivity of 100 mV/*g*. Therefore, the reference *g* unit corresponding to 1 V is 10 *g*.

To gain some familiarity with decibels, Table 4.2 shows voltage ratios compared to their decibel equivalents. Some useful values to remember are 6 dB is a ratio of 2 to 1; 10 dB is a little more than 3 to 1. This system of decibel volts requires considerable mental agility. It is better to use linear amplitude scaling and convert the vertical axis to a familiar unit like mils, inches per seconds, or *g*'s. FFT analyzers, with their digital processing and math capabilities, can do this very

TABLE 4.2 Decibel Values and Corresponding Voltage Ratios

Decibels	Ratio
1	1.12
2	1.26
3	1.41
4	1.58
5	1.78
6	1.99
7	2.24
8	2.50
9	2.82
10	3.16
15	5.63
20	10.00
40	100.00

easily. The vertical amplitude axis is redefined to read in engineering units. The operator needs to know the sensitivity constant for the transducer in use.

The frequency axis can also be displayed in logarithmic units. This expands the low-frequency part of the spectrum and compresses the high-frequency part. Figure 4.40 is the same vibration data as in Fig. 4.39, but here displayed with both log amplitude and log frequency

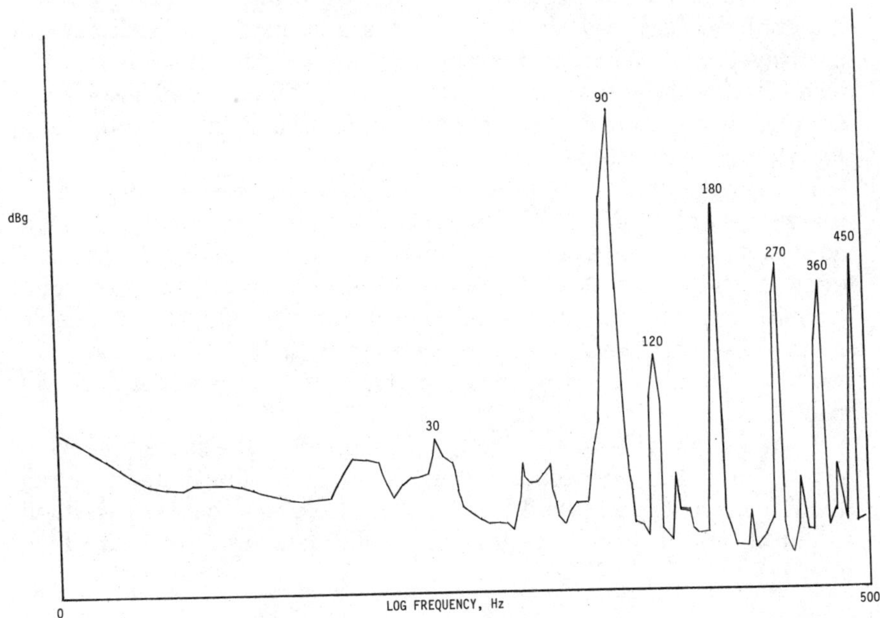

Figure 4.40 Logarithmic frequency axis of the same data of Fig. 4.39(*b*).

axis. Logarithmic frequency displays are favored by the military, and some old swept filter analyzers plot their data like this because that is how the filter sweeps—logarithmically. It is a clumsy display to use for machine monitoring.

Triggers

FFTs have the capability to start the sampling of a block of data at any time you select. This is done with an external or internal trigger. An external trigger is input from a separate sensor via a connection on the back or side of the instrument. This separate sensor is typically a Keyphasor and is useful for obtaining phase information. The internal trigger can be used to trigger on the signal itself, similar to the trigger on an oscilloscope. Figure 4.41*a* shows a transient signal that triggers on itself when the signal level rises to a value that the operator selects. Figures 4.41*b*, 4.41*c*, and 4.41*d* illustrate the use of pretrigger, normal timing, and posttriggering to set the delay timing. This feature is useful to position a transient event in a convenient location on the screen for viewing and recording.

Figure 4.42 is a time capture of the shock pulses of a plastic hammer impacting an aluminum block. The accelerometer was attached to the aluminum block with a stud mounting. An internal trigger was used with a level selected that would keep the analyzer in a hold state until an impact was detected. Every time the aluminum block was hit, the analyzer captured a new time block and displayed fresh data. A pretrigger was used to position the data to the right 2 msec so that the leading edge of the first impact was clearly visible. The FFT spectrum analyzer in the time domain is, and can be used as identical to, a digital storage oscilloscope.

The rectangular window is used to view transients. The window function is only useful in the frequency domain. The rectangular window is one with no windowing function at all. That is, all the data, even at the extremes of the time block, are processed with equal weighting. If the transient is positioned in the middle of the time domain view, with zero signal at the extremes, then the rectangular window will allow the frequency spectrum of this transient to be calculated with accuracy.

If the input signal is averaged and triggered at the same time while displaying in the time domain, then the result is time domain averaging. This is a very powerful technique for looking at gears. This application of time domain averaging will be discussed in Chapter 6, "Techniques."

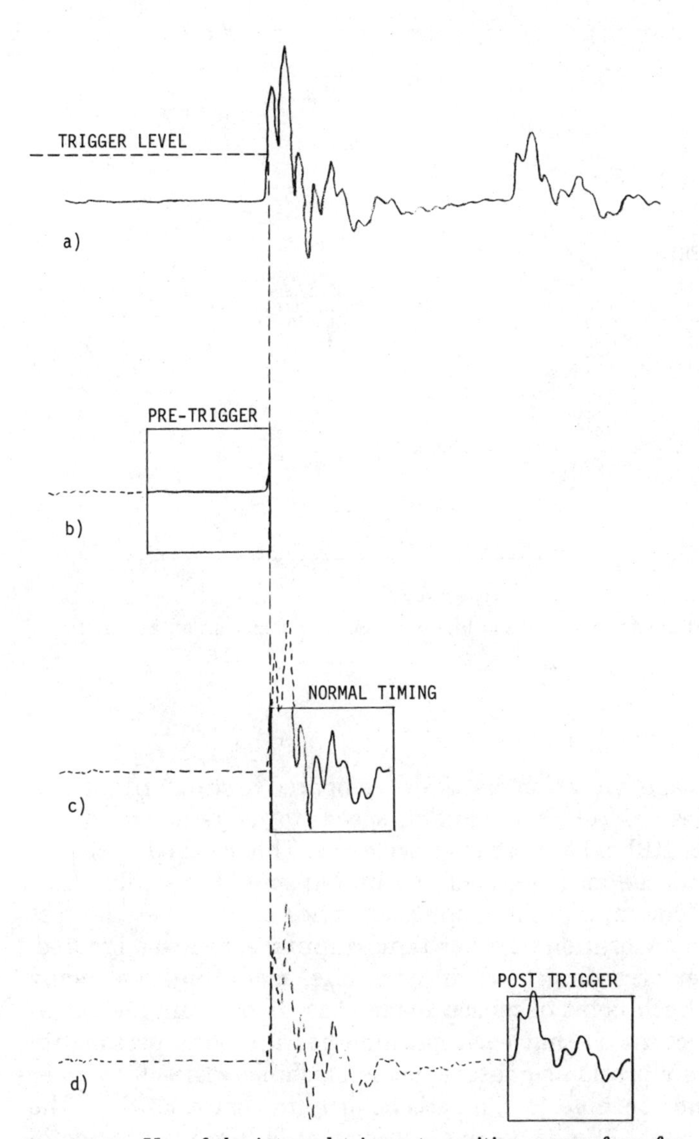

Figure 4.41 Use of the internal trigger to position a waveform for viewing.

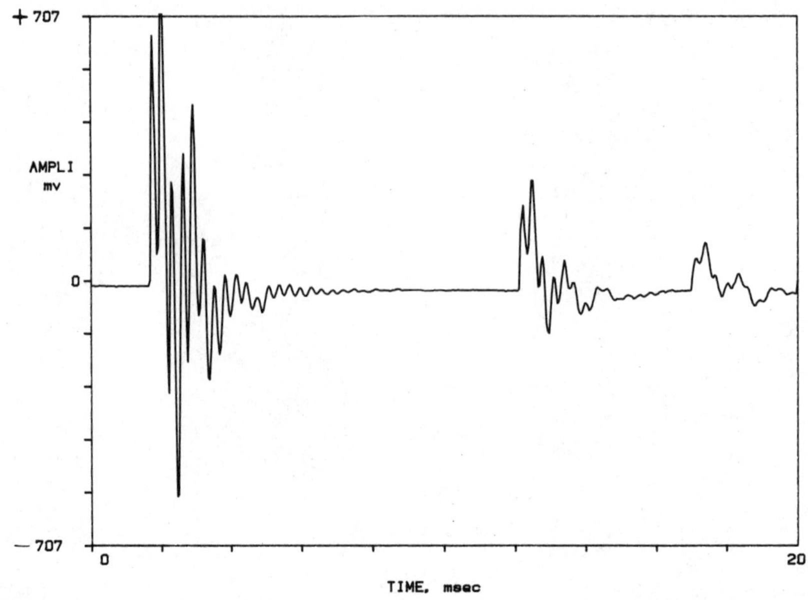

Figure 4.42 Initial impact and rebound hits of a plastic hammer on an aluminum block.

Comparator

Initially, years ago, vibration data were compared to standards at specific frequencies (usually the running speed) to judge acceptability. Such standards still exist and are in wide use. This method of single-frequency amplitude measurement at running speed is usually a fair criteria for judgment, but the amplitude varies locally over the machine, and every vibration at other than running speed was ignored. Machines differ according to their type, size, mounting, and many other factors. Each point of measurement has its own unique signature. It was recognized that each machine has a unique personality embodied in its vibration signature. Some machines start off rougher than others and continue to run satisfactorily in that condition. The important precursor of a defect and subsequent failure is a change in personality, or signature. Paper plots of these signatures were made, and the paper plots were overlaid on each other. The idea was that if the spectrums did not change, then neither did the machine. This was a wonderful idea and a good application of statistical process control to the maintenance of machinery. However, with many machines and/or frequent monitoring, an analyst could get buried in paper plots. To relieve this paper burden, the idea of defining a "good" envelope signature was started.

A "good" envelope was a bar graph of the maximum allowable vibration amplitude at each frequency. In simplified form, this is shown in Figs. 4.43 and 4.44 with the two spectrums of two separate gyroscopes of identical make and model. The measurement point on each gyro was also identical.

The running speed of both gyros is 11,400 rpm, or 190 Hz. Figure 4.43 is of a good, smooth-running gyro. The blocked area defines the maximum allowable vibration amplitude at running speed to judge imbalance. The smooth gyro would pass, or give a "go" indication, because the running-speed peak is below the limit. Figure 4.44 is of a rough-running gyro. The two gyros are from the same manufacturer, only a few months apart in production sequence. The rough gyro is obviously out of balance. Its amplitude at running speed punctures through the top of the block, and it fails, or gives a "no-go" indication. This example is of one frequency in the spectrum. Many other limits could be defined at other frequencies to "envelope" the entire spectrum if desired. The comparison can then be carried out electronically, or digitally. This relieves the analyst of overlaying paper plots looking for changes.

Some FFT spectrum analyzers have this comparator function built in, others implement it in a computer software that downloads data from the analyzer. This comparator function has tremendous potential for manufacturing quality control of mechanical equipment. Every de-

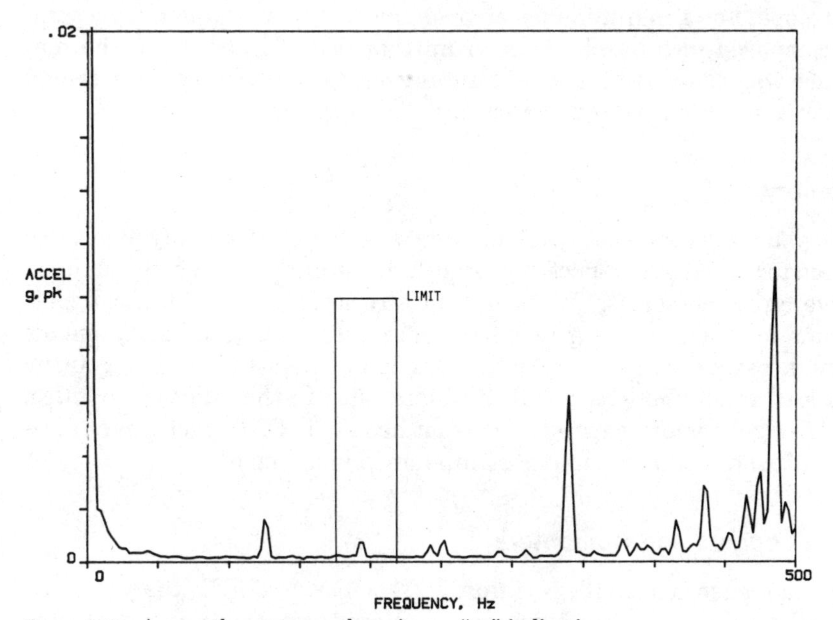

Figure 4.43 A smooth gyroscope that gives a "go" indication.

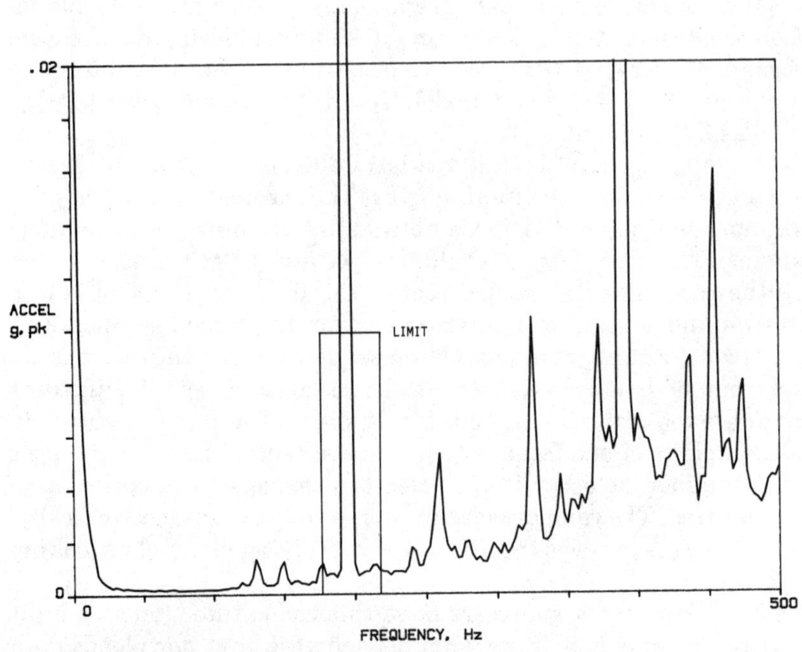

Figure 4.44 A rough gyroscope that gives a "no go" indication.

vice can be quickly checked and compared to a standard defined by the envelope. For a maintenance environment, each subsequent spectrum is compared to a baseline spectrum that was defined when the machine was in a known good running condition. Alarms are issued when a significant change occurs.

Memory

Ten years ago, the best spectrum analyzers could store only about two spectrums. All FFT spectrum analyzers recently put on the market have large memories. They can typically store several hundred spectrums for later plotting or downloading. This large memory means that a person can collect data from many points during a routine walk-through and look at all that data later in the quiet of an office. This large memory capability is what has made FFT spectrum analyzers so useful as a predictive maintenance instrument.

Data Loggers and Computers

A data logger is a small, portable, battery-powered FFT analyzer used in vibration monitoring programs. Sometimes there is no display screen. The data logger is typically strapped over the shoulder as a technician makes his or her rounds to collect vibration signatures. Attached to the data logger is a transducer on a coiled cord. The transducer is usually hand-held. Figure 4.45 shows a typical data logger.

The data logger is usually used with a computer. The computer programs the daily route to follow to gather vibration signatures. In this route download are the instrument's settings for each point—frequency span, sensitivity, window, number of averages, and every other setting that is required to use an FFT spectrum analyzer. The technician follows that route schedule and collects overalls or signatures from the required machines. At the end of the route, the technician returns the data logger to the computer. The computer then extracts the data from the logger, analyzes it, archives it, issues alarms if necessary, and prints a report.

The hazard of using the data logger in this way is that the man or woman taking the measurements is one step removed from the diagnostic process. He or she is actually in the best position to interpret the spectrum while standing next to that vibrating machine. With

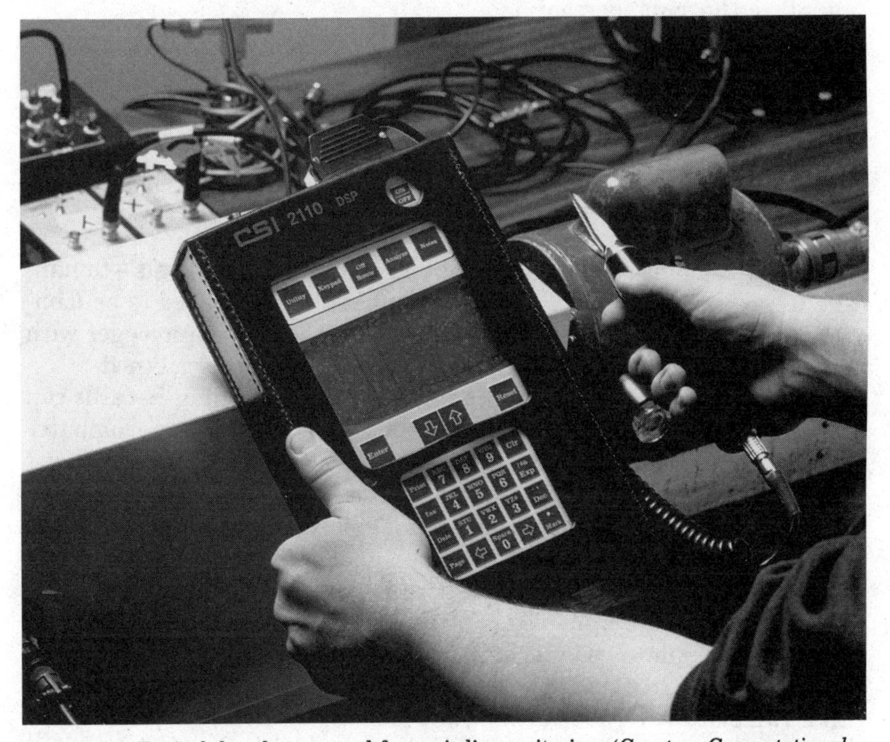

Figure 4.45 Typical data logger used for periodic monitoring. (*Courtesy Computational Systems, Inc.*)

some data loggers he or she is not allowed to even see the spectrum, and he or she feels like a vehicle used merely to transport the box around. Likewise, the analyst who sees the spectrum does not have the benefit of associating a sound and vibration with a spectrum. His or her task is more abstract. In some systems, no one sees a spectrum unless the computer analysis detects a change out of defined limits and issues a warning.

This use of data loggers and computers represents the most sophisticated level of vibration monitoring. Some manufacturers have been boasting larger memories and more data-handling capability to try to outclass their competition. Quantity does not mean a better monitoring program. These data loggers were developed to handle large amounts of data while allowing a maintenance person to operate it. In reality, it has alienated both maintenance person and analyst from the true task of analyzing vibration and relegated them both to routine data entry positions.

The most sophisticated monitoring system requires the most sophisticated type of operator. To use data loggers effectively, the vibration monitoring group must have the following skills, though not necessarily all in the same person:

Computers

Data acquisition

FFT spectrum analyzers

Machine diagnostics

This illustrates one use of computers in vibration analysis—to handle large amounts of data. If large numbers of points need to be monitored, i.e., more than 400 points per month, then the data logger with a computer and analysis software is the most practical method.

Another use for computers is as an FFT analyzer. Plug-in cards can be purchased that will convert a desktop, or even laptop, computer into an FFT spectrum analyzer. This is the least expensive way to get an FFT analyzer, but not the most trouble-free.

This completes the discussion of FFT spectrum analyzers. Figure 4.46 shows two types of reed vibrometers. These are two types of mechanical frequency analyzers. Figure 4.46*a* is a block with multiple reeds tuned to different frequencies by virtue of their lengths. The speed of variable-speed machines was regulated by keeping a specific reed vibrating in resonance. Figure 4.46*b* is a single-reed instrument that can be tuned by extending the reed out. In the 1930s prior to any electronic instruments being available, this was the state of the art in frequency analysis. Some are still in use.

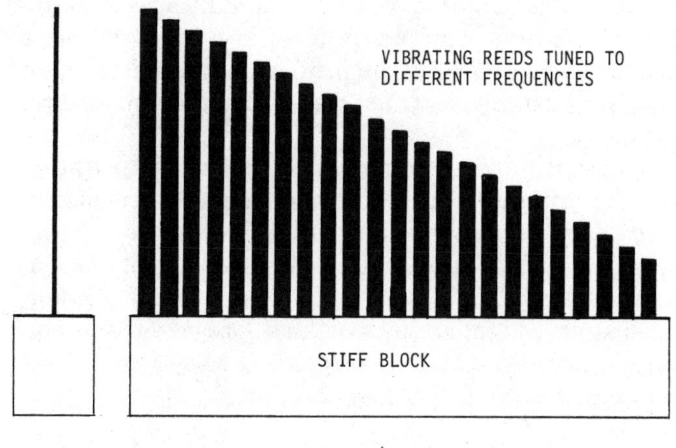

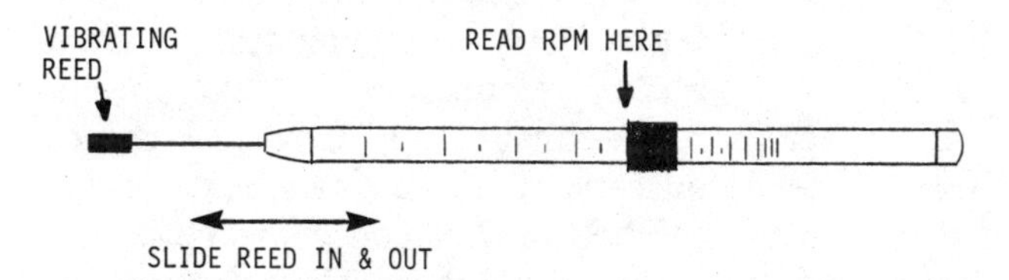

Figure 4.46 Reed vibrometers. (*a*) Multiple reed gage. (*b*) Variable-reed length gage.

Ancillary Instruments

Other instruments that have specialized purposes for vibration monitoring include:

- Laser vibrometer—a noncontacting instrument sensing reflected light
- Pressure transducers
- Force transducer
- Impact hammers
- Acoustic emission transducers and analyzers
- Seismological instrumentation

These instruments, not commonly used for machine health monitoring, will not be discussed further. The only purpose for mentioning them is to make the reader aware of their existence. There are also permanent monitoring systems that combine elements of instrumentation that have already been covered with multiplexer data acquisition hardware, shutdown controllers, and telecommunications equipment to transmit the data from remote locations. These systems raise the level of complexity and cost at least an order of magnitude above a portable analyzer system but are justified where the risks and consequences are serious.

In this section, I intend to cover some additional instrumentation that is useful in a machine monitoring environment. These instruments should be considered as supplements for any good, long-term, vibration monitoring program. They will be discussed under the following headings: stroboscope, oscilloscope, audio signal generator, small shaker, multimeter, weight scale or balance, programmable calculator, computer and printer, plotter, microphone, two-channel FFT spectrum analyzer, tape recorder, and tracking filter.

Stroboscope

As previously mentioned, the speed of rotation is the first and most fundamental number that must be known to do any meaningful vibration analysis of machines. An approximate number can be gotten from motor nameplates. For other than direct coupled ac-driven machines, it is best to measure the speed yourself. This includes variable-speed machines, dc-driven machines, and belt-driven and gear-driven machines. Mechanical tachometers can be used where it is safe to couple to the shaft. Photoelectric tachometers are both unreliable and inaccurate. A stroboscope, with an accurate digital readout, is the most accurate and reliable method to measure speed. The strobe is also useful for two other purposes. First to freeze the motion and examine the rotor without stopping the machine. By changing the flash rate slightly, it can be viewed in slow motion rotation—forward or backward. Second, the strobe can be used for measuring phase for balancing or analysis purposes. For these reasons, specify a strobe with an external trigger.

For high speeds, it is possible to measure a speed above the strobe light's flashing range. The technique is to stop the motion at two lower speeds and perform a calculation. For high-speed strobing:

$$\text{True speed} = \frac{A \times B}{A - B} \tag{4.2}$$

where A and B are adjacent flash settings below the true speed where motion is stopped with a single image.

Oscilloscope

The oscilloscope was one of the first electronic instruments used for vibration analysis, and it is still widely used today. With the oscilloscope, you can see the complex vibration signal directly from the transducer and measure the amplitude and period. With a tuneable bandpass filter and an oscilloscope, you can do frequency analysis. This instrument should be high on the priority list.

Audio signal generator

An audio signal generator is useful to generate pure sine waves, and perhaps square or triangular waves also. These signals can be input to your readout instrumentation as functional checks.

A pure sine wave can be used to drive a shaker table for calibration checks. A random vibration from a signal generator can be used for resonance testing and for a number of acoustic testing applications. A simple sine wave generator is sufficient and its cost is reasonable, but if more sophisticated testing is anticipated, then look for one with a random signal output.

Small shaker

A small shaker table is useful to calibrate your instruments from end to end, i.e., from the transducer through the readout device. At some time you will question the validity of your measurements, or worse, someone else will question them. Commercial testing laboratories and transducer manufacturers will calibrate your transducers for a fee, but this does not check your cables or readout instruments. For complete confidence it is best to calibrate your entire instrumentation system as a unit with a known vibration input. To move a shaker table, you must also have a signal generator and a power amplifier. A small portable shaker table can also be used to generate vibrations in the field at different frequencies for testing purposes.

A small shaker with a power amplifier is rather expensive. In addition, for calibration purposes you must also acquire some other instruments to measure the vibration table output—like a standard reference accelerometer, or some optical or mechanical means to measure table amplitude and frequency. This is all necessary for absolute accuracy, but not critical for machine monitoring situations. If it is recognized that a shaker table generates a known and repeatable output to compare your instruments against, perhaps you can find a known and repeatable vibration in your plant to compare against. This is not approved by American National Standards Institute (ANSI) Standards, but the purpose for machine monitoring is to detect relative changes in machinery and to have confidence that those changes are actually in the machine under test and not changes in your instruments. If you can find a machine tool, motor, or transformer that is stable in vibration, then drill and tap a place to mount your transducer, and measure its vibration under known conditions. Then, whenever you doubt your instruments, or just at regular monthly intervals, plug into that machine point and measure its vibration again. This is one way, albeit approximate, to check your instruments without spending huge dollars for a calibration system. If, however, you need to measure absolute amplitudes, such as in an equipment start-up inspection and acceptance, then your instruments need to be calibrated in an absolute manner. And again, your entire instrumentation chain should be calibrated, not just the transducer.

Multimeter

A small portable multimeter should be in your toolbox for continuity and battery checks in the field.

Scale

If you will be doing balancing, then you will need access to a precision scale. Many plants have precision scales in their laboratories, nurse's office, or mailroom. If not, then you will need to purchase one, because the numbers stamped on commercial balance weights are not accurate enough. A triple-beam balance with 1-gram resolution is sufficient for large fans. If small fans are to be balanced, then you will need a scale that can measure to fractions of a gram, that is, 0.1-gram resolution.

Programmable calculator

A calculator of some kind is necessary for a vibration analyst. A programmable calculator is extremely useful for balance calculations. Few balancers still use the graphical methods. The vast majority of balance calculations in the field are done with either a programmable, hand-held calculator, or on-board software resident in the readout instrument.

Computer and printer

Most vibration analysis and trending software is written to run on an IBM PC or compatible. Be sure to also purchase large mass storage capability if you expect to archive years of data.

The next most important reason to have a computer is for report generation. The purpose for vibration monitoring is not just for the analyst's benefit but for the entire plant. Keep people informed about your findings, and give them the satisfaction of knowing the condition of their machinery. A computer is useful for this routine administrative duty.

Plotter

Some FFT spectrum analyzers will plot their screen display directly to a Hewlett-Packard Graphics Language (HPGL) plotter. A multicolored plot on 8½- by 11-in paper makes an impact that a small 2-in strip plot on thermal paper cannot do. Most of the spectrums presented in this book were first plotted on an HPGL plotter on A size paper from data recorded in an FFT spectrum analyzer.

Microphone

It is almost a certainty that a need will arise for sound analysis; if not around machinery, then near some executive's office when the word gets around that a vibration analyst is now on staff. Any vibration measuring instrument is capable of sound measurements by replacing the vibration transducer with a microphone.

There are two basic types of microphones, electrodynamic and condenser. The condenser microphone, with a preamplifier, is superior because of its superior frequency response, especially in the low-frequency region below 100 Hz.

Two-channel FFT spectrum analyzer

More than 95 percent of all machine vibration measurements can be done with a single-channel FFT spectrum analyzer. The usefulness of a two-channel instrument for routine monitoring is almost nonexistent. However, once a problem machine has been identified and further analysis is required, then some of the features that a two-channel instrument provides can be put to good use if there is someone around who knows how to use it. These useful features are coherence, transfer function, frequency-response function, and more accurate phase measurements. This latter feature, more accurate phase measurements, has not been verified by the author, but others have reported this benefit by putting the Keyphasor in one channel and the vibration signal in the second channel. This feature holds the promise of being able to balance with a spectrum analyzer to a finer level than any other instrument.

The two-channel capability is also convenient for obtaining phase information during analysis. Two vibration transducers are attached at two separate locations on a machine. The data from one transducer is input to channel A, and the second transducer's data is input to channel B. The relative phase difference between the two is calculated by the spectrum analyzer. One transducer remains fixed, while the other is moved about on the machine to measure the phase at different points. Using this technique with a two-channel spectrum analyzer, phase information can be obtained quickly without stopping the machine to attach a reference mark.

Aside from this phase measuring and higher-level analysis functions, the two-channel spectrum analyzer has no advantage over a single-channel unit. It does have a disadvantage in term of complexity and cost. For routine machine monitoring and analysis, a better investment is two single-channel analyzers, one serving as a backup.

Tape recorder

Portable tape recorders are used to collect raw vibration data, which is later analyzed in the laboratory or office. This is useful if you must collect vibration at a place where you don't want to drag all your instruments with you. Some applications are:

- On aircraft in flight
- Inaccessible places (like on a tall structure or deep in a wet mine)
- Dirty or dusty environments
- Data collection from several transducers simultaneously (many tape recorders have multiple channels)
- Transient data acquisition, e.g., during machine start-up
- Short-term data acquisition, as when the machine cannot stay on for very long

The use of tape recorders requires considerable skill in monitoring tape speed, amplitude calibration, and transducer interfacing to acquire valid data. As a starting point, use an instrumentation quality tape recorder that has a high signal-to-noise ratio. These higher-quality tape recorders also have a better tape drive mechanism that controls the tape speed precisely.

Tracking filter

A tracking filter, sometimes called a *signal ratio adapter,* is a bandpass filter that locks onto the running speed of a rotating machine (via a trigger signal) and follows it as it changes speed. In this way, the tracking filter always admits the vibration at running speed, even if the speed changes. This is useful during machine run-up or coast-down to see how the rotor responds at various speeds. A tracking filter is needed to produce Bodé and polar plots which will be covered later.

In summary, Table 4.3 is a chart of the instruments discussed and their typical price ranges. With the information in this book, you should be able to decide what kind of vibration program you want and what instruments to purchase. This chart gives an idea of how much to budget for.

Those who have vibration programs typically have a need for one or all three of the following instruments:

- Analyzer for diagnostics
- Balancer
- Data collector for predictive maintenance (PM) programs

Some instruments can do all three (like a portable FFT analyzer with memory). With single-channel FFT spectrum analyzers costing as little as $4000, I do not see a reason to buy any other type of readout instrument.

TABLE 4.3 Range of Instrument Costs in 1990 U.S. Dollars

Instrument	Typical cost range
Vibration switches	$ 200–400
Proximity probes	200–400
Velocity pickups	200–400
Accelerometers	200–600
Cables	20 plus $1/ft
Vibration meters	300–2,000
Tuneable filters	1,500–4,000
Tuneable filter balancers/analyzers	4,000–16,000
Oscilloscope	600–2,000
Strobe light	800–2,000
FFT analyzer/single channel	4,000–14,000
FFT analyzer/two channel	8,000–25,000
Tape recorder	4,000–12,000
Data logger	3,000–12,000
Plotter	1,500
Computer	2,000
FFT card with software	2,000
Software for predictive maintenance	5,000+
Audio signal generator	600
Small shaker table	3,000–6,000
Multimeter	100
Scale for balancing	150–500
Printer for reports	400
Microphone with preamplifier	50–500
Programmable calculator	200

Chapter

5

Typical Vibration Problems

When Do You Know that You Have a Vibration Problem?

- When you find machine parts on the mechanical equipment room floor.
- When the machine starts walking out the building. This is an exaggeration, of course, but a tendency toward mobility is a good indicator that the level of vibration has gone too far.
- When no one can stand to be around the machine for more than 10 min. While the machine is running, people find excuses for being somewhere else.
- When someone complains, especially someone of importance. There is a correlation between "perceived" importance and sensitivity to noise and vibration.
- When there is a noise problem. All noise problems have as their origin a mechanical vibration.
- When machine wear parts are changed too frequently, i.e., coupling, bearings, or seals require changes at less than 6-month intervals.
- When you find cracks—these are fatigue failures that almost certainly point to excessive vibration.
- When the process produces bad product.
- When machinery does not live up to its expected life. Life depends on the designed-for life, load condition, maintenance (especially lubrication), and level of abuse. Typical rotating equipment, fans, pumps, and motors have a life of 10 to 20 years. Some are going beyond 30 years. If a piece of equipment does not achieve at least 10 years of operating life, and it can be established that it is not being abused or overstressed, and that the lubrication system is adequate,

then only two possible causes remain—excessive vibration or design fault. More guidance will be given later for when to question the design.

The U.S. Navy specified very low vibration levels on electric motors going aboard submarines. This was done years ago to have a quiet sub to avoid detection. The motors were designed to run quiet. The same design and manufacturing details that delivered a quiet motor also delivered a smooth motor. The pleasing result was that these low-vibration motors ran trouble-free long after their expected life was to have expired.

Life expectancy can be predicted by the level of vibration. This is routinely done in predictive maintenance programs using vibration data. The level of vibration depends on the total energy input to a machine. Part of the input energy is converted to vibration and noise. This is because of windage, bearings, motors, couplings, and other mechanical contact points. A large machine with a greater energy input will have a greater absolute amount of vibration. A small machine of less than 1 hp will have less vibration. There are international standards for the level of vibration versus a machine's size. For now, keep in mind that it is more difficult, and often impractical, to achieve a specified low vibration level on a larger horsepower machine, simply because of its large energy input.

Another factor that affects the level of vibration sensed on the outside casing of a machine is the stiffness of the supports. Larger machines tend to have massive and rigid casings. A large vibratory force on the rotor may be very damaging to the bearings but not manifest itself as a large vibration on the outside. By contrast, a small ¼-hp motor can be swinging wildly on a flexible support bracket, but not causing any serious damage except fatigue damage to the bracket. Remember, that the vibration sensed at the measurement location is a combination of the source and the path. The path introduces some impedance due to the mass that must move and the stiffness.

These considerations of machine size and support stiffness should enter into all judgments of vibration severity based on amplitude measurements. In addition, the frequency could be a factor depending on the units of measurement. High amplitudes at high frequencies in acceleration units are normal for most fluid-handling machines. In displacement units, a serious high-frequency vibration may not even register on your readout instrument. Likewise, a high amplitude at very low frequency, i.e., less than 20 Hz, may be of no concern because it may be normal building sway or traffic motion. In acceleration, however, almost anything below 60 Hz that is measurable is usually significant. These considerations of size, mass, stiffness, amplitude, and frequency are cognitive functions for the analyst. There are charts and standards that assist the novice, and also the experienced, in this process. But when any of the scenarios described at the beginning of this chapter occurs, then there may be a vibration problem at much lower amplitudes than the standards indicate.

The good news is that the majority of vibration problems can be corrected without ever disassembling the machine. In this chapter, we will cover the more typical vibration problems and their solutions. You will find that all these problems can be diagnosed without disassembly, and most of them can be corrected without a complete disassembly.

The two typical entry paths into a vibration problem are:

1. The machine does not live up to expectations, in terms of life.
2. It bothers someone.

These two criteria typically originate external to the analyst, and his or her services are called upon for a solution. The situation is usually an urgent one, and the diagnosis is the first step in a repair procedure. Another, less common, entry path into a vibration problem is a desire for excellence. In this case, a solution is first visualized, and dynamic analysis is used as a tool in search of a problem. An example is the recent emphasis on quietness and smoothness for appliances and passenger vehicles. This demonstrates the potential for improvement when someone expects something better.

Sources of Vibration

The source of all machine vibration is less than perfect design and less than perfect manufacturing. In other words, defects are the sources of vibration. A perfect machine would generate no vibration when operating. Perfection is not achievable at any cost, and just approaching perfection becomes astronomically expensive. Therefore, we will need to always coexist on this planet with defects. The remaining question is how severe a defect are we willing to tolerate. The answer to that question is rooted in human perception and expectations of machine longevity. General guidelines for severity criteria are contained in standards charts and will be presented in a later chapter.

Every vibration (and noise) problem is first a problem in identifying and locating the source. Identifying the source means to perform a frequency analysis to tag the offensive frequency, and then locate the source by tracking this frequency to its origin. The amplitude is then measured to judge the severity. This sounds simple enough, and it is straightforward for typical problems as described in this chapter. These typical problems have accounted for 98 percent of all past and known vibration problems. However, I have spent enough time troubleshooting field vibration problems to know that the source cannot always be clearly identified. Each vibration problem is a new situation, and it is possible that that particular set of parameters has never been encountered before or described in the literature. These situations serve to humble us and to sharpen our vibration analysis skills. I have walked away from some problems without identifying the source with confidence. It is best in such cases to admit to a less than 100 percent confidence to the owner and let him decide on a remedial course.

I also realize that every conceivable configuration for machines has not yet been built. One hundred years from now, new kinds of bearings may be in use with totally different vibration characteristics. When presented with an unfamiliar problem, the analyst must return to fundamentals. First, remember that forced vibration is caused by a repetitive force. What can cause a force to repeat at that frequency? Second, the phenomena of natural vibrations is unlikely to change in the foreseeable future. The source must fall into one of these two categories: forced vibrations or natural vibrations.

It is usually best to identify the source and fix it. Sometimes this is not practical. Situations are described in this book and in a subsequent book where it is more appropriate to treat the path of vibration or the symptoms, rather than the source.

Imbalance

Mass imbalance is at the top of the list because it is the most common cause of vibration and the easiest to diagnose. Imbalance is a condition where the center of mass is not coincident with the center of rotation (Fig. 5.1*a*). The reason for this is a nonuniform mass distribution about the center of rotation. The vector sum of all the density variables can be combined into a single vector, or single weight at one location (for a thin disk where single-plane balancing applies). This can be viewed as an imaginary heavy spot on the rotor. The heavy spot pulls the rotor and shaft around with it causing a deflection that is felt at the bearings. The task for the balancer is to find the amount and location of the heavy spot and apply an equal and opposite weight (180°) to compensate. This will bring the center of mass to be coaxial with the center of rotation, and the result is a smooth running rotor (Fig 5.1*b*).

The amount of runout, as measured with a dial indicator, is irrele-

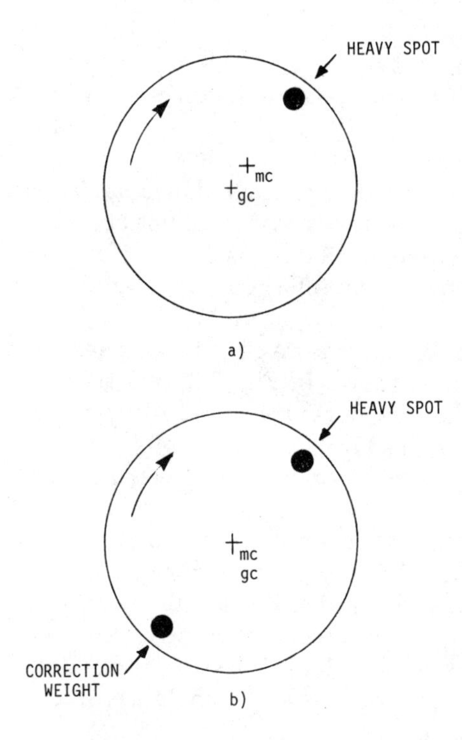

Figure 5.1 (*a*) Out of balance: the mass center *mc* is not coincident with the geometric center *gc* or center of rotation. (*b*) In balance: a correction weight has been applied opposite the imaginary heavy spot.

vant to mass balancing. Runout and mass imbalance are independent quantities. A well-balanced rotor can have significant runout. If the rotor's mass distribution is compensated for by mass balancing, then a runout condition will not cause significant forces on the bearings, and it can run in this condition indefinitely if there is clearance.

Following are the causes of imbalance:

- Porosity in casting
- Nonuniform density of material
- Manufacturing tolerances
- Gain or loss of material during operation
- Maintenance actions, like changing bearings, or cleaning
- Changing bolts
- Machining
- Loose material moving around, like water in cavities
- Keys
- Couplings
- Anything else that affects the rotational mass distribution

Unbalance shows up as a vibration frequency exactly equal to the rotational speed with an amplitude proportional to the amount of imbalance. A spectrum of an out-of-balance rooftop fan is shown in Fig. 5.2.

A question can be posed at this point. How can imbalance be diagnosed with only a hand-held meter? The answer is, it cannot because frequency information is not available. In fact, this is true for the analysis of any problem. Without frequency information, no analysis with confidence is possible.

After balancing, this rooftop fan displays a much smaller peak at 875 rpm—its operating speed. This is shown in Fig. 5.3, and the reduction in the 875-rpm amplitude is a measure of the improvement. In this case, the vibration amplitude dropped from 0.045 to 0.007 *g*, an 85 percent improvement. Notice that the remainder of the spectrum was not changed much by the balancing. Balancing procedures are presented in a separate chapter of a subsequent book. For analysis purposes, imbalance always shows up as a high-amplitude vibration at 1X rpm. There may be other peaks in the spectrum associated with imbalance, like harmonics of running speed, if the imbalance is severe. Also, other defects can cause 1X-rpm vibrations. These compounding indications sometimes complicate the diagnosis of imbalance, but I

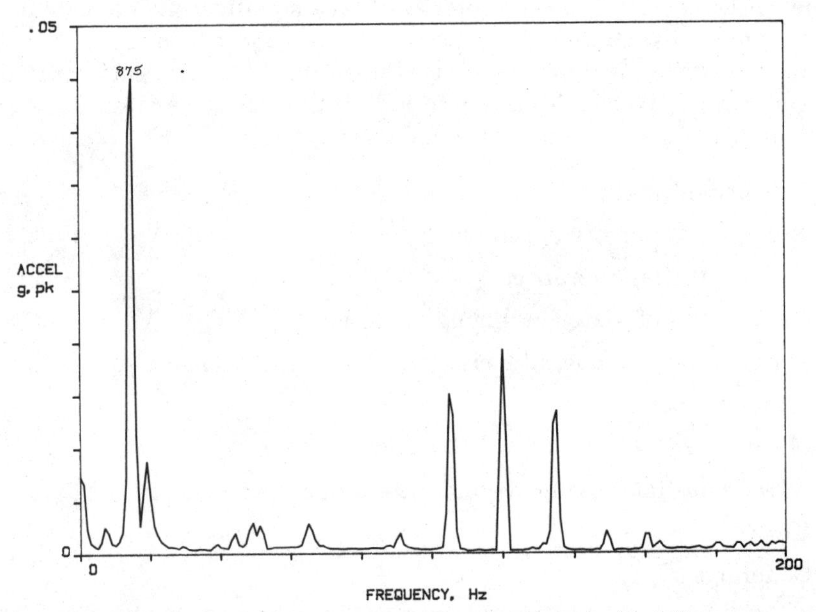

Figure 5.2 Spectrum of an out-of-balance rooftop fan. Fan speed is 875 rpm.

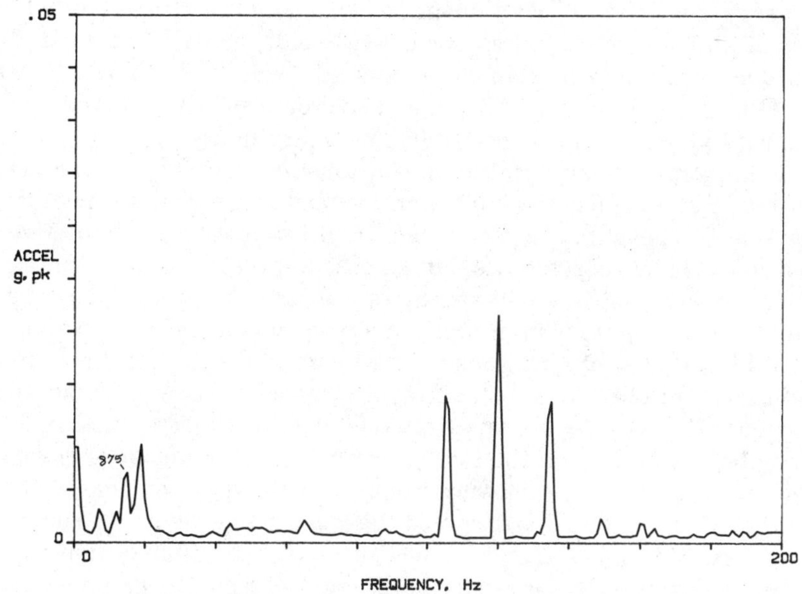

Figure 5.3 Spectrum after balancing fan of Fig. 5.2.

have never encountered an imbalance condition with the absence of the 1X-rpm vibration. This is a certainty. If a 1X-rpm vibration is present, then imbalance must be on the list of possible causes, and assigned a high priority on that list.

Perfect balance is a zero quantity and cannot be measured. What is measured is the centrifugal force on the rotor due to the heavy spot. The centrifugal force formula is

$$F = mr\omega^2 \tag{5.1}$$

where F = centrifugal force
m = mass of heavy spot
r = radius of heavy spot
ω = speed of rotor in radians/second

A more convenient formula, in English engineering units, is

$$F = 1.77 \left(\frac{\text{rpm}}{1000}\right)^2 \times mr \tag{5.2}$$

where F = centrifugal force, lb
rpm = speed, cycles/min
m = mass of heavy spot, oz
r = radius of heavy spot, in

The heavy spot is actually an imaginary quantity. It is the vector addition of all the mass nuances; inclusions, voids, actual weights, and other defects that are combined into a single weight at a single location. This is valid for a rigid rotor, and the correction weight is a single mass 180° opposite this invisible, imaginary heavy spot.

This imaginary heavy spot creates a constant centrifugal force on the rotor. But as the rotor spins, a stationary transducer senses this force as a cyclic event once per revolution. It is a matter of perspective based on our frame of reference. For an observer on the rotor, spinning with it, the heavy spot would appear as a weight that got heavier the faster the rotor spun. This weight creates a constant direction force outward from the center that distorts the rotor and shaft. To this must be added the gravitational force. This is a constant force in a stationary frame of reference, but a variable force on a rotating disk. This gravity force also acts on the heavy spot with a constant force downward, but this force pulls outward from the shaft when the heavy spot is on the bottom, and inward toward the shaft when the heavy spot is on top (see Fig. 5.4). Thus, in addition to the constant centrifugal force, a cyclical gravity force acts on the rotor and shaft, when viewed from the rotor frame of reference. Thus, an observer sitting on the heavy spot and spinning with the rotor would feel a very strong constant centrifugal force outward, and on top of this, a cyclical gravity force. This gravity force creates an alternating stress on the shaft.

With a stationary piezoelectric or electrodynamic transducer on the

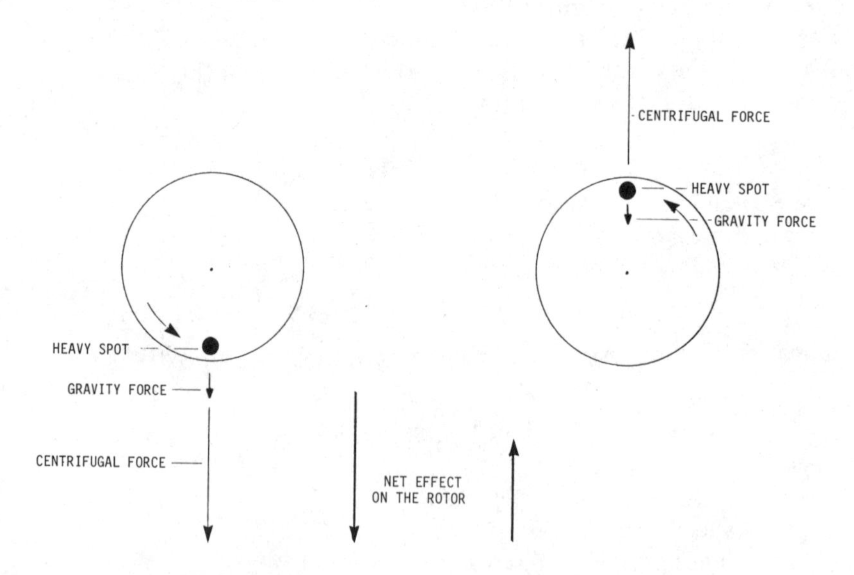

Figure 5.4 The gravity force creates a cyclical force on the heavy spot for an observer on the rotating disk.

bearing casing, the gravity force is constant and not measurable. The passage of the centrifugal force due to the heavy spot registers as an oscillation at the frequency of rotation. The effect of the gravity force can be measured if the transducer is moved around the circumference of the bearing. A larger total motion would be detected in the downward direction than in the upward direction. This has the effect of displacing the zero position of the sine wave and, in effect, creating small shifts in phase readings. For a transducer mounted in the horizontal direction, the gravity force is zero and no error occurs. The transducer senses the effect of the centrifugal force only.

For normal machine speeds of 600 rpm and higher, the centrifugal force is at least 100 times more than the gravity force, and the gravity force can be virtually ignored. This analysis illustrates how to visualize rotating forces from a stationary frame of reference. It also shows how balancing at higher speeds makes the detection of the heavy spot easier. The heavy spot of a few ounces when stationary as a gravity force can turn into a few hundred pounds as a centrifugal force when rotating.

As a final example of diagnosing imbalance, Fig. 5.5 is a spectrum of an imbalance that was created in a tabletop demonstration machine by adding a bolt and nut. The speed of rotation was 1500 rpm (25 Hz), and a single tall peak is present at this frequency. The amplitude of

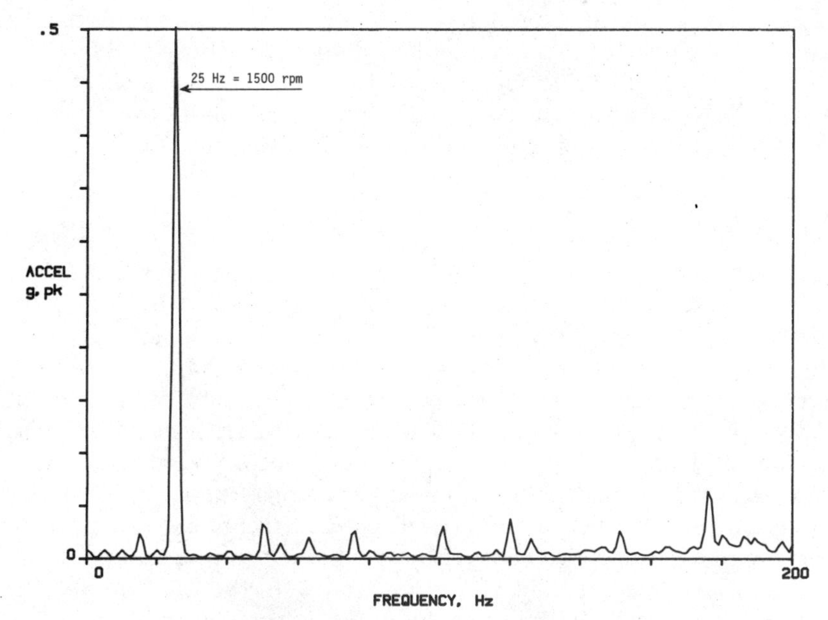

Figure 5.5 Spectrum of imbalance created in a tabletop demonstration machine turning at 1500 rpm (25 Hz).

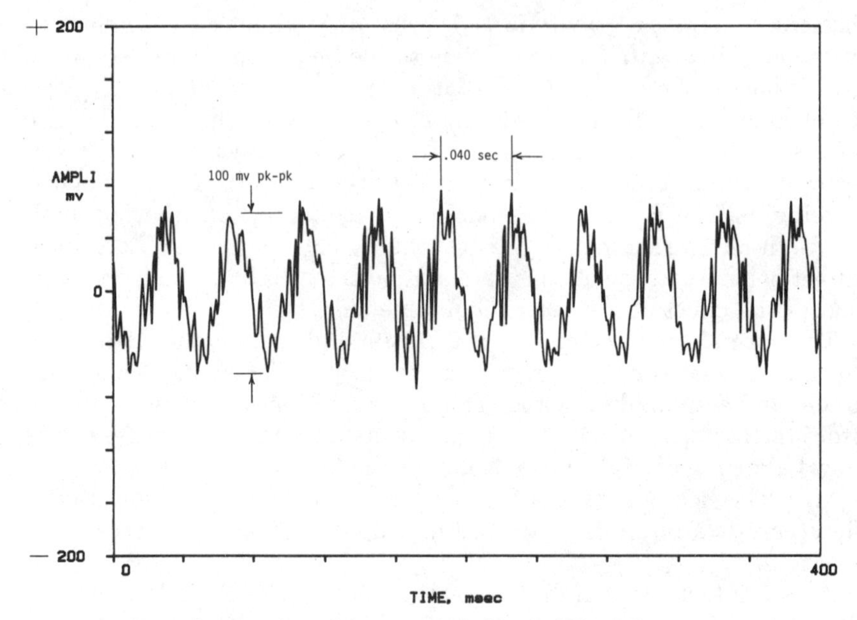

Figure 5.6 Time domain view of the same imbalance of Fig. 5.5.

this peak is 0.5 *g* peak. Figure 5.6 is a time domain view of this same imbalance. The period of the most significant vibration is 0.040 sec, which corresponds to a frequency of 25 Hz. The peak-to-peak amplitude is 100 mV. The accelerometer used had a sensitivity of 100 mV/*g*. Therefore, this vibration is 1-*g* peak to peak, or 0.5-*g* peak, as expected. This example illustrates how simple problems can be diagnosed with an oscilloscope by performing frequency analysis.

Misalignment

Coupling misalignment is a condition where the shafts of the driver machine and the driven machine are not on the same centerline. For discussion purposes, the noncoaxial condition can be parallel misalignment or angular misalignment as depicted in Fig. 5.7. The more common condition is a combination of the two in both the horizontal and vertical direction (Fig. 5.8). This compounding of misalignment, parallel and angular in both vertical and horizontal directions, is what has made the correction of misalignment so frustrating. Correcting for one affects another, and the alignment technician hunts his or her way to a better alignment with repeated measurement and move, then another measurement and move, and so forth. The illustrations in Fig. 5.7 suggest coupling misalignment is the subject. This is the more common situation, but the analysis method is applicable to gear and bearing misalignment also.

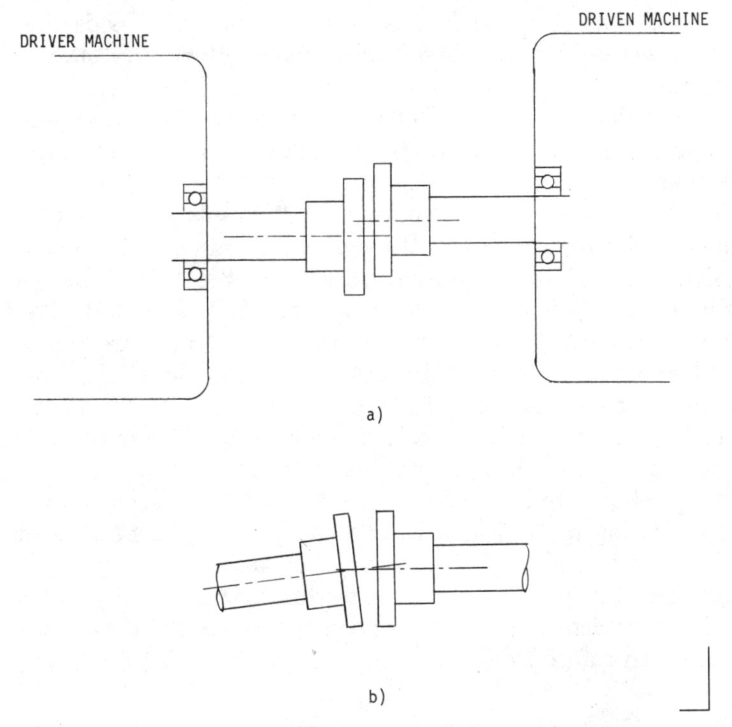

Figure 5.7 (*a*) Parallel misalignment. (*b*) Angular misalignment.

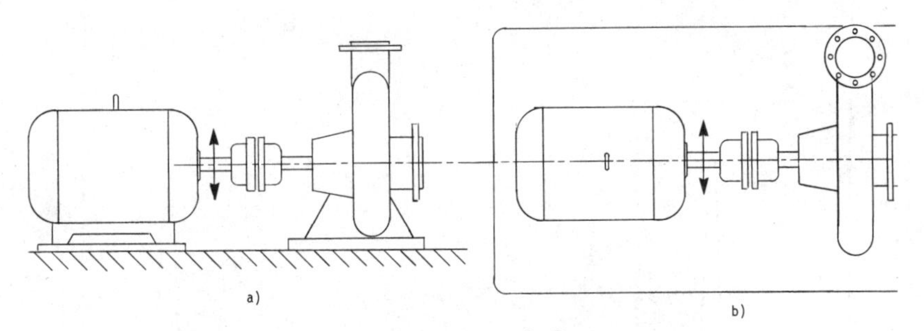

Figure 5.8 (*a*) Vertical alignment. (*b*) Horizontal alignment.

We are forced into this situation of coupling alignment because equipment from different suppliers must be mated together. Motor manufacturers do not make pumps, and fan manufacturers do not make motors. Some equipment, like vaneaxial fans and some centrifugal chillers, is built with an integral motor and the impeller is mounted directly on the motor shaft. In this case, there is no shaft or coupling misalignment problem because there is only one shaft. We are also forced into this situation because fluid-handling machines can produce different flows and pressures at different speeds and horsepower ratings. It is desirable to maintain this flexibility of mating different motors to fans and pumps to achieve the desired flow and pressure. For this latter reason, it looks like industry has a problem that is not going to go away soon, and there is job security in learning precision alignment procedures.

Since it is impossible to get the output shaft of one machine perfectly aligned to the input shaft of the driven machine, flexible couplings are available to take up the misalignment. Flexible couplings allow the two machines to operate, but not necessarily smoothly. The further away from perfect alignment the shafts are, the more strain there is on the couplings. This strain causes a higher level of vibration. It also causes bearings, seals, and couplings to wear faster. If you are changing couplings, bearings, or seals more frequently than at 5-year intervals, it is likely that you have a misalignment problem.

Misalignment shows up in the frequency domain as a series of harmonics of the running speed (Fig. 5.9). The harmonics occur because of the strain induced in the shaft. The harmonics are not really vibrations at those frequencies, but a fallout of the digital signal process when motion is restricted. If two shafts are not aligned *and* not coupled, then they can rotate freely on their own axis (Fig. 5.10*a*). When

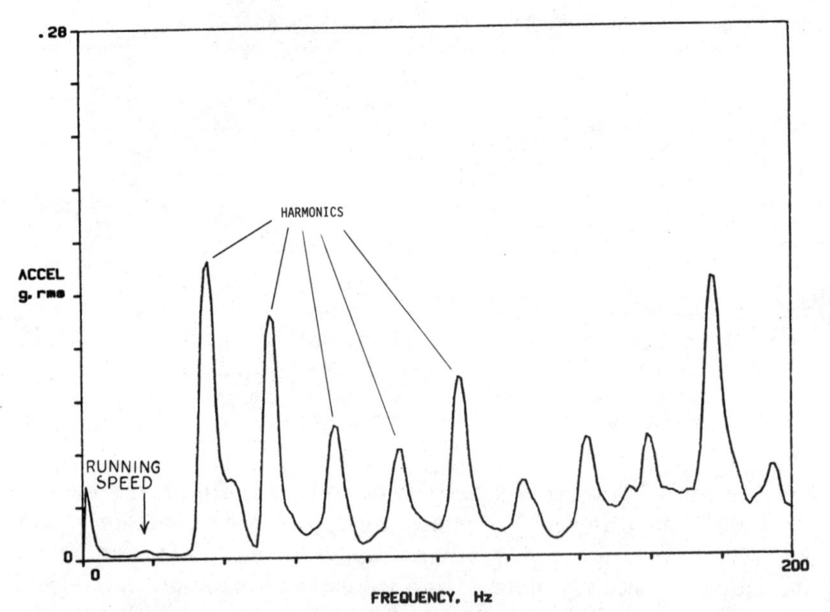

Figure 5.9 Spectrum of a 150-hp motor coupled to a boiler draft fan. The running speed is 1094 rpm (18.2 Hz). These machines are well balanced, but the harmonics of running speed indicate misalignment.

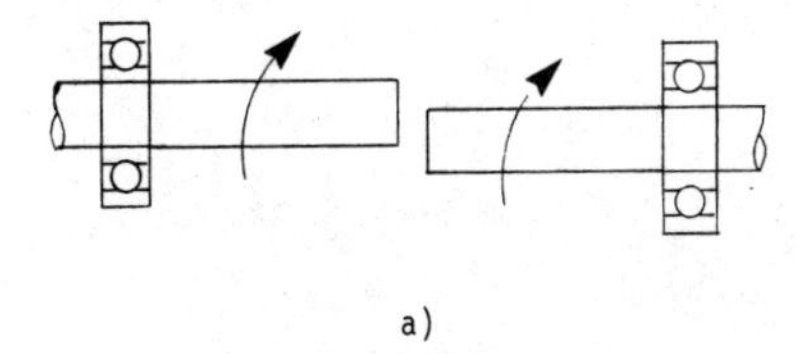

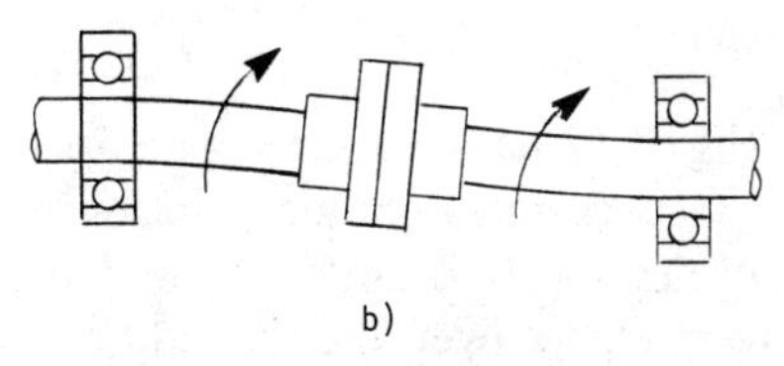

Figure 5.10 (*a*) Uncoupled shafts free to rotate without binding. (*b*) Shafts coupled and strained.

the two shafts are coupled together, they are now strained toward each other (Fig. 5.10*b*). When coupled and rotated together, the two shafts are cyclically strained at running speed. Steel shafts are very stiff, but they still deflect a small amount. This deflection creates forces on the nearby bearings and sets the entire housings of both machines into cyclical motion. The housings and bearings create reactionary forces that prevent the shaft from moving as much as it would tend to. These restrictions prevent the normal sine wave motion from achieving its full excursion in amplitude. In other words, the sine wave motion of shaft deflection is distorted at the extremes (Fig. 5.11). It is this distortion that generates the harmonics. The conservation of energy principle requires the generation of harmonics for distortion when the sine wave is clipped off. The missing part of the waveform is displaced to a different part of the spectrum, i.e., to higher frequencies. The unwanted shaft deflection also causes accelerated wear of the coupling and adjacent bearings and seals.

This gives you a clue to the problem on machines that have frequent inboard bearing changes. The misaligned shafts also cause a strain on the coupling and cause the internal parts of the coupling to press against each other every revolution. The parts touch, press together, come apart, then touch again on the next rotation, and continue this cycle. This cyclical touching causes a real vibration at the frequency of the coupling elements contacting. This frequency is the number of elements times the rotational speed.

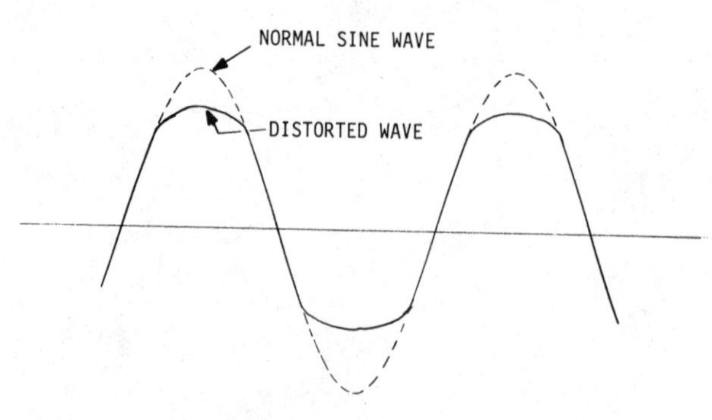

Figure 5.11 Distorted sine wave generates harmonics.

If you saw the pattern of vibration in Fig. 5.12 on a rotating machine, with flexible couplings, you could strongly suspect misalignment. Notice in Figs. 5.9 and 5.12 that the amplitude is in acceleration units. Remember that acceleration emphasizes the higher frequencies over velocity and displacement. The acceleration display is the best place to see the harmonics of misalignment.

This same pattern of harmonics appears for bearing misalignment also. Some further analysis needs to be done to separate the two. For instance, harmonics in the spectrum of a machine without a flexible coupling means that coupling misalignment is out of the question. The harmonics could be caused by bearing misalignment or some other distortion. Likewise, the disappearance of harmonics on a motor run solo is a good indication of coupling misalignment. If they remain, then the problem is in the motor and not in the alignment of the two machines.

Misalignment is temperature dependent. All materials grow with increasing temperature, and metal is no exception. Motors warm up several degrees, and the driven machine may warm up or cool down from ambient depending on the fluid it is handling. This temperature change causes very slow movement as the machines reach operating temperature. This movement changes the alignment condition. Therefore, a change in vibration, and specifically in the harmonics, during temperature changes is a strong indication of misalignment.

Total harmonic distortion is an effective measure of misalignment. Some analyzers have this feature and it can be used as a trending tool. Some analyzers also have a histogram display. This is a probability diagram of the chances of measuring the voltage around a mean of zero. In the time domain for a pure sine wave, the voltage is most likely to be at zero, with decreasing probability farther from zero in either the positive or negative direction. By contrast, a square waveform has a greater probability at the extremes and a lesser probability of being at zero. When some harmonic distortion is taking place, as with misalignment or looseness when the waveform is being clipped, then the histogram display shows a decrease in probability and a broadening at the base as in Fig. 5.13*b*.

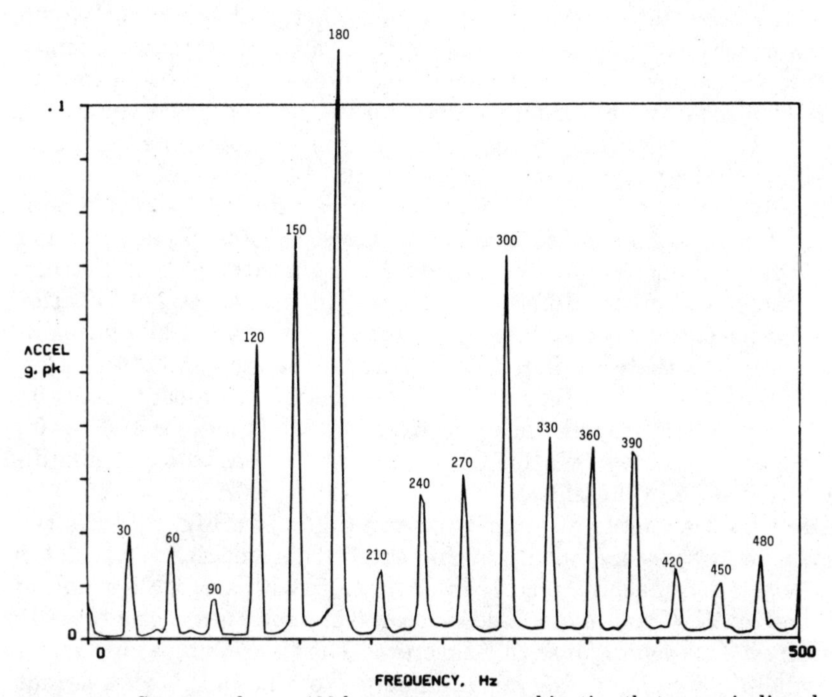

Figure 5.12 Spectrum from a 400-hp motor-pump combination that was misaligned. The rotational speed was 1800 rpm (30 Hz), and the machine emitted the audible "growl" of misalignment.

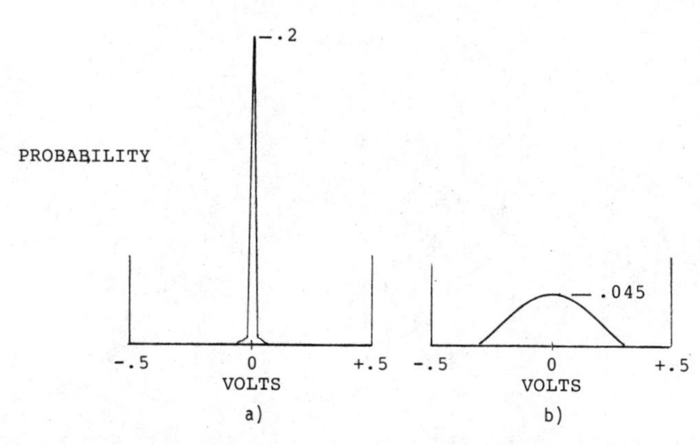

Figure 5.13 Histogram displays of the voltage from an accelerometer on: (*a*) a machine well aligned, and (*b*) the same machine misaligned.

As the two shafts roll around with each other, both shafts rock in an angular motion. Some bending takes place in the shaft, but the angular motion causes the machine housing to also trace out this same motion. The result is that the machine vibrates up and down vertically, and back and forth horizontally, but also in and out axially. If you have an analyzer that can measure phase, then this rocking motion (actually the shaft traces out a cone) can be detected by 180° out-of-phase readings. The machine casing will rock 180° out of phase up and down vertically, or in and out horizontally from front to back. You can also detect 180° phase difference axially. This measurement is better taken adjacent to the coupling. You can also detect the 180° phase difference across the coupling; one reading on the motor and the other reading on the pump (be careful of the pickup orientation, see Fig. 6.11). The difference will rarely be exactly 180°. It may be 160°, 140°, or maybe even only 110°. But the point is that there will be a significant difference in the phase.

As a final example of diagnosing misalignment, Fig. 5.14*a* is the time domain view of a misaligned machine. This machine is a tabletop demonstration machine pictured in Fig. 5.15, with a flexible coupling between a 1790-rpm motor and a rotor. The motor was intentionally misaligned by moving it with jackscrews. The time domain view is not very helpful for diagnosing misalignment. Nothing in this view stands out as characteristic of alignment.

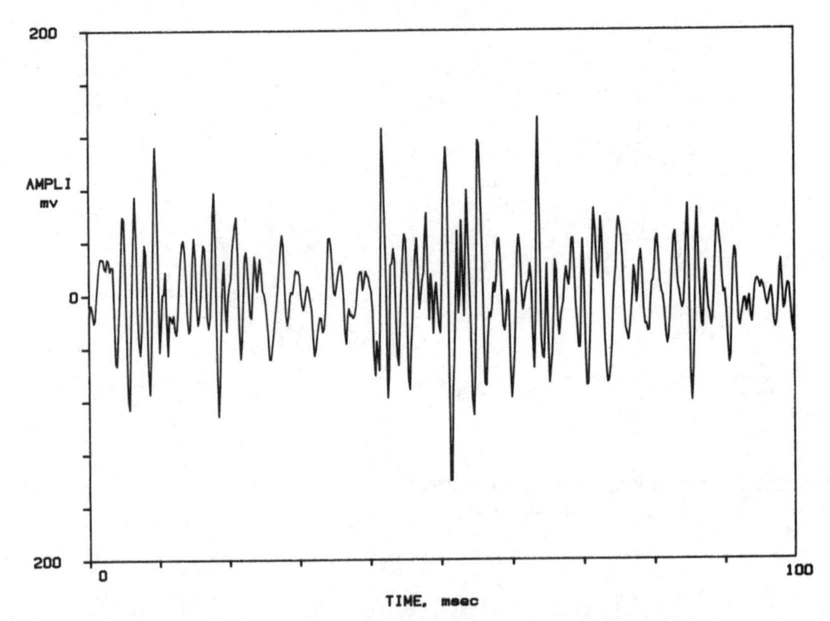

Figure 5.14(*a*) Time domain view of a misaligned machine.

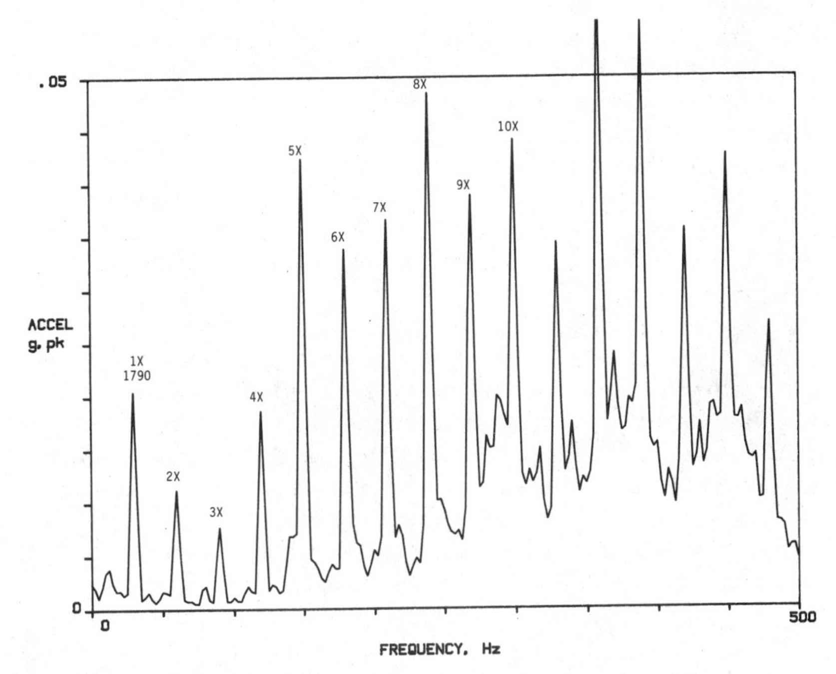

Figure 5.14(*b*) Frequency domain view of the same misaligned machine.

Figure 5.15 Photograph of misalignment demonstrator.

By contrast, Fig. 5.14*b* is a frequency domain view of the same misaligned tabletop demonstration machine. The harmonics of running speed indicate misalignment, loud and clear. The conclusion is that the frequency domain view, especially with the acceleration display, is the best way to detect misalignment.

You now know the two main causes of vibration—imbalance and misalignment. How can you tell the difference from vibration readings? Table 5.1 offers information to help separate imbalance from misalignment.

The question occasionally comes up on which corrective technique should be applied first, balancing or alignment. At times, it may not be clear which is the dominant cause of vibration. The answer is to perform alignment first because there may be erroneous 1X-rpm vibration readings if a large misalignment is present. This 1X-rpm vibration due to misalignment must be removed first for balancing to be successful. Also, aligning is a procedure that usually can be completed in a shorter time than balancing. If a high 1X-rpm vibration remains after aligning, then it is time to balance.

Resonance

Resonance is a condition whereby the driving force applied to a structural part is close to its natural frequency and amplification occurs. The source of the driving force is most likely residual imbalance in a rotating machine attached to the structure (Fig 5.16), or broadband turbulence due to fluid motion. This small imbalance is transmitted throughout the machine, and attached parts, as a vibratory force. If that force encounters a structural part that is tuned to that frequency

TABLE 5.1 Recognizing Imbalance from Misalignment

Imbalance	Misalignment
High 1X rpm	High harmonics of 1X rpm
Low axial readings	High axial readings
In phase	About 180° out of phase
Temperature independent	Temperature dependent; therefore, vibration changes on warm-up
Speed dependent due to centrifugal force. Vibration at 1X rpm increases as the square of the speed.	Less sensitive to speed changes. Forces due to misalignment remain constant with speed.

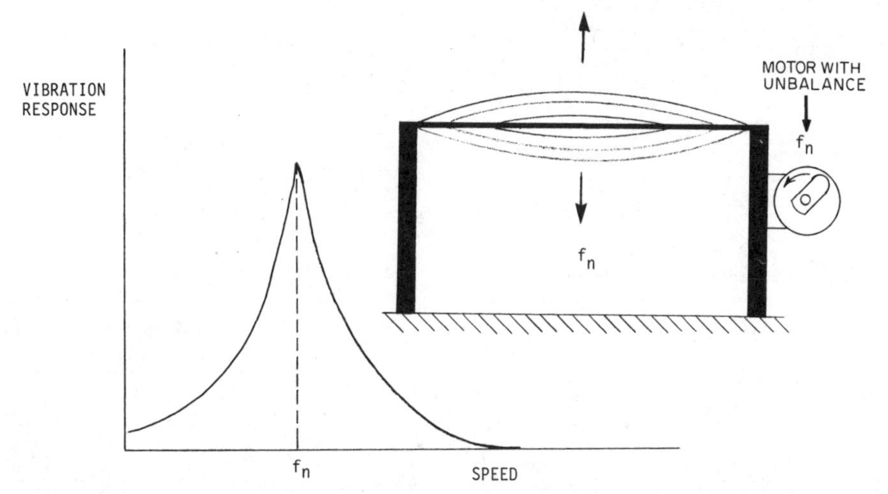

Figure 5.16 Resonance occurs when the speed of a driving force equals the natural frequency of a structural part.

by virtue of its mass and stiffness, then that part will be excited into resonance. Its vibration amplitude can be 10 to 100 times the amplitude of the input force, depending on damping.

The equation for resonance is $\omega_n = \sqrt{k/m}$. This says that the resonant frequency, in radians per second, is equal to the square root of the stiffness divided by the mass. (2π rad = 360° or one cycle.) This equation is for a single-degree-of-freedom system, i.e., a mass supported by a massless spring.

Beams, plates, and other objects also have resonant frequencies. The formulas for calculating their resonances are more complex and can be found in the *Shock and Vibration Handbook,*[1] or any other good mechanical engineering reference book. All objects have more than one natural mode of vibration and hence more than one natural frequency.

Figure 5.17 shows the first, second, and third modes of vibration of simple beams. The points where no motion is taking place are called *nodes*. For the first mode of vibration, the beam supports are nodes. Each successive higher mode of vibration has one additional node.

Resonances are highly speed sensitive, depending to some extent on damping. Damping not only decreases the maximum amplitude of vibration but also broadens out the response curve. With light damping,

[1]C. M. Harris and C. E. Crede, *Shock and Vibration Handbook,* 3d ed., McGraw-Hill, New York, 1988.

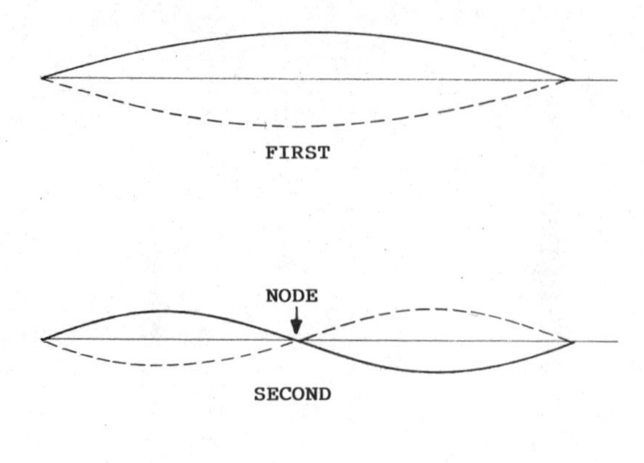

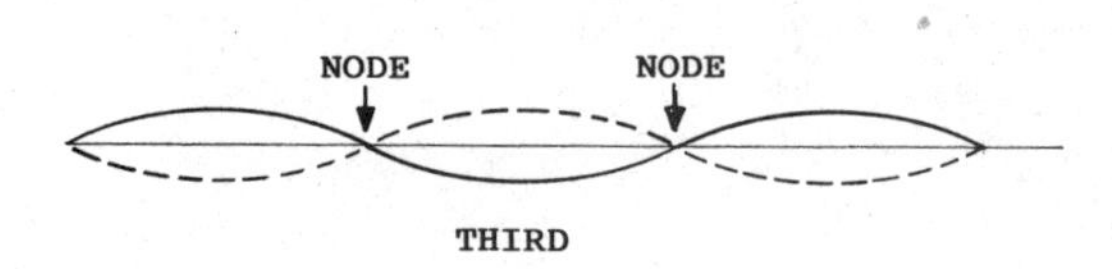

Figure 5.17 Resonance modes of simple beams.

as is usually the case with metals, the resonance curve is a sharp peak. A slight change in speed will dramatically reduce the resonance response.

Rotors also have resonances, called critical speeds. According to the formula for centrifugal force, $F = mr\omega^2$, the vibratory force should increase as the square of the speed. This is true in the low-speed range. When approaching resonance, or critical speed, the vibration increases much more than expected by the centrifugal force formula. It peaks at the critical speed; then, at higher speeds, it smooths out. The reason for this is because of the 180° phase shift at resonance. The response of the rotor to the residual imbalance is delayed 180° and causes the mass center to be pulled in closer to the center of rotation. Rotors run smoother above the first critical speed than below it. However, they must pass through that first critical speed for every run-up and coast-down.

Resonances are more typical on stationary parts than on the rotating parts. Rotating parts are in the fluid stream, and this usually provides sufficient damping to keep amplification down. Except for critical speeds of high-speed rotors, resonances on rotating parts are rare. Sheet metal resonance on fan cabinets is probably the most common case of resonance.

A single impulse will excite a system to vibrate at its natural frequency (Fig. 5.18). The vibration amplitude dies away because of

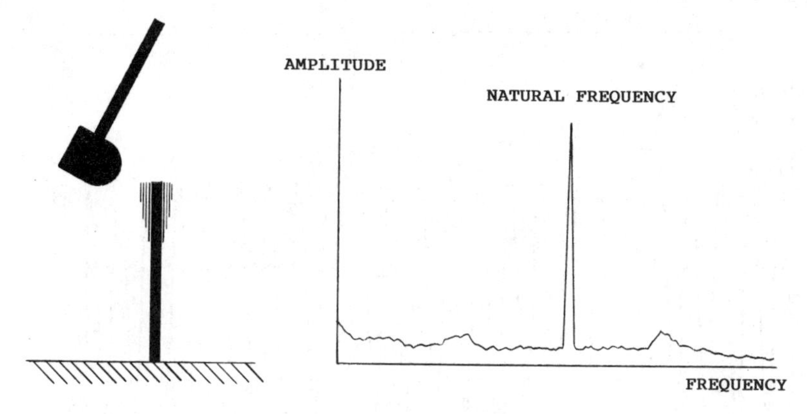

Figure 5.18 An impulse excites natural frequencies.

damping, but the frequency of vibration is the same—its natural, or resonant frequency.

Another example of resonance is a squeaking in an automobile at highway speeds. The squeaking is metal or plastic edges rubbing when one of the parts goes into resonances. It usually goes away with a change in speed. This is the key to identifying a resonance—it is highly speed sensitive. Notice in Fig. 5.16 that a slight change in speed, either faster or slower, will dramatically reduce the vibration response.

Directional vibrations suggest resonance. When the vibration amplitude in one direction is significantly greater, i.e., more than 5 times the vibration in any other direction, at the same frequency, then this is a clue to resonance. Machines vibrate typically in all three directions with similar amplitudes and frequency content. If, however, a motor shows a very strong vibration at 30 Hz (1800 rpm) in the vertical direction which is 10 times larger than the amplitude in the horizontal direction, then a vertical resonance of the motor mount is likely, and not imbalance.

In the frequency domain, a resonance appears as a discrete peak unrelated to the running speed (unless the running speed or a harmonic coincides with the resonance). A resonance coincident with a driving frequency will have a relatively constant amplitude. A resonance near a driving frequency, not exactly coincident but close enough to be excited, will rise and fall rapidly in amplitude. It looks like a vibration that comes and goes. The rate of decay is an indication of the damping present. With very little damping, the vibration amplitude will hang in there a long time.

Figure 5.19 is a resonant vibration from an aluminum wheel. The wheel is pictured in Fig. 5.15 as one of the larger wheels. The wheel

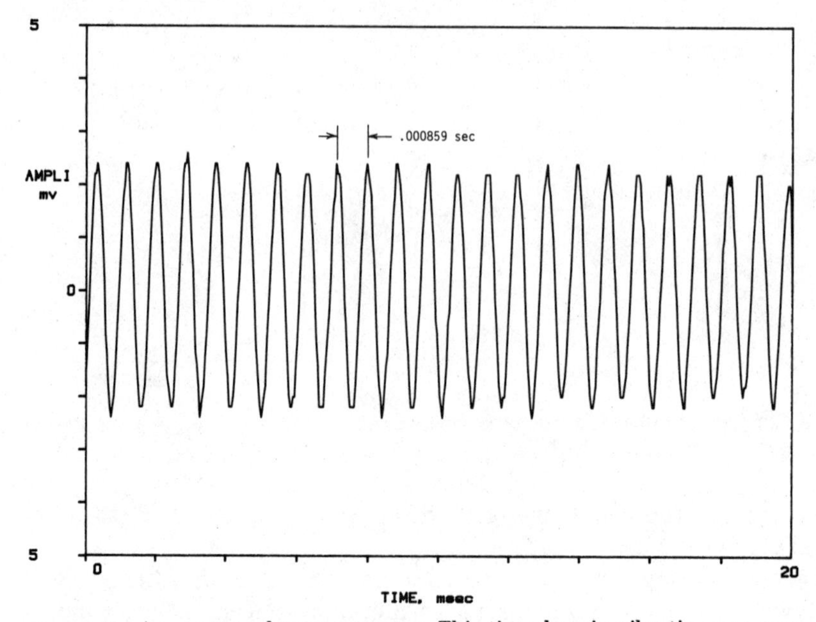

Figure 5.19 A pure tone due to resonance. This time domain vibration was measured with an accelerometer on a bearing block that supported a shaft on which was mounted an aluminum rotor. The rotor was 8 in in diameter and 3⁄16 in thick. Excitation was with a wooden dowel.

was not rotating but was excited into resonance by striking it with a wooden dowel. The vibration was measured on the adjacent bearing block. It "rang" like a bell and was clearly audible. The "ringing" continued for about 10 sec before finally fading away. Very little damping was present here. The period of this sine wave was 0.000 859 sec, which corresponds to 1164 Hz.

Figure 5.20 is the frequency domain view of this same aluminum wheel vibrating in resonance. The resonant frequency is 1170 Hz, as measured in the frequency spectrum. The key indicators of resonance are

1. An audible pure tone
2. A clean sine wave in the time domain
3. A single tall peak in the frequency domain

There are two ways to positively identify a resonance. One is to stop the machine and perform a bump test on the suspect parts. This method is fully described in Chapter 6, "Techniques." This measures the natural frequencies of the suspect parts. If these frequencies then appear in the spectrum when the machine is running, then you have

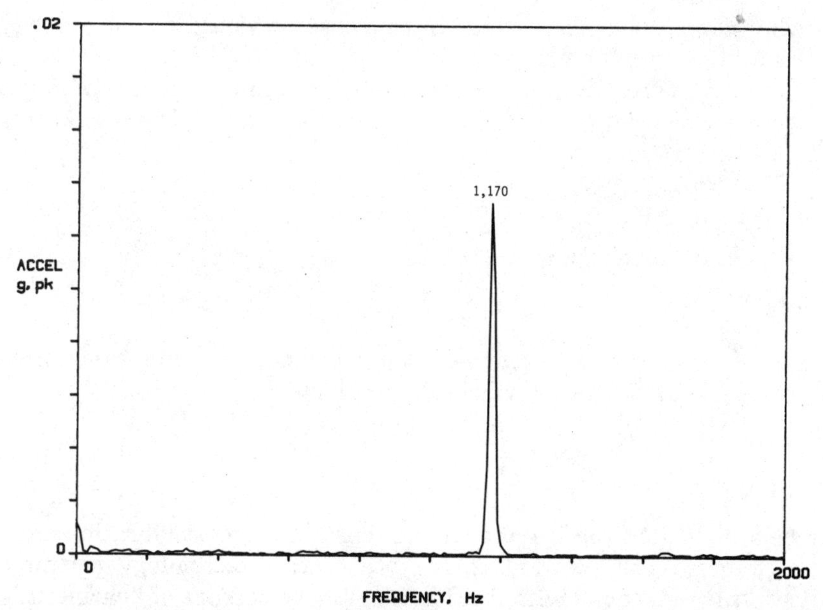

Figure 5.20 A pure tone due to resonance in the frequency domain. This spectrum corresponds to the same vibration as Fig. 5.19.

confirmed a resonance. The second method is to watch the spectrum as the machine changes speed, possibly a coast-down. Resonances will rise as another vibration peak comes into coincidence with it. The key is that resonances do not change frequency as the machine speed changes. A peak hold averaging during coast-down is a handy way to record all the resonances.

To conserve energy nowadays, variable-speed drives are being added to machines that were designed to run at constant speed. The hazard here is finding a structural resonance somewhere within the speed range of the machine, which was not active when the machine ran at a constant speed. The key to recognizing a structural resonance is a pure tone that is speed sensitive.

Valves on automatic controls will sometimes "hunt"—that is, oscillate between extremes. This is a resonance at a system natural frequency. Technically, this is a self-excited vibration sustained by a constant force—the flow in the pipe. The solution is to change the system natural frequency with a spring or mass change. Remember:

$$\omega_n = \sqrt{\frac{k}{m}}$$

In this case, the spring is whatever is providing the restoring force. It may be a physical spring or it may be a pneumatic cylinder. The mass is what compresses the spring. In this case, it would be the velocity pressure impinging on the valve.

The above three sources, imbalance, misalignment, and resonance, account for over 90 percent of all vibration problems. The breakdown is approximately:

40 percent imbalance

30 percent misalignment

20 percent resonance

If past experience is a teacher, then we can expect this same probability to occur for future vibration problems.

Bearings

To be able to monitor bearings is the reason that most vibration analysis programs are started. The state of this technology certainly allows this to be done with confidence. Ninety percent of bearing failures can be predicted months beforehand. There are still approximately 10 percent of bearing failures that are abrupt and unforeseen. Being able to predict the 90 percent majority is a good enough reason to invest in a bearing monitoring program for many companies. However, if this is the only use of the vibration instrumentation, then it is underutilized.

There are very few "bad" bearings coming out of bearing factories. The state of quality control at these facilities is of the highest caliber of any manufactured goods. Bearings fail for several reasons, the least of which is a manufactured-in defect. All bearings have some defects, and they are graded accordingly. It is only a matter of degree of defects that separates out the highest-quality bearings from the lowest-quality ones. The presence of these defects is not the primary cause of bearing failure. The primary causes of bearing failures are:

1. Contamination, including moisture (Some sources claim that 40 percent of bearing failures are caused by contamination. This is certainly believable based on my field experience.)
2. Overstress
3. Lack of lubrication
4. Defects created after manufacturing

Bearings typically achieve only about 10 percent of their rated life. Tests of bearing life under laboratory conditions yield lives of 100 to

1000 years. Clearly, the design and manufacturing do not present deficiencies that limit their life. So why don't bearings under service conditions achieve those running times? The answer is that in the laboratory, there is no contamination of dirt or water, there is little imbalance or misalignment to cause overstress, the lubrication is the best, and the bearing is handled as if it were a delicate instrument, which it is. Under service conditions, these factors are not all optimum as they were during the laboratory tests. The tests prove that long life is achievable with some care. If a bearing is not achieving its rated life, the appropriate question to ask is, "Why not?" The prediction of bearing failure is useful information to the production manager, but the root cause of the failure is the key information that can prevent a recurrence. The point of this discussion is that very few root-cause failure modes originate at the factory, the majority of failures are worked into bearings once they get into our hands. This is good because we can't do much about factory defects. We can do a lot with preventing the defects that we, as users, create. The defects that will cause failure are probably present the instant the machine is energized, and modern vibration instruments can detect them. The goal of the vibration analyst should be to not only warn the production manager of impending bearing failure (a computer can do that) but to analyze the data, look for patterns, and apply reasoning to find the root cause. A computer has some difficulty with this latter function.

If the above causes—contamination, overstress, lubrication, and defects—can be ruled out and continued bearing failures occur at the same machine, then it is appropriate to question the design. Is this the right bearing for this application? Some bearing problems have been solved by redesigning the bearing, or specifying a bearing that is bigger or rated for a higher speed, than the original one installed. Before attacking the bearing design or application, however, the analyst should rule out the above factors by establishing vibration baseline and patterns of vibration leading to failure.

All vibration occurs at some frequency. Knowing the frequency of the vibration is paramount in diagnosing the problem. This is especially true for bearings. All roller bearings give off specific vibration frequencies, or tones, that are unique. The amplitude of these tones is an indication of their condition. Ball bearings give off four distinct tones. These are:

Fundamental train frequency (FTF)

Ball spin (BS) frequency

Outer race (OR)

Inner race (IR)

The formulas for calculating these specific frequencies are

$$\mathrm{FTF} = \frac{\mathrm{rps}}{2}\left[1 - \frac{Bd}{Pd}\cos\phi\right] \tag{5.3}$$

$$\mathrm{BS} = \frac{Pd}{2Bd}(\mathrm{rps})\left[1 - \left(\frac{Bd}{Pd}\right)^2\cos^2\phi\right] \tag{5.4}$$

$$\mathrm{OR} = N(\mathrm{FTF}) \tag{5.5}$$

$$\mathrm{IR} = N(rps - \mathrm{FTF}) \tag{5.6}$$

where rps = revolutions per second of inner race
Bd = ball diameter (refer to Fig. 5.21)
Pd = pitch diameter
N = number of balls
ϕ = contact angle

Ball bearings that carry no thrust load are assumed to have zero contact angle.

Be advised that these formulas are theoretical, and the difference between calculated and measured bearing frequencies can be as much as several hertz. The discrepancies arise when ball bearings have significant thrust loads and internal preloads. This changes the contact angle and causes the outer-race frequency to be higher than calculated.[2]

Tables of tabulated bearing frequencies are available. As an example, Fig. 5.22 contains a vibration spectrum of an NSK 6311 bearing turning at 1715 rpm. The bearing, its mounting, and the drive motor

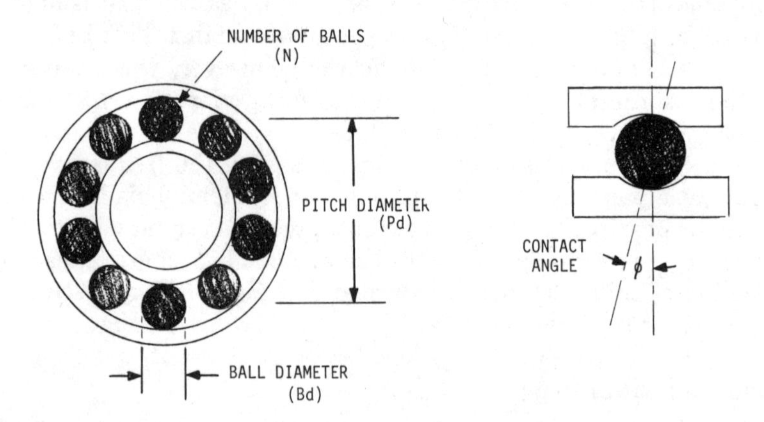

Figure 5.21 Parameters for calculating bearing frequencies.

[2]For more information on this, see *Vibrations,* Vol. 6, No. 1, March 1990, pp. 3–7.

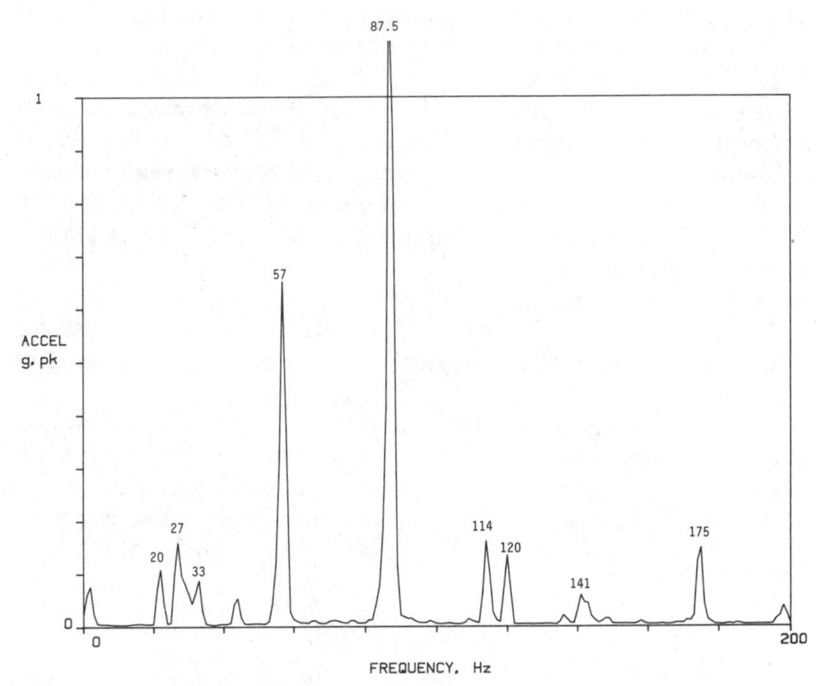

Figure 5.22 Vibration spectrum of an NSK 6311 bearing pictured in Fig. 5.24. Rotational speed is 1715 rpm (28.58 Hz).

is pictured in Fig. 5.24. The accelerometer was stud mounted in the vertical direction. The motor has sleeve bearings. The bearing tables I have access to do not list an NSK 6311 bearing. This points out one problem with bearing tables. The bearing you are interested in may not be listed. But 6311 bearings are listed of FAG, SKF, and NTN brands. Their data is tabulated in Table 5.2. This table points out another feature of bearing tables, i.e., the same bearing number from different manufacturers may produce different frequencies. The discrepancies may be larger than this. Bearing tables assume that the shaft rotates and that the outer race is stationary. If these conditions do not exist, then the tabulated numbers do not apply. One final point: Bearing manufacturers reserve the right to change the internal con-

TABLE 5.2 Bearing Frequency Factors*

	FTF	BS	OR	IR
FAG 6311	0.378	1.928	3.024	4.976
SKF 6311	0.382	2.003	3.057	4.943
NTN 6311	0.384	2.040	3.072	4.928

*These factors are multiplied by the rotational speed to obtain the vibration frequency.

struction of their bearings without changing their number designations. That is, they can change the number of balls, size of balls, or pitch diameter, and as long as it does not affect form, fit, or function, the new bearing will retain the old number. Their only obligation is that it be interchangeable on the machine, not that it generate the same frequencies. I have found some tabulated bearing frequencies to be grossly in error with measured frequencies. The reason for the gross error is believed to be old tabular data which did not match the present bearing geometry.

The bearing frequencies are found by multiplying the numbers in Table 5.2 by the rotational speed of the shaft. If we take the data for the SKF 6311 bearing, as an average, we will come up with the frequencies in Table 5.3.

In Fig. 5.22, the higher three frequencies do appear. The fundamental train frequency (FTF) does not show, and it rarely does, unless a serious cage defect exists. The vibration peaks at 57 and 87.5 Hz are likely the ball spin (BS) and outer-race (OR) frequencies, but they could also be harmonics of running speed. The harmonics of running speed are

$$2 \times 28.58 = 57.16 \text{ Hz}$$

$$3 \times 28.58 = 85.74 \text{ Hz}$$

One way to sort this out is to zoom in closer to these peaks. This has been done in Fig. 5.23. Here we see the 57-Hz peak divide into two prominent vibrations at 55.5 and 57.25 Hz. The 57.25 Hz is most likely a combination of ball spin plus a harmonic of running speed. The two frequencies are too close and overlap. Further zooming will separate the two. This illustrates a problem with monitoring bearing frequencies that the analyst should be aware of. Whenever the bearing frequency factors in Table 5.2 are very close to a whole number, then the calculated bearing frequency will overlap a harmonic of running speed. The 87.5-Hz peak divides into a vibration at 85.75, which is certainly the 3X harmonic of running speed, and a peak at 87.5 Hz which is clearly the outer-race frequency.

This bearing was run dry with no lubrication, in a test setup shown in Fig. 5.24. The motor rotating the NSK 6311 bearing had sleeve

TABLE 5.3 Bearing Frequencies for SKF 6311 Turning at 1715 rpm (28.58 Hz)

Fundamental train frequency = 0.382 × 28.58 = 10.92 Hz
Ball spin frequency = 2.003 × 28.58 = 57.25 Hz
Outer-race frequency = 3.057 × 28.58 = 87.37 Hz
Inner-race frequency = 4.943 × 28.58 = 141.27 Hz

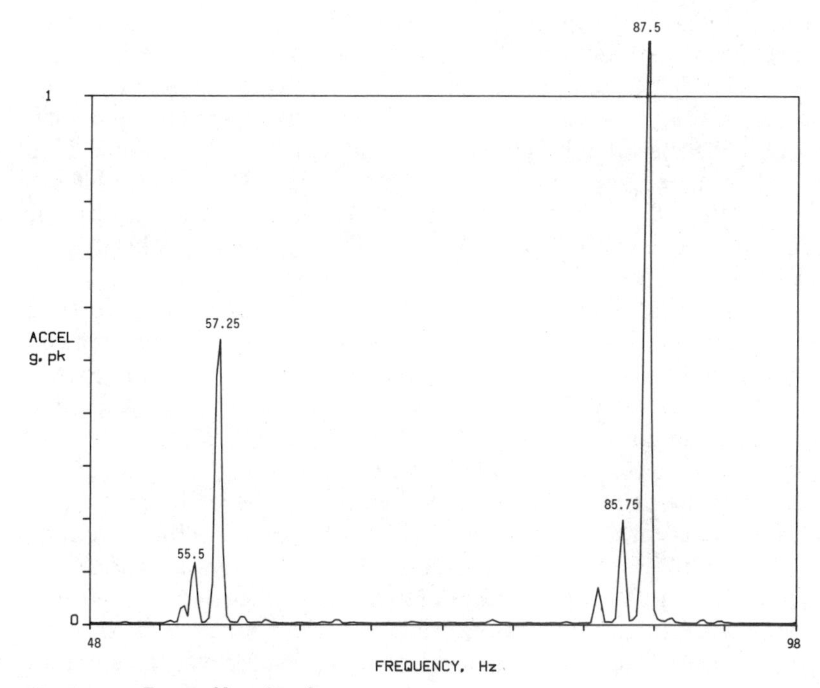

Figure 5.23 Zoom of bearing demonstrator.

Figure 5.24 Test setup to measure bearing vibration.

bearings, which generate no discrete vibrations. This NSK 6311 bearing came from a 20-hp motor that had run for several years driving a fan via a belt. This bearing was on the drive end of the motor. It was removed because it made a noise that caused the maintenance person concern, so he installed a new bearing and gave me the old one. This NSK 6311 bearing was actually okay, that is, there was no visible defects on either of the races or the balls, but it generated significant vibration when run dry, as evidenced by Fig. 5.23. An acceleration of 1 *g* at a bearing frequency is cause for concern.

Notice in Fig. 5.24 that the accelerometer is mounted in a vertical direction. This is the best orientation to monitor bearing vibrations on horizontal shaft machines because the accelerometer-sensitive axis is in line with the bearing load zone. The bearing load zone is typically a 45° arc centered on the bottom of the bearing (Fig. 5.25).

This bottom-centered load zone is valid where the gravity weight of a machine is the primary force on the bearings. Where different directional forces act on a bearing, such as belt-driven fans, then the load zone will be different. Thrust bearings obviously have an axial load zone, and they are better monitored in the axial direction. For normal machines with a dominant gravity force on the bearings, monitoring in the vertical direction is best to see bearing frequencies. Imbalance, however, is best detected by mounting a transducer horizontally because most machines are more flexible in that direction. If only one transducer is to be mounted on a bearing, then a compromise needs to be reached based on the purpose of monitoring. For bearing frequencies, it is actually best to monitor directly in the load zone. For most bearings, this means on the bottom, which is impractical. If that is the case, then it is sufficient to just monitor in the same direction as the dominant bearing forces, i.e., vertical.

Figures 5.26*a* and 5.26*b* are comparable plots of the same NSK 6311 bearing tests as in Fig. 5.22. Figure 5.26*a* is with the accelerometer

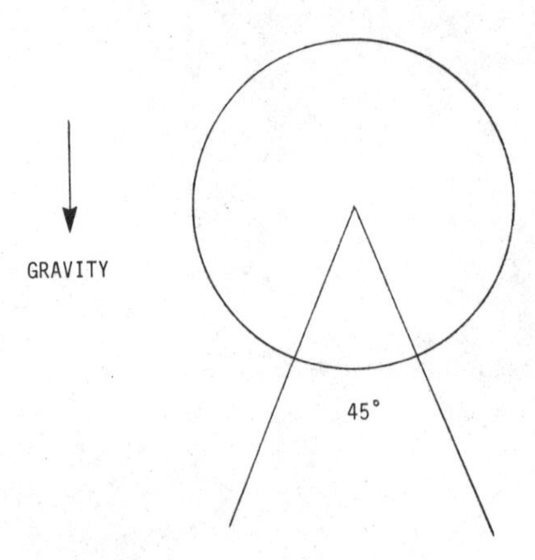

Figure 5.25 Bearing load zone.

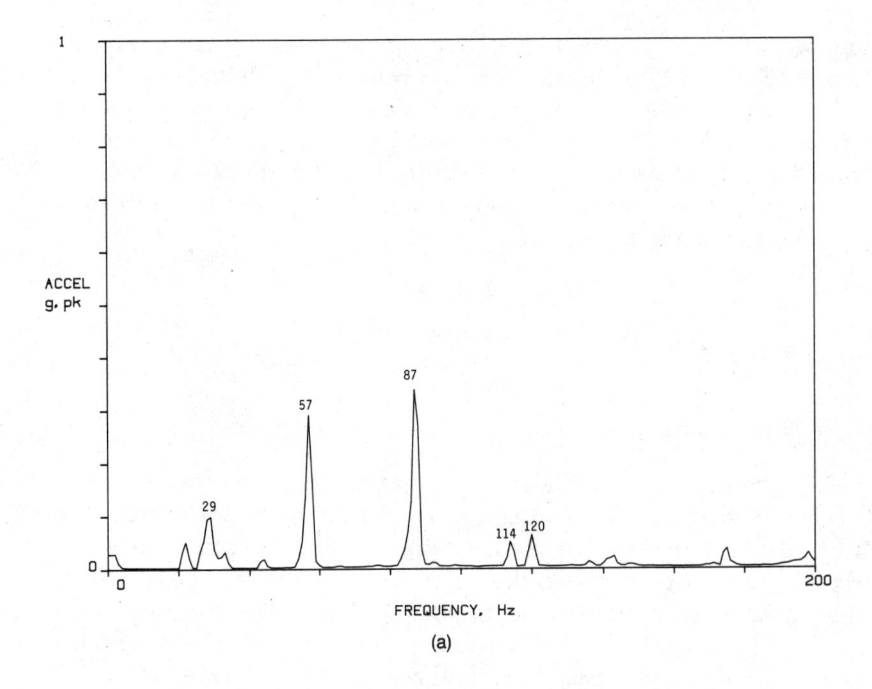

(a)

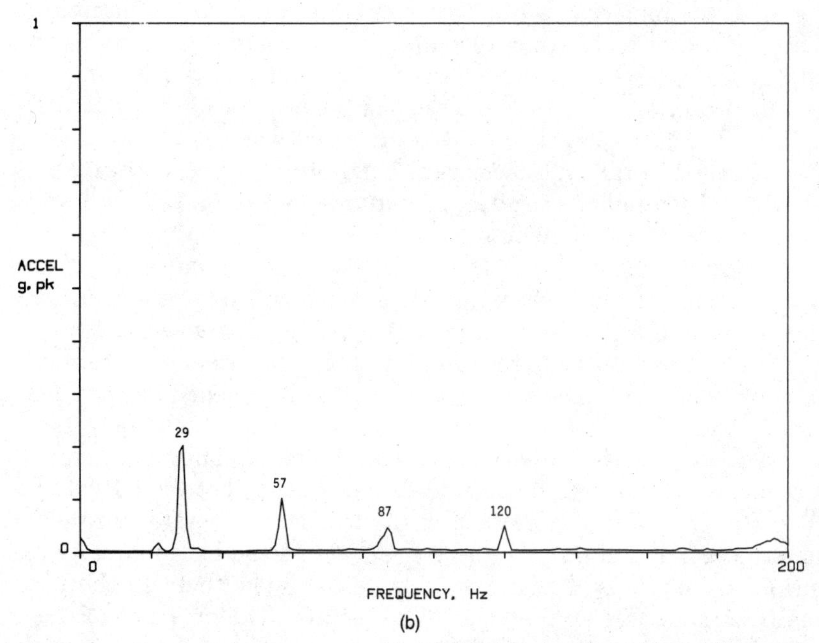

(b)

Figure 5.26 Bearing demonstrator with NSK 6311 bearing. (*a*) Accelerometer mounted at 45°. (*b*) Accelerometer mounted horizontal (see Fig. 5.22 for the vertical direction).

mounted at a 45° angle, and Fig. 5.26*b* is with the accelerometer mounted horizontally. Notice how the bearing frequencies at 57 and 87 Hz decrease dramatically when moving the accelerometer away from the vertical direction. This example emphasizes the need for proper transducer orientation to monitor bearing vibrations.

If bearing tables are unavailable, then bearing frequencies can be approximated with these simple rules of thumb:

$$\text{FTF} = 0.4\,(\text{speed}) \tag{5.7}$$

$$\text{OR} = 0.4\,(\text{speed}) \times N \tag{5.8}$$

$$\text{IR} = 0.6\,(\text{speed}) \times N \tag{5.9}$$

where N = number of balls

These formulas are based on the fact that the cage rotates 40 percent as fast as the shaft. This can be easily verified by hand rotating any ball bearing after marking the cage and turning the inner race 360°. According to these formulas, the outer-race frequency is 40 percent of the rotational speed times the number of balls. The inner-race frequency is 60 percent of the rotational speed times the number of balls. The speed can be measured. Most ball bearings have between 7 and 15 balls, with the majority falling between 9 and 11. So if the number of balls is not known, assume 10 balls, then a rough estimate can be made of the outer- and inner-race frequencies by multiplying the speed by 4 and 6, respectively. This has proven useful to tentatively identify bearing frequencies while taking field measurements.

Equations (5.7) through (5.9) have been proven to be within 2 percent when the number of roller elements is less than 12.[3] The error increases with more elements.

The relative amplitudes of these vibrations is an indication of the bearing condition. These bearing frequencies will be present on new bearings, even if lightly loaded. As the bearing parts wear, these specific tones get louder, or higher in amplitude. The frequency remains the same. The very first indications of bearing wear are metal-to-metal impact shocks. These metal impacts are the balls, or rollers, making contact with the races or cage assembly. They show up as shock pulses in the time domain with a frequency between 1000 and 10,000 Hz. They can also be seen in the frequency domain as random peaks that come and go. They average out to some broadband vibration in the frequency domain, so it is best to view them in the time domain with an accelerometer as a transducer. The presence of these

[3] Richard L. Schiltz, "Forcing Frequency Identification of Rolling Element Bearings," *Sound and Vibration,* Vol. 24, No. 5, May 1990, pp. 16–19.

shock pulses do not necessarily indicate a bad bearing. What they do indicate is metal-to-metal impacts due to defects, high loads, or lack of lubrication. If this condition is left to continue, it will most certainly lead to accelerated wear and premature failure.

Figure 5.27 shows a time plot of a "bad" bearing. Notice the shock pulses that swing between the top and bottom extremes of the amplitude. This bearing came from an automotive transmission that was found lying in the road. There was significant spalling and pitting on both the outer race and inner race that was clearly visible. I once viewed a bearing on a 100-hp motor that displayed even more severe shock pulses than this. I thought it was due to heavy load and did not notify the owner. Four days later it failed during a holiday weekend.

When viewing bearings for shock pulses, it is best to set up a consistent display window for comparison purposes. With a 100 mV/g accelerometer, I have found a time length of about 40 msec and an amplitude scale of about ± 0.5 V to be good. For comparison, Fig. 5.28 is a new bearing, and it is well lubricated. The 8.3-msec period is the 120-Hz vibration from the motor. If the 120 Hz from the motor were absent, then the time plot of this good bearing would have a small amplitude excursion. No severe shock pulses are present.

Even without any instruments, these shock pulses due to metal impacts can be heard with your ear to the bearing via a screwdriver. It is

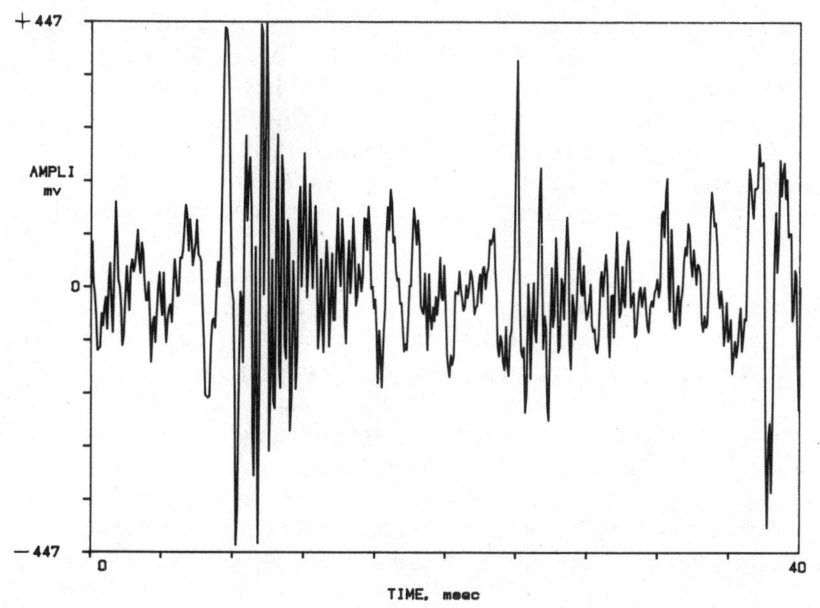

Figure 5.27 A "bad" bearing. Shock pulses indicate metal-to-metal contact. Accelerometer sensitivity 100 mV/g.

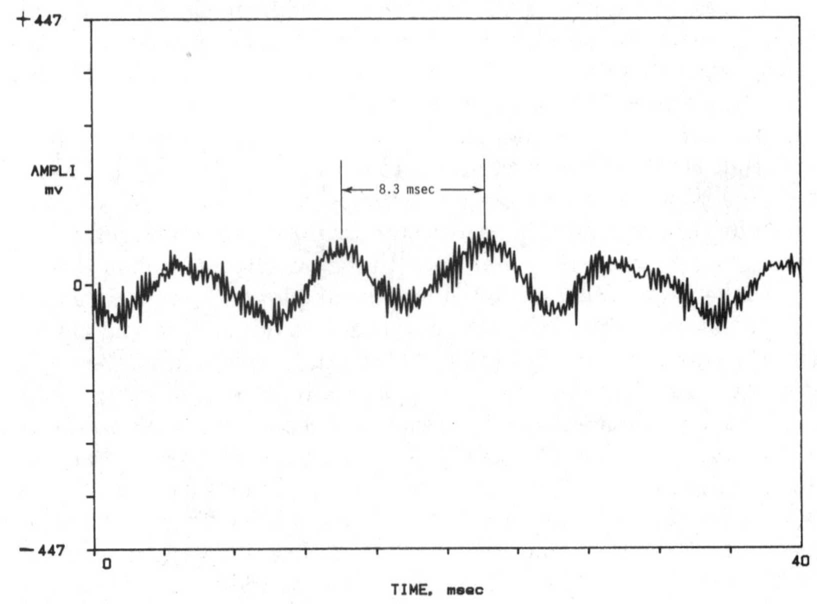

Figure 5.28 A "good" bearing. No shock pulses.

best to listen to this "ticking" during coast-down when power has been removed and electrical and turbulence noise is absent.

As wear progresses, these shock pulses increase in number and severity. Bearing analyzers typically look for and count these high-frequency shock pulses. They display their result as "Spike Energy"[4] or "BCU," bearing condition units. Be aware that other, nearby high frequencies can be detected also by these instruments, i.e., belt squeal, coupling guard rubs, pump cavitation, control valve noise, steam leaks, or other impacts. These other sources get counted in as bearing noise.

It is important, at this point, to summarize the two types of vibrations emanating from bearings. The first discussed was the repetition rate of balls encountering a defect. These are the low-frequency vibrations from 4 to 10 times running speed. These are shown in Fig. 5.29 as the bearing defect recurrence rate. The period of this defect is the time interval between balls encountering the same defect. When the ball encounters this defect, it is a metal crash, and a high-frequency shock pulse travels through the bearing housing as a pressure wave. Both of these vibrations can be detected by the accelerometer. The short-duration transient nature of the shock pulses makes them not

[4]Spike energy is a trade name of IRD Mechanalysis.

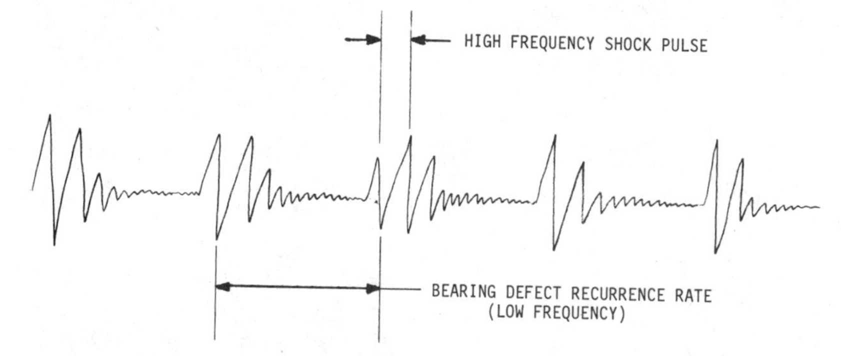

Figure 5.29 A hypothetical bearing vibration to illustrate low-frequency periodic vibration and high-frequency transients.

very visible in the frequency domain, but they can be clearly seen in the time domain. Therefore, the earliest detector of bearing degradation are these high-frequency shock pulses.

As bearing wear progresses, defects show up first on the outer race as spalling. The amplitude of the outer-race frequency increases as the spalling gets deeper. In the final stages of failure, the rollers transfer the spalling to the inner race, and inner-race frequencies increase. During this process, the shock pulses get bigger and more frequent. The relative amplitude of outer-race frequencies and the presence of inner-race frequencies will indicate the time to failure.

From experience, 0.1 in/sec at a specific bearing frequency indicates a significant visible bearing defect. This corresponds to 0.1 to 0.3 *g* in the range of normal bearing frequencies from 60 to 200 Hz. That is, if the bearing were disassembled at that point and visually inspected, a visible defect with the naked eye could be seen, probably on the outer race. This does not mean, however, that the bearing needs to be changed. That decision is up to the responsible maintenance professional with other factors being considered, such as production requirements, adverse consequences of continued operation, availability of spare parts, and the next scheduled maintenance downtime.

Bearings are load-carrying members and will continue to carry the load even into the final stages of failure. Some companies continue to operate bearings beyond the first signs of degradation and well into the failure cycle. Sometimes this is due to necessity to get beyond a production cycle, and other times it is a matter of choice based on past experience. The responsibility of the vibration analyst is first to notify the production manager of a developing defect, and then to keep him or her advised with more frequent monitoring as the bearing progresses toward failure.

One final note concerning bearing vibrations. Bearings can be in-

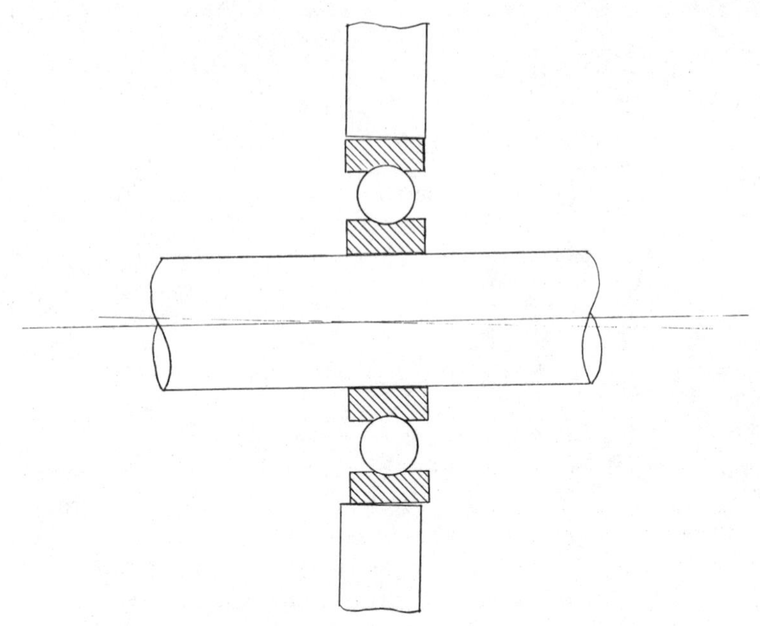

Figure 5.30 Misaligned antifriction bearing.

stalled in a misaligned condition, i.e., their axis not being perpendicular to the housing bore (Fig. 5.30). The shaft and housing are then both distorted cyclically during rotation. The vibration spectrum of a misaligned bearing looks just like a misaligned coupling, with harmonics in the spectrum. The other diagnostic indicators of misalignment are also present; high axial vibration at 1X rpm and 180° phase differences. It is difficult sometimes to separate the two, coupling misalignment versus bearing misalignment. A bowed rotor can also display the same symptoms of harmonics. Some additional analysis techniques need to be used to separate out the true fault, such as:

- Run driver solo.
- Measure shaft runout with a dial indicator while hand turning.
- Carefully inspect shaft and bearing surfaces visually for contact points and wear patterns.

Gears

Gears transmit power from one rotating shaft to another. The full power generated by the prime mover, less minor losses, is transmitted by the gear teeth in contact. Consequently, significant forces are present at the surfaces of the mating teeth. These forces cause the teeth to deflect under load, and then rebound when unloaded. The local stress level is high and fatigue damage accumulates. If the teeth were of perfect form and without defects, this cyclical loading and unloading would cause very little vibration. The presence of nonperfect gears is what gives rise to vibration.

The spectral pattern of gear vibration is the crucial information to understanding their defects. A single defect on a single gear tooth will cause a force perturbation at 1X running speed. Figure 5.31*a* is a time domain view of a 30-tooth gear with 1 defective tooth. The defective tooth was created by filing flat a tooth face. A shock pulse clearly arises every time this defective tooth comes in contact. The speed of the gear was 1200 rpm (20 Hz), and the period of one rotation is 50 msec. However, unless the defect is large, the energy generated by this defect is of short duration and looks like a transient. It may not show up at all at 1X running speed. And, in fact, this defective tooth caused no noticeable change in the vibration at running speed. Figure 5.31*b* is the frequency spectrum of this same gear. The vibration at running speed is very small at 20 Hz. It will, however, modulate the gear-mesh frequency and appear as a 1X sideband of gear mesh. Needless to say, gear-mesh frequencies are high, typically above 1000 Hz, so an accelerometer is necessary to detect them, and the acceleration display is necessary to see them. The gears of Fig. 5.31 generated a

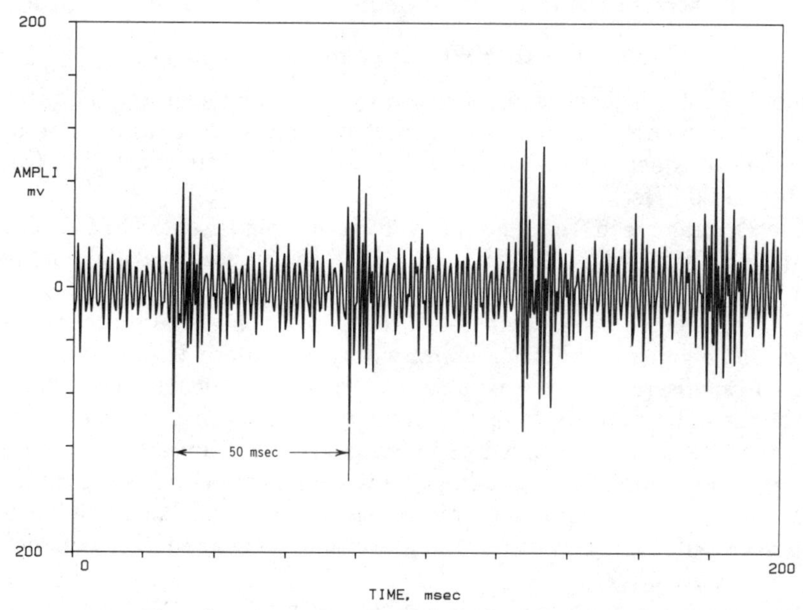

Figure 5.31(*a*) Time domain pattern of one defective gear tooth. Speed: 1200 rpm (20 Hz), 30 teeth.

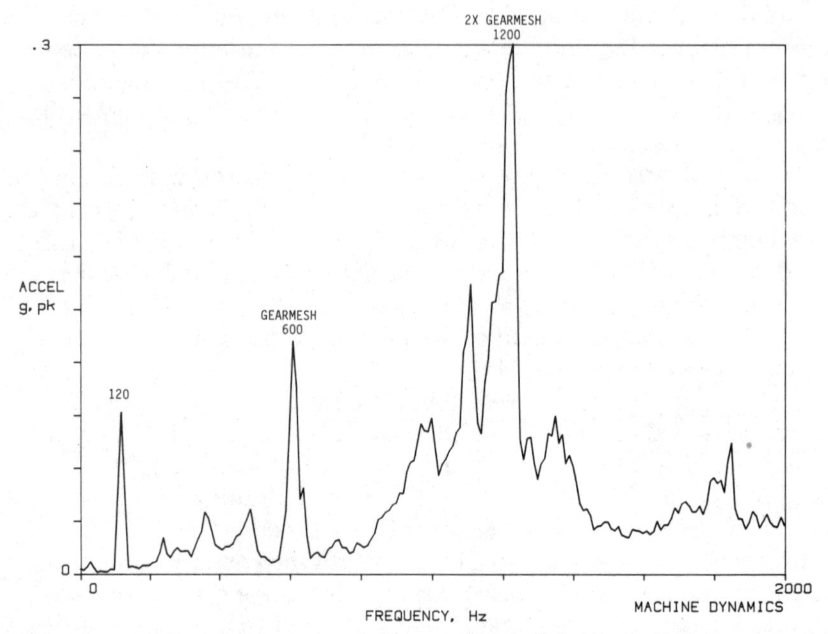

Figure 5.31(*b*) Frequency spectrum of gear with one defective tooth.

gear-mesh frequency of 600 Hz. This is calculated by multiplying the rotational speed of the gear times the number of teeth. In this case

$$20 \text{ Hz} \times 30 \text{ teeth} = 600 \text{ Hz gear-mesh frequency}$$

It should be made clear that there is only one gear-mesh frequency for two mating gears. The high-speed shaft has fewer teeth, and its product of speed times number of teeth will yield the same gear-mesh frequency of 600 Hz.

The 2X gear-mesh frequency of 1200 Hz is also present. The 2X gear mesh usually shows up larger than the 1X in the acceleration display and is a better indicator of developing gear problems. As mentioned earlier, sidebands of the gear-mesh frequency contain significant information about the condition of individual gear teeth. Zooming on the gear-mesh frequency is usually necessary to clearly see these sidebands. The amplitude of vibration at gear-mesh frequency does vary with load. Therefore, when judging the condition of gears using the vibration amplitude, it is important to take measurements under the same load conditions to obtain comparative data. Excessive backlash does cause an increase in the amplitude at gear-mesh frequency when unloaded.

A broken gear tooth obviously cannot carry any load. This momentary lapse in torque transmission causes a 1X-rpm abnormality that shows up as sidebands of the gear-mesh frequency. The sideband spacing is the running speed. These kinds of defects, broken, cracked, or chipped gear teeth, show up best in the time domain as in Fig. 5.31*a*. There is metal-to-metal contact either at this defective tooth or when the next mating abruptly takes up the load. These metal crashes appear as shock pulses in the time domain. The time interval between shock pulses is the time interval between defect encounters. For example, Fig. 5.32 is a hypothetical time plot of a mating spur gear set. The rotational speed is 1800 rpm (30 Hz) which corresponds to a period of 33.3 milliseconds. This is the time interval between shock pulses of similar patterns. There are two separate patterns, therefore, two teeth have abnormalities. The time interval x is proportional to the arc distance along the circumference between the bad teeth. Serious gear problems will cause accelerated wear of metal particles and should be confirmed with a wear particle analysis.

Finally, the metal crash at defective gear teeth is equivalent to striking the gearbox with a hammer because of the significant forces involved. These "blows" will excite structural resonances of the gears, shafts, and even housing parts. High-quality gearboxes should be relatively free of serious resonances. This is achieved by building massive and stiff housings, and supporting them in a like manner. The consequences of having serious resonances on gearboxes is accelerated gear wear due to relative motion between operating parts, affecting the alignment in a dynamic environment.

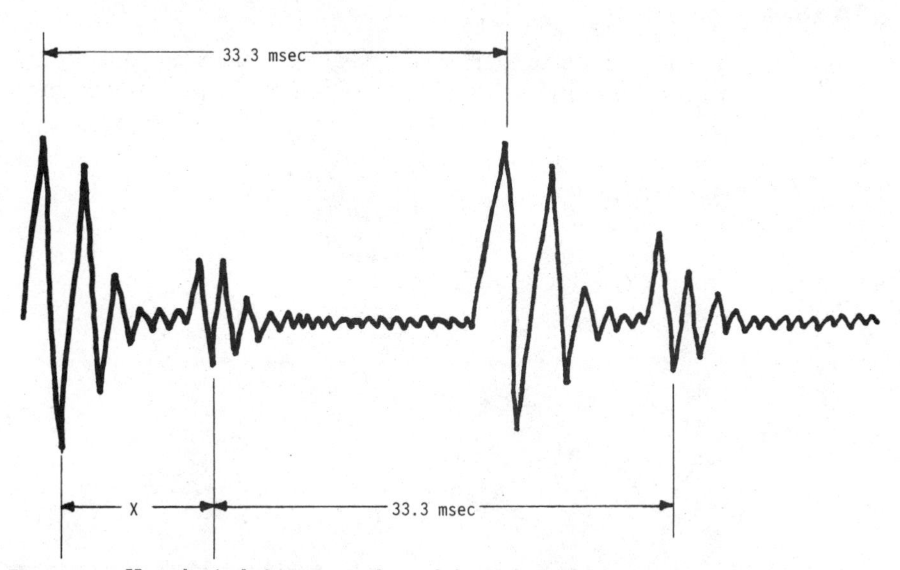

Figure 5.32 Hypothetical time trace of two defective gear teeth on a shaft spinning at 1800 rpm (period = 33.3 msec).

Gears, when produced, are not perfect with respect to tooth profile. During the early few hours of operation, the teeth wear in and the gear-mesh frequency can be expected to decrease. As gears wear, the frequencies remain the same, but the amplitudes increase. The peaks at gear-mesh frequencies broaden and develop sidebands of the primary rotational speed.

It is important to mention two facts concerning gear faults. The first one is that as these load-bearing members wear, they get louder and louder (increased amplitude), but they will continue to carry the load. At some point cracks begin to develop, and these are the first signs leading to a catastrophic failure. It is sufficient, usually, to monitor the gear frequencies and watch their amplitudes grow due to normal wear. With experience, and a previous history to rely on, you will be able to predict at what level a gear will fail. A sudden change is very significant. The change may be an abrupt decrease to a lower plateau. Figure 5.33 shows such a change. The abrupt drop is not good news. It indicates that a dramatic decrease in stiffness has attenuated the force transmission path. The crack has propagated much more, leaving the gear or shaft very flexible and able to absorb the forces by bending more. Catastrophic failure is eminent. The second fact concerning gears is that they are designed to last the life of the machine. If they do not, then suspect some other cause, such as imbalance, misalignment, or improper lubrication.

Vane Passing

A phenomenon similar to gear-mesh frequency is vane passing frequency on fluid-handling machines. This occurs on pumps and fans.

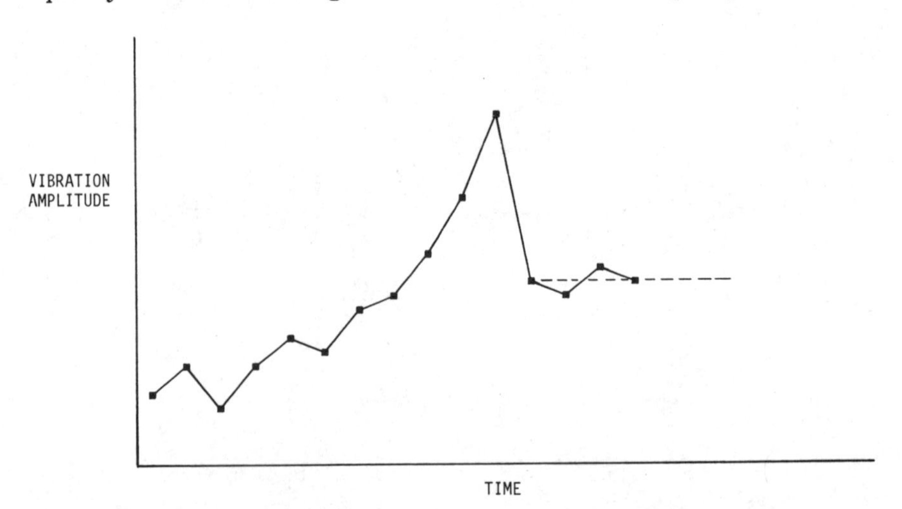

Figure 5.33 An abrupt drop in vibration amplitude signals a significant change.

The source of the vane passing frequency is a pressure fluctuation as a vane passes a discontinuity within its chamber. This discontinuity is usually the edge of the discharge opening, but it could also be a structural support on a propeller fan, or the trailing edge wake of an airfoil fan. The vane passing frequency is always the product of the rotational speed times the number of vanes. Figure 5.34 is a vibration spectrum of a small centrifugal fan with 24 blades. The speed of the fan was measured with a strobe light to be 3355 rpm, or 56 Hz. The vane passing frequency is therefore calculated to be

$$56 \text{ Hz} \times 24 \text{ blades} = 1344 \text{ Hz}$$

The existence of vane passing frequencies can be expected as a fact of operation of all pumps and fans. They are usually not a problem unless they excite a structural resonance or cause an acoustical problem on fans.

Fans

Fans are the cause of more field vibration problems than any other category of machines. Motors easily outnumber every other kind of machine and are a significant source of vibration on their own, but fans require more attention to keep running. The reasons are their function and their construction.

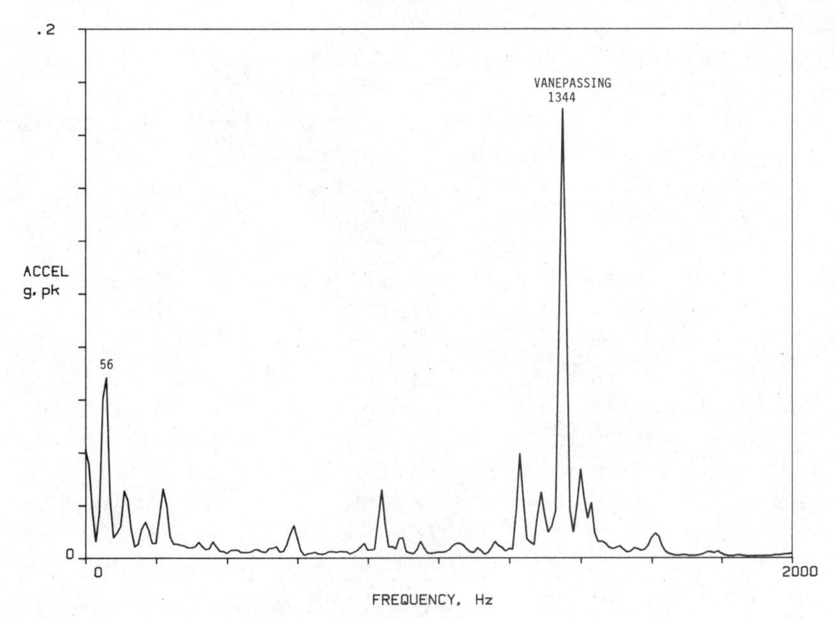

Figure 5.34 Spectrum of a small centrifugal fan.

Fans handle air, and if that were all that they handled, then they would operate clean. In reality, they handle grease-laden air, they exhaust corrosive gases, and they move erosive particles like sand, fly ash, or cement dust. Under these conditions, they gain or lose material and require frequent rebalancing. The level of balance must also be very fine because of their large wheel diameters compared to other machines. Fans are usually mounted on springs, and this added flexibility demands better balancing to keep down motion. During final trim balancing, it is not uncommon to add only 1 or 2 g of weight to a fan. Fans that handle material that tends to stick to the blades, like grease, must be cleaned periodically. After cleaning, the weight distribution can change, either because a previous balance weight washes off or previous balancing was done on a dirty fan to compensate for an accumulation which is now gone. Fans require frequent rebalancing. When approaching a fan vibration problem, the first task of analysis should always be to measure the fan speed and then measure the vibration amplitude at that speed to judge the balance condition.

The second reason that fans cause so much trouble in the field are their flimsy construction. This is not meant as a degradation of their quality, but just to point out the realities of physics and affordable construction. To move large volumes of air without compression, large chambers and passageways are required. Sheet metal construction is the standard today, with fiberglass also being used. These large panels are flexible and easily set into motion at the driving frequency. With low damping, large resonant vibrations on sheet metal panels are easily sustained. If fan housings and ductwork were made of concrete, then we would have fewer vibration problems on them. But concrete construction is not economical, and we are left with sheet metal construction that is affordable but prone to vibration.

These sheet metal vibrations are not a concern in terms of life or damage to the fan or its bearings. They are a problem when they transmit to a sensitive receiver. These vibrations are typically described as duct rumble. These are separate from air noise through room delivery registers or vane passing frequency which is a pure tone. The rumble is very low frequency, less than 100 Hz, and can typically be felt in the walls or floors. The path of this vibration is along the ducts and into the structure. It is many times a combination of airborne noise and structural vibration. These vibrations are not usually caused by a "bad" machine. They are usually caused by a failure in architecture and space planning by locating sensitive receivers in close proximity to an air handler. They are difficult to correct in the field. I have had little success in solving such problems when the owner is unwilling to relocate either the air handler or occupants.

Other solutions are possible, but they are even more expensive. Balancing of the fan usually has little effect. These duct vibrations can be recognized by the low-frequency broadband vibration below 100 Hz in the structure or in the airborne sound.

Figure 5.35 is a sound spectrum in the lobby of a reception area. This sound spectrum was obtained with a condenser microphone that had good low-frequency sensitivity. Notice the broadband nature of this sound field out to 100 Hz. Some discrete peaks are visible, which are the speeds of the supply and return fans immediately above the lobby on the roof. A vibration measurement on a column shows a very similar spectrum. Balancing of the fans was done with no improvement in the problem. Improvement in this spectrum, and in the perceived noise and vibration, could only be achieved when the fan speeds were slowed down 40 percent. This was unacceptable to the owner and the occupants because of reduced air conditioning performance. Various other fixes were tried over a 5-year period, such as stiffening, enclosing the ductwork in sheetrock, and sound baffles, with little or no improvement. This situation was brought up at this time to show you how to recognize a common and difficult fan problem—low-frequency broadband vibration below 100 Hz due to duct rumble.

Fans move air and create turbulence within ducts when the air velocity is too great. This turbulence is what sets the duct walls into mo-

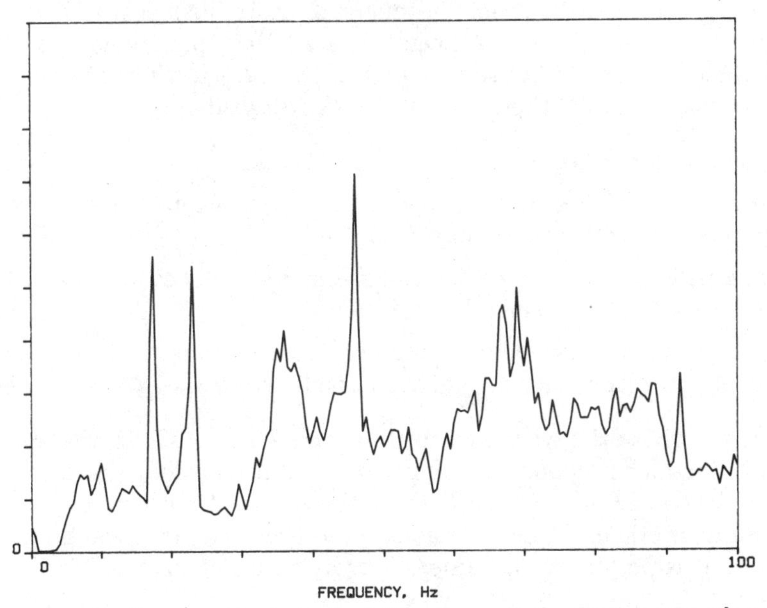

Figure 5.35 Sound spectrum in the lobby of a medical reception area where noise was objectionable.

tion. On the duct walls, this vibration is either a broadband vibration or a vibration at the duct's resonant frequency, or more commonly, both. Other aspects of fan vibration, such as belts, pulleys, and bearings, are covered in other sections.

Fans can also generate pure tones at vane passing frequency which transmit into the occupied spaces through the ductwork. This is most common on airfoil-type blade fans. This is really an acoustical problem, rather than a mechanical vibration. Three case histories of fans generating a pure tone are presented in Chapter 9, "Acoustics."

Motor Vibrations

Every electrical machine in the United States that is powered by 60 Hz line frequency gives off a mechanical vibration at 120 Hz. This includes all ac motors and power transformers. This is normal and is called *magnetostriction*. It is caused by the magnetic field expanding and collapsing in the iron laminations of transformers and motors. A second source of 120 Hz is the variation of air gap in ac motors. The presence of 120-Hz vibration is normal. The relative amplitude is a measure of the quality of construction. Better electrical machines generate a smaller 120-Hz hum.

Figure 5.36 is a spectrum of a typical restaurant exhaust fan. This was a belt-driven centrifugal fan. This measurement was taken on the fan bearing. The 120 Hz from the motor is quite strong on the fan bearing. Also notice the imbalance at 34 Hz (2040 rpm). This imbalance was the reason for the service call. If the 120-Hz vibration from motors is objectionable, then several things can be done:

1. Isolate the motor better.
2. Change to a higher-quality motor. Energy-efficient motors are manufactured with better quality control.
3. Go to a three-phase motor. Three-phase motors are better balanced electrically to deliver smoother rotation.
4. Replace with a sleeve-bearing motor. Ball bearings transmit the 120 Hz better from the rotor to the outside world.

In other parts of the world, where line frequency is 50 Hz, the mechanical vibrations emanating from motors and transformers are at 100 Hz. Electric motors also produce torsional vibrations that depend on the number of poles. These torsional vibrations couple to the linear mode, and it is possible to get linear vibrations at 60 Hz or multiples thereof.

Large motors generate high-frequency vibrations in the range of 500 to 5000 Hz. These higher frequencies are also related to manufac-

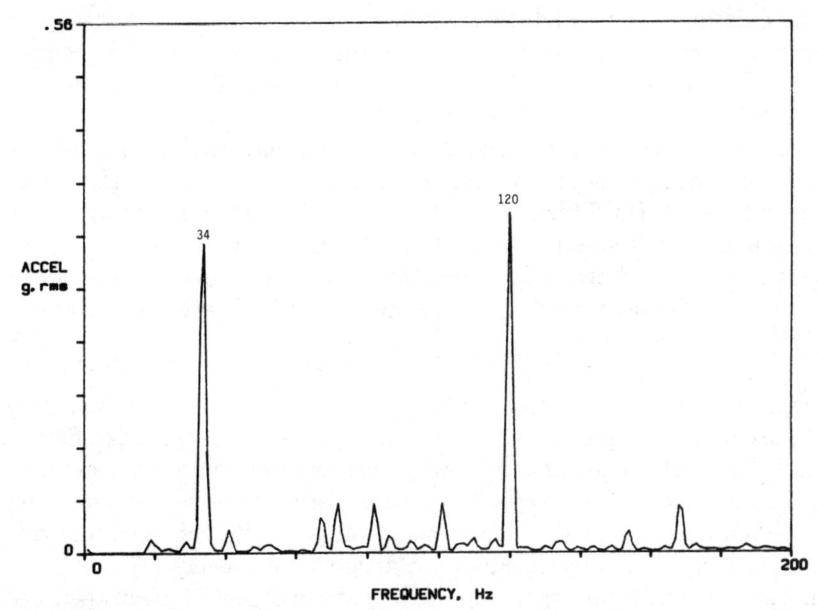

Figure 5.36 Vibration spectrum of a restaurant exhaust fan rotating at 2040 rpm (34 Hz) with a 120-Hz motor hum.

turing variability but are not well correlated to motor reliability. That is, many large motors have been observed to generate these high frequencies for many years with no detectable damage or decreased life. The primary complaint is the objectionable audible noise of high-pitch tones. The best solution is to apply some acoustical treatment around the motor and monitor the lower frequencies of less than 200 Hz for developing defects.

Vibrations due to electrical problems can be easily determined by disconnecting electric power to the machine and seeing if the vibration (and noise) immediately disappears or gradually winds down. An abrupt decrease in vibration upon power disconnection indicates an electrically related vibration. If the vibration remains high and gradually decreases as the rotor spins down, a mechanical cause due to rotation is indicated, possibly imbalance, misalignment, or bearings.

DC motors do not generate a 120 Hz, nor are they plagued with 60-Hz torsional vibrations. They are smoother in terms of vibration. Since they are variable-speed machines, there is always a potential for exciting resonances within their speed range. DC motors are a good choice where variable speed and smoothness of operation are requirements. The worst situation, in terms of vibration, is a variable-speed ac motor.

It has become common recently to install variable-speed drives on ac motors. The reason is to conserve energy by slowing down a motor rather than throttle the output of a fan or pump. The problem is twofold. First, many existing motors, and the machines that they drive, were designed to run at a constant speed, which is how they were tested. To retrofit an existing machine in the field to run at other than its tested constant speed is to expose the possibility of finding a resonance.

Second, many variable-speed drives for ac motors pulse the motor windings. These pulses have higher rise times and are at a higher frequency than 60 Hz. This aggravates the motor laminations and generates a higher-frequency hum than 120 Hz. For example, a pulse-width-modulated drive that generates 20 pulses for every cycle of the incoming 60-Hz power will create a mechanical vibration at 1200 Hz:

$$60 \text{ Hz} \times 20 = 1200 \text{ Hz}$$

This high frequency may not only cause degradation at the bearings but may also be unpleasant to be around. Other variable-speed drives (other than pulse-width-modulated types) are smoother. The best protection for the owner is to specify vibration and acoustical testing to be performed throughout the entire speed range whenever machines are to be purchased or retrofitted to run at variable speed.

Large motors, greater than 100 hp, have been observed to abruptly develop an imbalance. I have seen two of these in the past 6 months. One had run for 13 months and the other for 8 years, when a large imbalance came about almost overnight. The identification of this imbalance is easy if the motor is periodically monitored. The precise cause of gain or loss of material in the motors I examined was not identified, but the correction was easy before serious bearing damage occurred.

I must tell you a story about a motor that burned up its supply wiring when single phasing. Single phasing on a three-phase motor is when the current in one leg is deficient or missing. In this case it was caused by a bad electrical connection. This was a 20-hp motor driving a vaneaxial fan. The motor was changed for a planned upgrade from 15 to 20 hp. The 15-hp motor was removed and replaced with a new 20-hp motor. I trim-balanced this fan after the motor change, and everything looked normal. The vibration spectrum out to 200 Hz was below 0.003 *g*. It was a smooth-running fan.

Two weeks later, while doing some follow-up work, this fan had a strong 120-Hz hum with sidebands of running speed (Fig. 5.37*a*). The fan turned at 1200 rpm or 20 Hz. Something had obviously changed in the 2 weeks since it was first energized. The owner was advised to check the motor for single phasing before more elaborate checks were ordered for soft foot or loose field windings. The owner wrote a work order to check for single phasing, but the motor failed 5 days later before the work order floated to the top of the priority pile. The wiring leading to the fuses was burned and the insulation melted. The fuses were also burned open. After repair, the fan was again checked 25 days later, and its vibration signature is shown in Fig. 5.37*b*. The 120 Hz

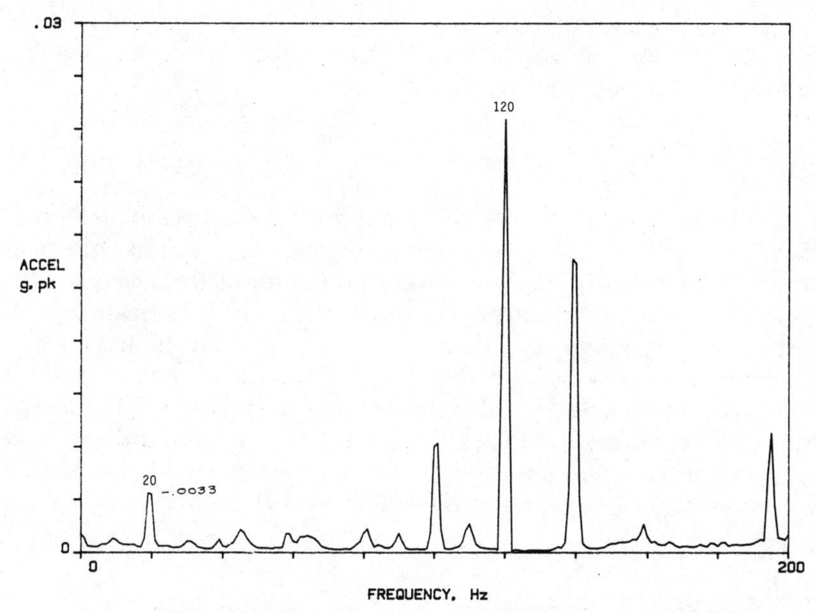

Figure 5.37(*a*) Spectrum from a vaneaxial fan with a single-phasing motor.

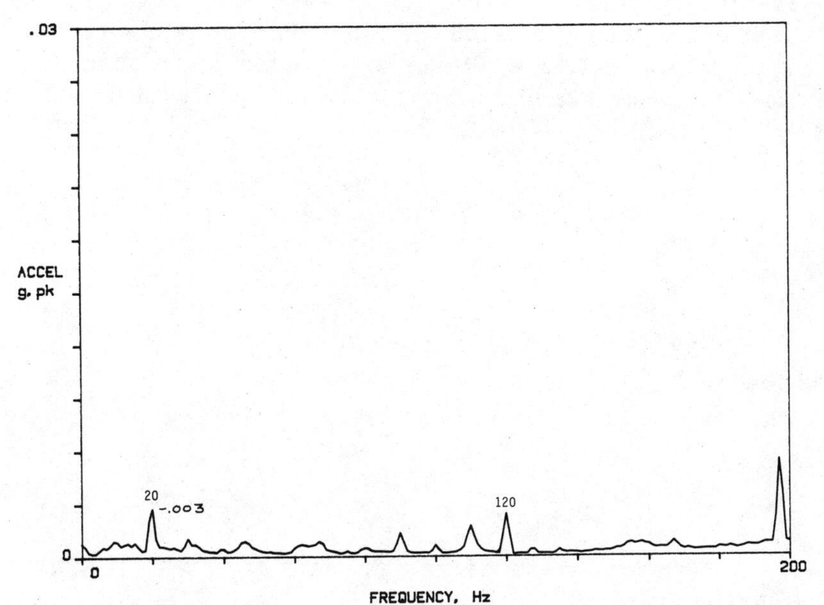

Figure 5.37(*b*) Spectrum after single phasing was corrected. (Same motor, new power wiring to motor.)

is now back to normal, and the remainder of the spectrum is unchanged. This story illustrates that faulty motor wiring can manifest itself as a 120-Hz mechanical vibration.

Cavitation

Cavitation is the vaporization of fluid within the pump. It occurs when the fluid pressure is less than the vapor pressure at that temperature. Technically, cavitation is the boiling of fluid at ambient temperature due to reduced pressure. Prolonged cavitation will cause erosion damage to a pump impeller. The causes and correction of cavitation is beyond the scope of this book, but cavitation can be recognized as a broadband vibration in the 3- to 5-kHz range (Fig. 5.38). With a spectrum analyzer monitoring the pump vibration, the controls and operating parameters can be varied to find the conditions of least cavitation and minimum vibration.

Oil Whirl

Oil whirl is a condition peculiar to journal bearings. It manifests itself as a vibration of less than one-half rotational speed. It is caused by a lightly loaded bearing riding up on its high-pressure wedge and going "up over the top and around." The journal is actually revolving around inside its bearing opposite the direction of rotation at about 45 percent of the rotating speed (Fig. 5.39). The 45 percent, or less than one-half rotation speed, comes from the average fluid velocity with some slippage.

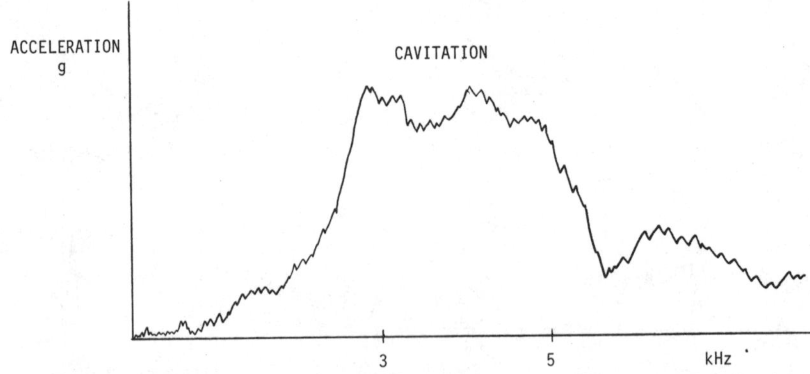

Figure 5.38 Pump cavitation symptoms are broadband vibrations typically from 3000 to 5000 Hz.

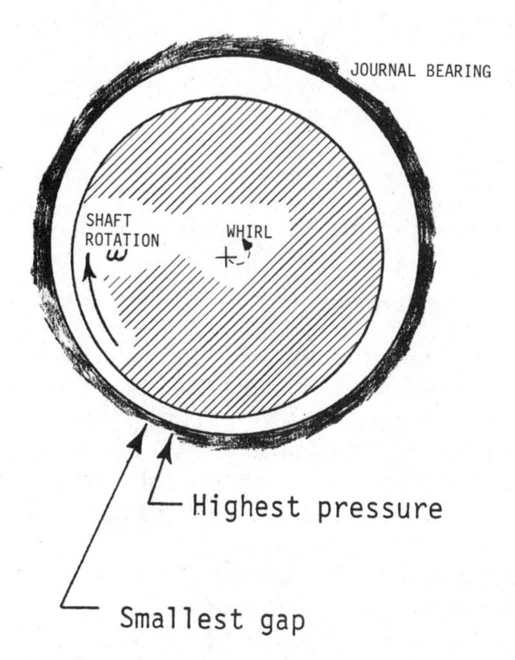

Figure 5.39 Oil whirl.

The cure is to increase the load on the bearing. This usually requires a redesign of the bearing and is best handled by the bearing or machine supplier. Some redesigns are narrower bearings, axial grooves, pressure dams, lobed journals, or tilting pad bearings. The narrower bearing increases the load on the journal. The other redesigns break up the symmetrical oil flow pattern.

There are some temporary measures that can be taken to alleviate oil whirl. This is a destructive condition, and temporary relief is usually required until a permanent fix can be redesigned. The temporary measures are to change the oil viscosity, by adjusting the oil temperature or a different oil. Another temporary measure is to run the machine in a more loaded condition. Introducing some small misalignment to load the bearings and reduce oil whirl has been applied successfully as a temporary measure.

Oil whirl is aggravated by excessive bearing clearance. This should be recognized as a condition of looseness. The motion of the shaft within its journal can be reduced by increasing the load on the machine. This pins the shaft against one side and limits its freedom of motion. Therefore, a test for oil whirl is to increase the load on the machine and observe a decrease in the subsynchronous vibration. Another test is to measure the bearing clearance. This can be done by lifting the shaft with a pry bar or jack and measuring the movement with a dial indicator. A general guideline for journal bearing clearance is:

$$0.002 + 0.001(d) \text{ in} \tag{5.10}$$

where d = shaft diameter, in

For example, a 6-in-diameter shaft should have a diametral clearance of

$$0.002 + 0.001(6) = 0.008 \text{ in}$$

Figure 5.40 shows a condition of oil whirl on a 400-hp motor hermetically sealed in a centrifugal compressor. The motor turned at 3,600 rpm (60 Hz). The 22.5-Hz vibration is 38 percent of the running speed of 60 Hz. This is due to oil whirl in the journal bearings caused by excessive clearance. At higher loads this 22.5-Hz amplitude decreases. This 22.5 Hz also decreased when these worn bearings were replaced with new ones that had a smaller clearance.

This section on oil whirl has been brought up so that you can recognize the symptoms as a subsynchronous vibration at less than one-half rotation speed. This condition appears only with journal bearings, typically on larger machines.

When oil whirl becomes severe, there is a potential for the shaft to

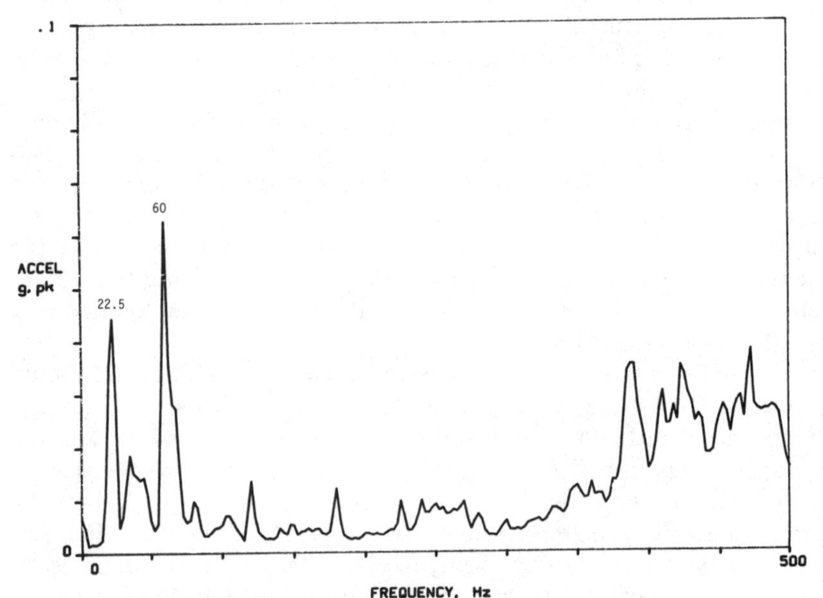

Figure 5.40 Spectrum from a 400-hp motor showing oil whirl at 22.5 Hz.

rub the inside of the journal. This causes friction and subsequent localized heating. A rub is a symptom and can be caused by other factors, most notably heavy imbalance or severe misalignment. The rub is a metal-to-metal contact and as such, shows up best in the time domain. In the frequency domain it is not so clear. The friction is a broadband high-frequency vibration that can excite resonances. The most notable effect of rubbing is a localized temperature rise followed by metal particles in the oil. Both of these should be used as supporting evidence to confirm a suspected rub. The analysis of rubbing based on vibration spectral data is not defined well enough to analyze this condition with confidence. The best indicator, in terms of vibration, is an increase in the overall vibration level, followed by observing the time domain view for evidence of metal-to-metal contact.

Piping

Pipes are passive elements and cannot produce any vibrations of their own. They are driven to vibration by some other vibrating machine that supplies the vibratory forces. Usually these vibration sources are the machines connected to the piping.

The piping will vibrate at one of two frequencies:

1. The frequency of the source
2. Its own natural frequency

It is relatively easy to determine which by measuring the frequency of vibration on the pipe. Resonance can be verified by turning off the machine and performing a bump test on the pipe.

The adverse consequences of allowing pipes to continue to vibrate is fatigue cracking at high-stress points (like flanges). Leaks then develop. Another consequence is inducing vibrations into the structure. Pipes are attached to buildings and become highways for the transportation of vibration energy. I have a story to relate that will make this concept clear.

A newly constructed four-story bank building had unacceptable vibrations on the fourth floor. Actually, the vibrations were worse on some parts of the second and third floors, but the executive offices were on the fourth floor. The building owner refused to occupy the building, and hence, pay the contractor, until these vibrations were removed.

Also on the fourth floor was computer equipment. The computers were cooled by 4-in cooling water lines passing down a pipe chase to two 10-hp pumps on the ground floor. On the fourth floor, the largest floor vibration was measured to be at a frequency of 147.5 Hz. This frequency was larger in amplitude on the third and second floors, respectively. On the ground floor, the investigation led into the mechanical rooms and directly to one of the pump motors. Figure 5.41 shows the vibration spectrum on the offensive pump motor from the bearing adjacent to the coupling. The 147.5 Hz was not the largest vibration on the motor, but it was the one that passed through the piping to the fourth floor, virtually unattenuated.

It is interesting to note that the adjacent pump "threw" a coupling within the first month of operation. In addition to replacing the coupling, the adjacent pump was aligned. This troublesome one was not. The 147.5-Hz vibration was a bearing frequency that overlaid a harmonic of running speed. The bearing was complaining due to the misalignment. The recommendation was to align this pump; however, the contractor elected a far more expensive option, to install flexible pipe connectors in all the lines leading to and from both pumps. This story illustrates how moderate vibration can be transported along a pipe to become troublesome elsewhere. It also illustrates how common sense and analysis can be overruled by other considerations, namely, cost-plus contracts.

The best solution to piping vibration problems is to find the machine that is the source of the vibration and balance or align it first. If

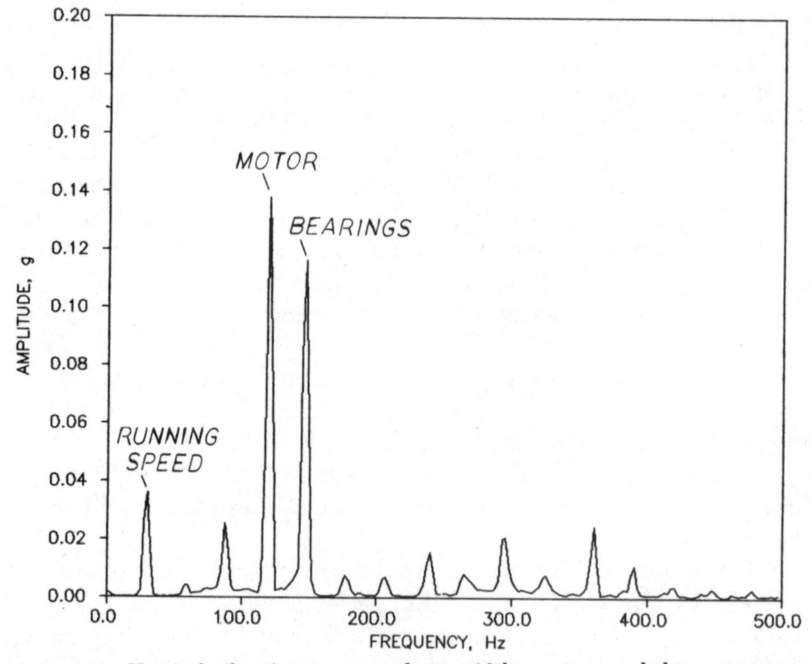

Figure 5.41 Vertical vibration spectrum from a 10-hp motor coupled to a pump.

that results in insufficient improvement, then the next step is to install flexible pipe connectors in the pipe to isolate the pipe from the machine. A third option is to detune the pipe if it is resonating, by stiffening. This is effective on small pipes (less than 6 in in diameter). It is unlikely to be successful on larger pipes. The reason will be discussed later in this chapter under the heading, "Fix the Source or the Symptom." The last alternative is to isolate the pipe from the structure if building vibrations are the complaint. Isolation is a more expensive option and should be considered if none of the other fixes work. Isolation of pipes and ducts after building construction is especially painful.

Bent Shaft and Bowed Rotor

A bent shaft and a bowed rotor are actually the same phenomena. The bent shaft is measurable outside the machine housing while the bowed rotor is the same condition inside. These defects sometimes develop on a motor that has been allowed to sit stationary for a long time. When sitting stationary, the weight of the rotor causes the shaft to deflect. After a period of time (about 6 months), the deflection takes a permanent set (Fig. 5.42).

When running, the vibration spectrum appears identical to imbalance, and in fact it is an imbalance condition. If a perfectly balanced rotor (such as a fan impeller) is attached to a bent shaft, it will run out of balance. It can be balanced back to a smooth-running condition. But then this rotor and this shaft are matched. Any other rotor on this shaft (even a new, well-balanced one) will run out of balance.

A bent shaft can be detected by measuring runout on the shaft with a dial indicator. A TIR (total indicator reading) of 0.001 or more on the end of the shaft is unacceptable. As indicated, a bent shaft can be com-

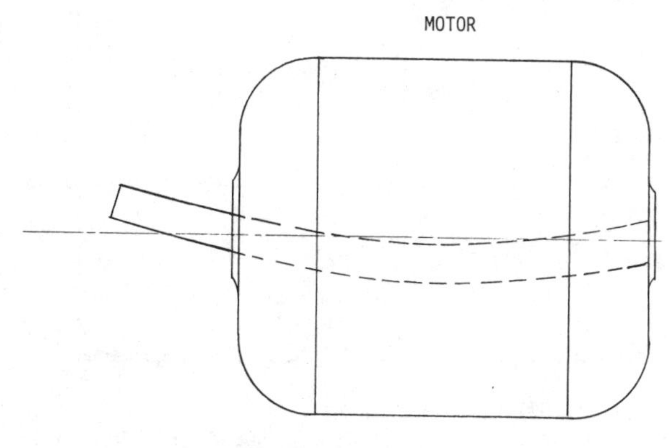

Figure 5.42 Bent shaft. Bowed rotor.

pensated for by balancing. Some motor shops claim that they can correct a bent shaft, usually by a big press or some application of heat. The best protection against a bent shaft or bowed rotor is not to let a motor, or any machine with significant rotor weight, sit stationary for any length of time. For motors in storage, it is a good idea to store them with the shaft vertical. Otherwise, in a horizontal position, have the shafts rotated on a weekly basis. The next best protection is to measure for a bent shaft before installation and not accept those motors with more than 0.001-in runout. After the fact, it is worthwhile attempting to compensate for these conditions by balancing in-place. A bent shaft or bowed rotor is an out-of-balance condition and shows up as a 1X-rpm vibration. The correction by balancing is also a standard balancing procedure. The analyst, or balancer, will rarely be aware of these conditions beforehand. His or her only indication will be an unusually large amount of balance weight required for correction.

Figure 5.43*a* shows the vibration spectrum of a cooler fan. It turned at 480 rpm and showed a large peak at 480 rpm (8.0 Hz). This was a small fan, only 2 ft in diameter, and turned at a slow speed. It sat stationary, however, for about 7 months of the year. After balancing, the vibration came down nicely as shown in Fig. 5.43*b*. The unusual feature of this balancing job is that this small, slow-speed fan required

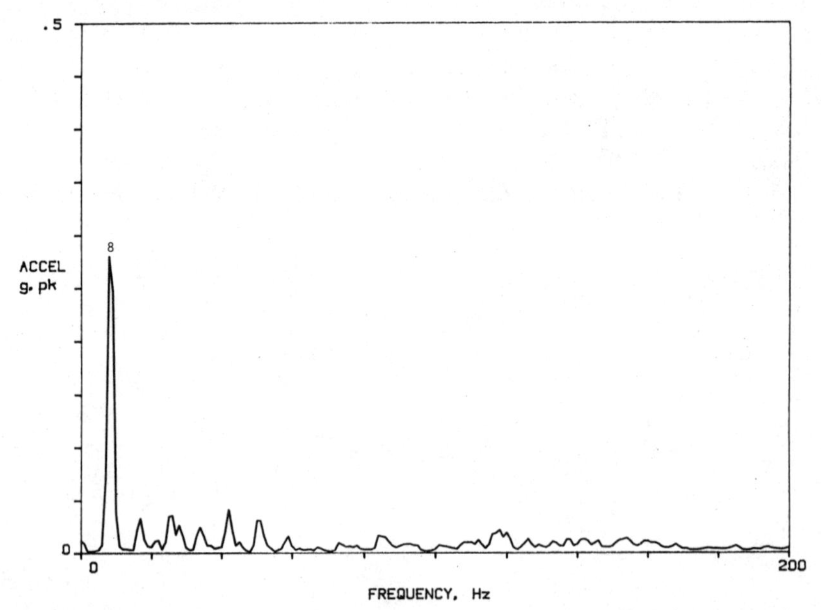

Figure 5.43(a) Spectrum of a cooler fan with a bent shaft. Fan speed was 480 rpm (8 Hz).

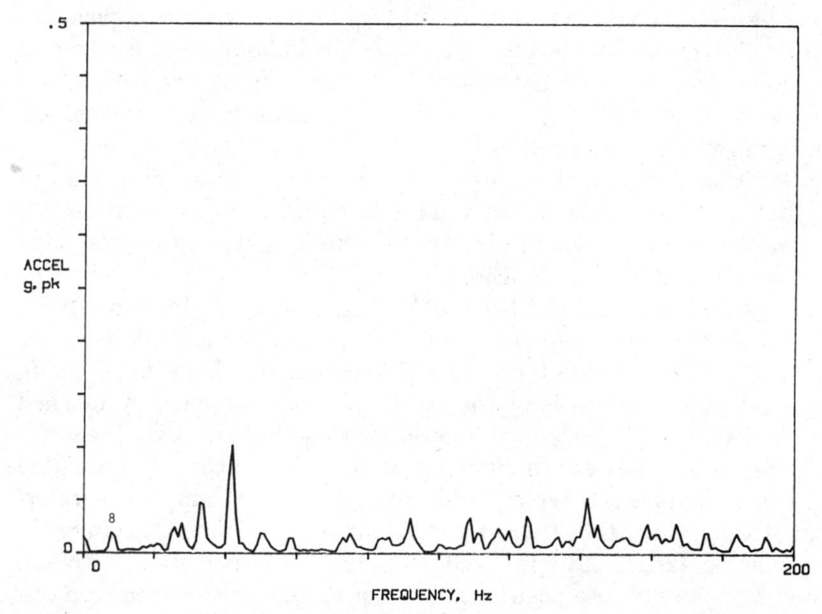

Figure 5.43(*b*) Spectrum after balancing. Ten ounces of correction weight was applied.

280 g, or 10 oz, of correction weight, all on one side. This is a case of a bent shaft corrected by massive balancing weights.

Looseness

Mechanical looseness shows up in the frequency domain as a large number of harmonics of running speed when lightly loaded. Typically ½ harmonics show up, as 1½X, 2½X, etc. The harmonics show up because of clipping of the waveform when the loose parts hit against their limits of movement. Figure 5.44 is an illustration of a sine wave that has been clipped. The vibrating part is not allowed to swing to the full extreme of motion that it would like because it hits up against physical stops. The sine wave of its motion is clipped and the resulting

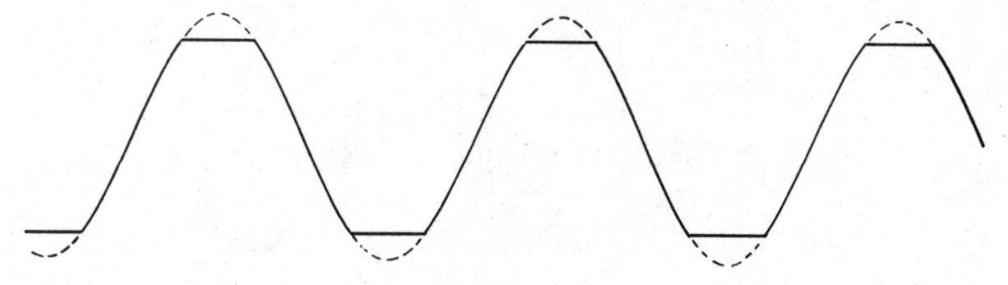

Figure 5.44 Clipping of waveform when loose part hits against physical stops.

spectrum of looseness is shown in Fig. 5.45*a*. This spectrum was created by clamping a 3600-rpm motor with an imbalance in a vise and then loosening the vise hold-down bolts until things rattled. Upon retightening, the 3600-rpm vibration dropped, as did the harmonics. The ½ harmonics disappeared (Fig. 5.45*b*).

In addition to harmonics, looseness can be detected by a decrease in amplitude at higher load. On fans and pumps, higher load causes greater thrust forces which pins the machine against its stops, thus limiting free motion due to looseness.

A large building fan had been vibrating excessively for some time. The "excessive" judgment was made by the maintenance supervisor who was comparing this fan to the three other quadrant fans. Being new to vibration analysis at that time, and having recently learned how to use an FFT analyzer, I was eager to go balance this one.

The fan wheel was 4 ft in diameter, turned at 650 rpm, and handled 60,000 cfm. This was a typical building supply fan. Using the acceleration display, I saw that there was not only a significant 1X-rpm peak, but an even larger 2X-rpm peak. The maintenance man with me changed the inlet vane positions as we observed on the spectrum analyzer. With the vanes fully open and the fan moving the maximum amount of air, the 2X rpm decreased dramatically. With the vanes fully closed, the 2X-rpm vibration was a maximum. This perplexed us at the time. We decided to balance the fan, and our first step in this

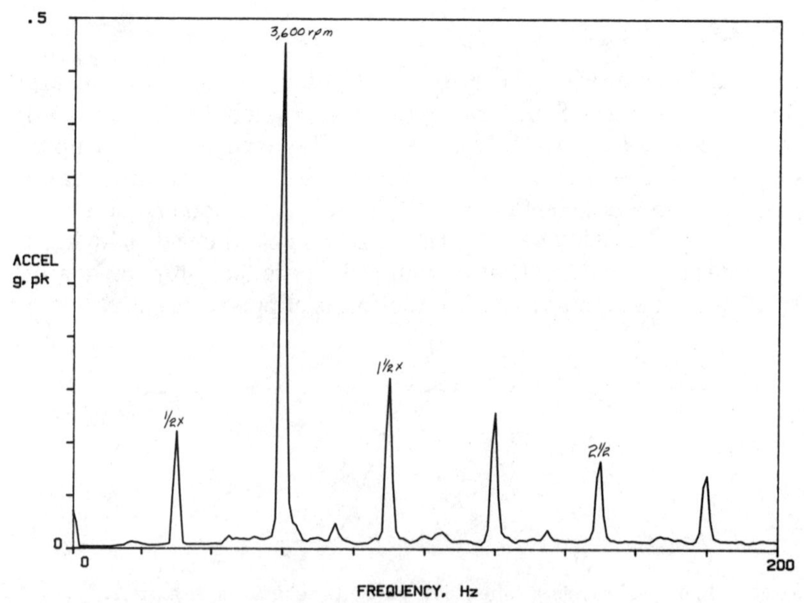

Figure 5.45(*a*) Spectrum of looseness of a 3600-rpm motor.

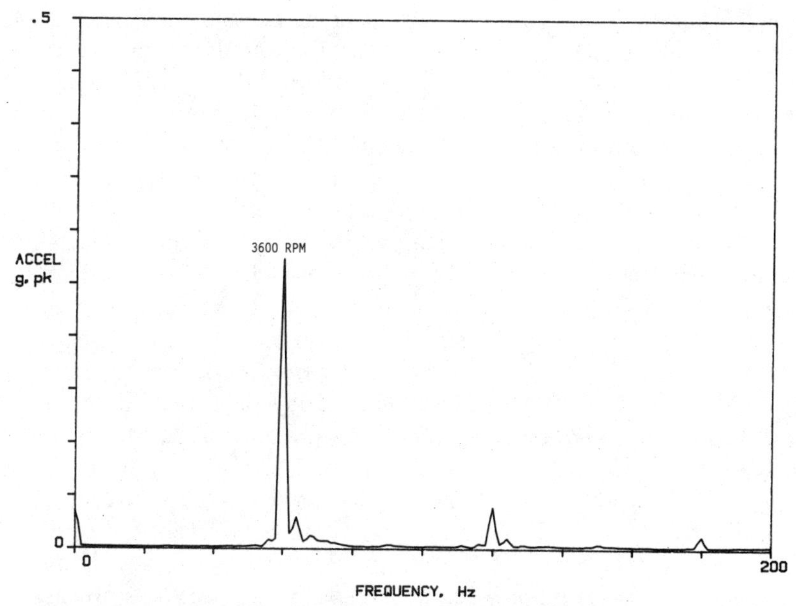

Figure 5.45(*b*) Looseness removed by tightening all bolts.

procedure was to check all the nuts and bolts for tightness. My partner found one of the four corner hold-down bolts that secured the fan housing to the concrete inertia base to be significantly loose. As he tightened it, we could both hear the vibration and noise decrease. We watched on the spectrum analyzer. As he loosened this one bolt, the 2X rpm increased dramatically. When he tightened it, the 2X rpm dropped to almost nothing and the 1X rpm also decreased about 70 percent.

We had fixed the problem with a wrench and packed up our instruments. On the way out, my astute colleague reasoned that this fan discharged air horizontally. With the vanes fully open, the thrust forces caused the fan housing to lean back and pressed it against its stops. Unloaded (with the vanes closed), it had more freedom of movement and vibrated more (high 2X rpm).

Some sources report ½ harmonics, and others report 1X harmonics, for looseness. I have observed both. It depends on the nature of the loose part and how it interacts with the rotating imbalance. Generally, the presence of ½ harmonics indicates a more severe condition of looseness. On all kinds of machines, looseness can be confirmed by changing the load and observing the vibration.

Belts and Pulleys

Some years ago in a semiconductor facility, I remember a woman worker who had difficulty aligning a photographic mask under a microscope because of vibrations. The vibration came and went about every 40 sec, building up to a maximum gradually and dying away gradually. She always had to do her aligning during a lull period. And of course, the problem would go into seclusion whenever engineers tried to investigate. The problem could not be duplicated, much less solved.

A few months later, a maintenance man was observing the belts on an air handler above this clean-room space. He saw two of the three belts build up to a resonance and die away on a cycle of about 40 sec. He changed the belts with a matched set, and we never heard any more about the microscope vibration. Evidently, the belts were slipping relative to each other. This is rather common on multiple-belt drives. When two of the belts slipped around to an unfavorable orientation and this spot went over the motor pulley, it jerked the motor more than at other times. This vibration transmitted to the microscope table 40 ft away.

In normal situations, this problem would have gone unnoticed, because the vibration on the microscope table was imperceptible to human touch. But in a sensitive environment where optics are in use, vibrations below the level of human perception can ruin the process, and the results.

Belts and pulleys need to be discussed together because they interact in a dynamic manner. They can have several problems with a corresponding variety of symptoms. Belt seams rolling over a pulley create an impulse at belt revolutions per minute. An eccentric pulley looks like an imbalance as it stretches the belt. Stretched belts can resonate like a plucked string. Pulleys that wobble in the axial direction create an axial vibration. Let's discuss these one at a time.

If there is only one discontinuity in a belt, like its joining seam, it will deliver one pulse each time it passes over a pulley. This vibration will show up at the belt speed, and its harmonics. The spectrum of a bad belt, in Fig. 5.46*a,* was created by wrapping some electrical tape around the belt and letting the machine run in that condition. The good belt spectrum, in Fig. 5.46*b,* is of the same machine with the tape removed. The two peaks at 24.5 and 30 Hz are the driven rotor and the motor speeds (1470 and 1800 rpm, respectively). The belt speed was 720 rpm. This vibration peak shows up on both spectrums with about the same amplitude. The difference between the two spectrums is the harmonic and ½ harmonic content, and the overall level.

An eccentric pulley is one that is egg shaped, or has a bore that is not concentric with the diameter of the pulley groove (Fig. 5.47). The pulley can be perfectly mass-balanced, but it will still produce a 1X-rpm vibration of that pulley speed as it stretches the belt when the high spot goes around. This looks just like imbalance with vibration analysis, but unfortunately, it cannot be balanced out. It is especially

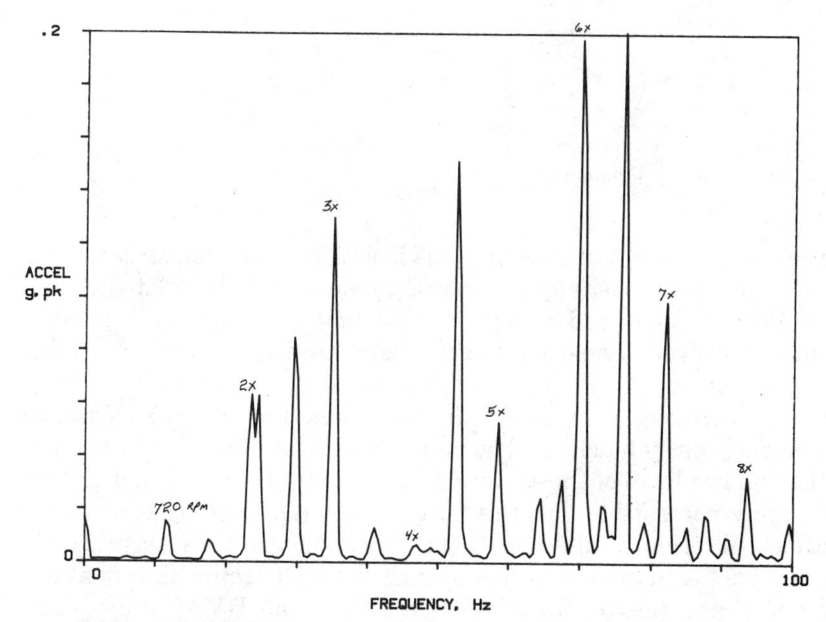

Figure 5.46(*a*) Spectrum of a bad belt seam created by wrapping electrical tape around the belt. Belt speed was 720 rpm. Overall was 0.41 *g*.

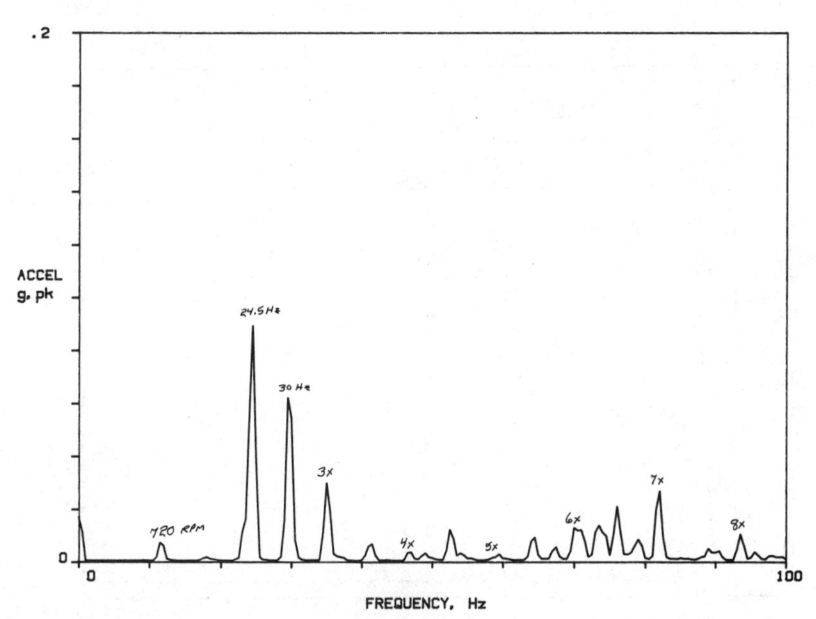

Figure 5.46(*b*) Spectrum of a good belt. The tape was removed. Overall was 0.13 *g*.

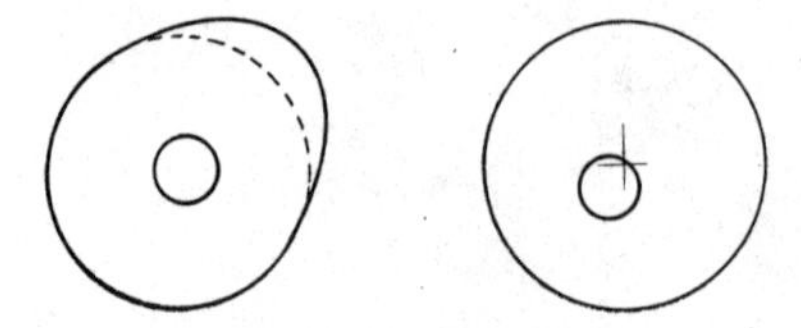

Figure 5.47 Eccentric pulleys.

aggravated by high belt tension. This is a rather common situation on HVAC equipment. Usually it is not a problem on well-isolated equipment. When the transmission path to the building is poorly isolated, a sensitive receiver is nearby, and the belts are tight, there is a setup for a problem.

Figure 5.48 is a spectrum of the vibration that was taken on the handle of an x-ray machine. The two peaks are at 1000 and 1740 rpm. Notice the low level of these vibrations at about 0.006 to 0.008 *g*. This was barely perceptible, but the x-ray technician knew that any perceptible vibration would create blurred images. This technician's hands were his instruments for sensing vibration. Immediately above this x-ray machine, on the roof, were eight small HVAC units, each with a small fan turning at 1000 rpm driven by a ½-hp motor at 1740 rpm. This looked like a simple imbalance problem. However, all attempts at balancing were futile. The solution was simply to loosen the

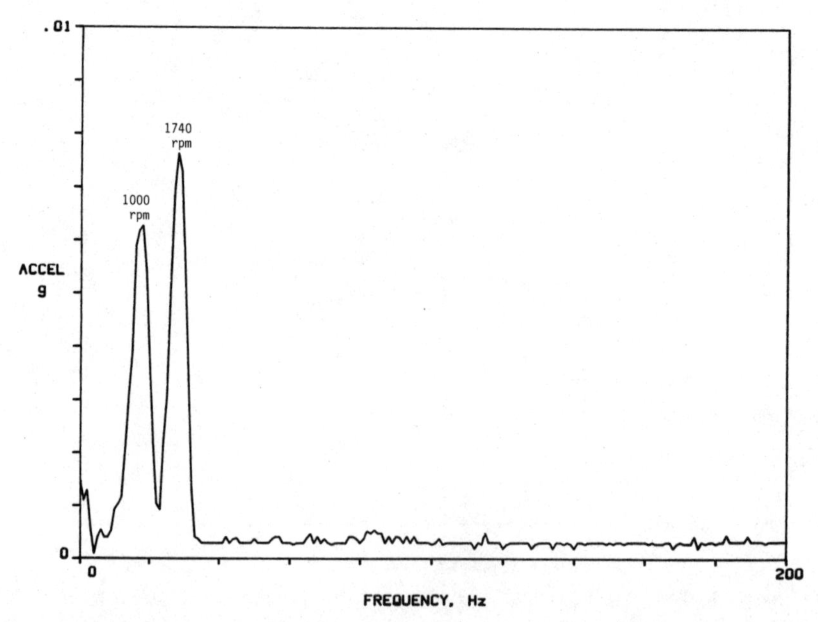

Figure 5.48 Vibration spectrum on the handle of an x-ray machine.

belts on five of the eight units, and the vibration on the handle of the x-ray machine dropped to about 0.003 *g*. This was acceptable.

Eccentric pulleys will also stretch the belt momentarily, which will then resonate at its own natural frequency. This resonant vibration of belts has been observed in both the radial and axial directions. This pulley eccentricity needs to be discussed along with wobbling pulleys because they are the same problem in different directions. These conditions can sometimes be observed under stroboscopic light, and even sometimes with the naked eye. The best check is to stop the machine and set up a dial indicator on the belt running surface of the pulley. The indicator stem should be perpendicular to this surface (Fig. 5.49).

A good pulley should have a TIR of 0.002 in or less while slowly turning the pulley around by hand. Usually this test can be done without removing the belts. A TIR of 0.005 in is acceptable. Anything more should be corrected if it is causing a vibration transmission problem. It is not uncommon to see pulley runout of 0.010 to 0.035 in. An extreme runout of 0.035 will not cause excessive damage or wear to the

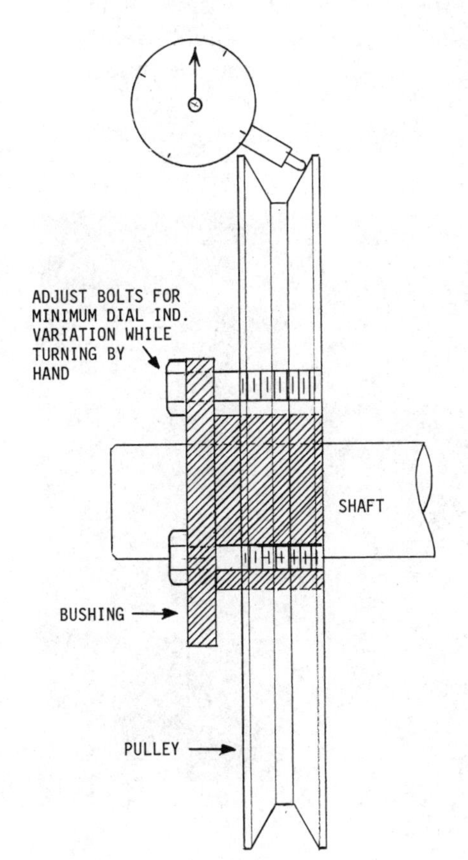

Figure 5.49 Checking pulley eccentricity with a dial indicator.

belts or the bearings. It will transmit this vibration elsewhere if the path is not well isolated, and it will complicate balancing because the eccentric pulley creates a 1X-rpm vibration.

The usual cure is to adjust the bushing bolts to get the pulley to run truer. If the pulley does not have an inside bushing, then remove the pulley, and either remachine it on a lathe or replace it with a better pulley. The bushing-type pulleys are recommended because they can be adjusted to run truer by varying the bolt tension.

In some situations, it is desirable to machine the pulley groove in place. This has been done successfully with a file and a small grinding tool on an extension handle. For multiple-belt pulleys, one belt can be removed and that groove ground, while the other belts drive the machine. This is a tedious process, but over time, the high spot of the pulley wears away faster than the low spot. No special fixturing is required, just hand holding the file or grinding tool, while keeping your hands away from moving parts. The technique is shown in Fig. 5.50. This may sound hazardous, but it is no more hazardous than filing burrs from a machined part on a lathe. The process is slow. It may take 1 or 2 h to remove a few thousandths of an inch of material, but this is still less time than it typically takes to exchange the pulley if one is not readily available.

A mass imbalance in a pulley is correctable by balancing. Eccentric-

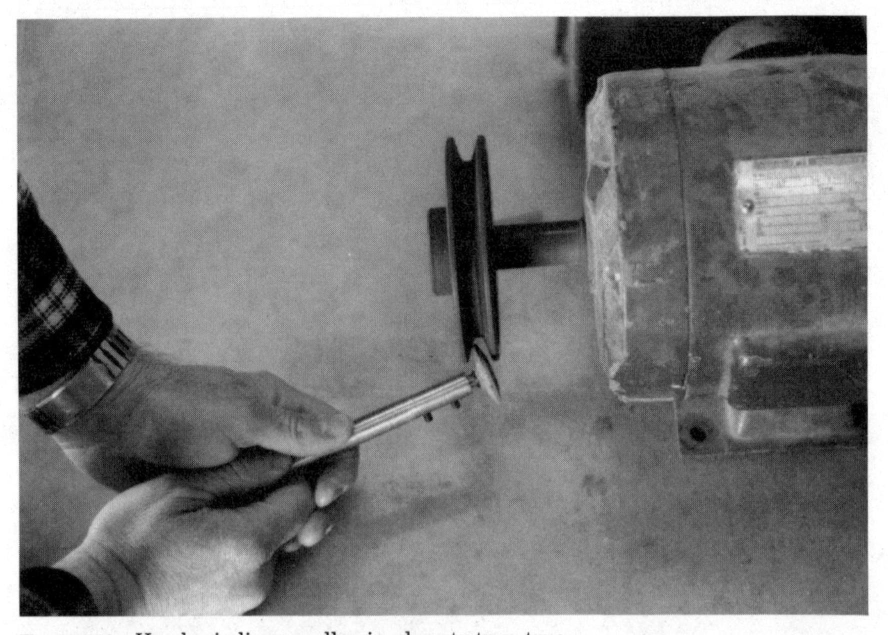

Figure 5.50 Hand grinding a pulley in-place to turn true.

ity in a pulley is *not* correctable by balancing. Even when well balanced, the eccentricity will stretch the pulley with each rotation and cause a 1X-rpm vibration.

When approaching a vibration problem on belt-driven equipment, I make it a standard practice to check the belts and pulleys before beginning balancing:

1. First, look at the belts and pulleys with the naked eye while operating.
2. Next, observe them with a strobe light.
3. Stop the machine and check the pulley runout with a dial indicator.
4. Start the machine and adjust the belt tension while running for minimum vibration as observed on a vibration measuring instrument.
5. If all is in order, then balancing can proceed with a fair chance of success.

Fix the Source or the Symptom

Since we have been discussing vibration energy, let's introduce the concept of a vibration sink. Just like water flows to and settles at the lowest depressions, vibration energy seeks a position of equilibrium. Its equilibrium is on a part where maximum motion is achieved. This could be a resonant part or just a very flexible part. Vibration energy flows through a structure seeking a place to settle, which is inevitably on a resonant or flexible part.

It is always better to reduce the vibration at its source. This means finding the offending machine and balancing or aligning it. But sometimes we are tempted by expediency to just detune the vibrating part by bracing. Many times this works and is a satisfactory fix. But when the vibratory source is putting out a large amount of energy, bracing can cause the vibration energy to travel farther down the structure until it finds another part to settle on.

I remember a problem that a colleague of mine from Colorado described.[5] The machine was a very large reciprocating compressor, 9 ft wide × 21 ft long. The output pipe was 18 in in diameter and vibrating terribly. This output pipe was braced (stiffened) with angle iron and clamps. After bracing, another section past several 90° elbows began vibrating. That section of pipe was also braced. After this second bracing, a large bottle attached to the piping took up the energy and

[5]Ralph T. Buscarello, Update International, Inc., Denver, Colo.

began vibrating badly, which it had not done before the bracing was applied.

At this point, my colleague stopped this futile bracing, scratched his head, and reasoned that the source of all this vibration energy had to be found and corrected. When the investigation was directed to the source, an incredible 2000 lb of unbalanced mass was found in the crankshaft. This was a design error. Rather than detune resonant parts, it would have been much better to balance the compressor. This stiffening could have resulted in a significant area of real estate braced, and the bearings would still be taking a beating. The true cause was identified and corrected when the investigation shifted focus from the symptom to the source. This story brings up the questions of when to stiffen the symptom or when to correct the source?

Fortunately there is a rule, based on the above example. The rule is called the *8-in rule.* When the resonant part is smaller than 8 in in cross-section, the vibrational energy is also small, and bracing may be successful in eliminating the problem. When the resonant part is greater than 8 in in cross-section, so much vibration energy is involved that bracing the resonant part will probably only displace the problem to another location. In this case with an 18-in-diameter pipe vibrating, it was better to fix the problem at its source. This rule has worked satisfactorily on piping and structural vibration problems.

Structural Vibrations

Structural vibrations range from less than 1 Hz to ultrasonic in frequency, and from many inches (or even feet) of sway to microinches at higher frequencies. As structures, we will consider not only buildings but bridges, towers, and movable structures like aircraft, ships, and trains.

The lower-frequency vibrations are due to earth movements and the building's natural frequencies. The natural frequencies of structures are typically from 0.1 to about 10 Hz. Figure 5.51 shows three structures of various heights and their associated natural frequencies in the horizontal directions. It should be recognized that buildings are cantilever beams with attachment to a stable foundation at the bottom. The top end is unrestrained and is free to sway horizontally. Horizontal earth tremors occur at about 1 Hz; it is possible to have a situation where the natural frequency of buildings is close to the driving frequency of earthquakes. Under these conditions of resonance, large amplitudes will build up, and it is these large deflections that cause failure. Earthquakes do not hurt people. Collapsing structures during earthquakes hurt people.

Another consideration during earthquakes is the restraining of mechanical equipment as the building floor moves. This involves hard

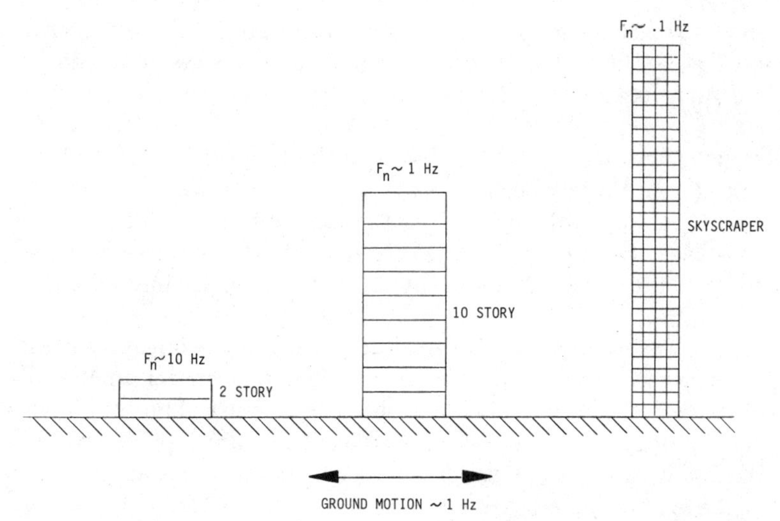

Figure 5.51 Various height structures and their typical, natural frequencies in the horizontal direction.

bolting the equipment to the floor, installing snubbers, restraint cables, or lockable hold-down devices. The level of restraint depends on whether the equipment must remain in operation during the earthquake or if it needs only to survive and not fall off its supports or break pipes, and be ready for a restart within 24 h. A diesel-powered generator in a hospital obviously needs to be operational during an earthquake, whereas a bathroom exhaust fan must only stay in place and not fall down on someone.

This entire subject of designing mechanical equipment to withstand earthquakes is a separate field of structural engineering that has implications for the mechanical engineer. Earthquake restraining devices allow less freedom of movement of the equipment and make vibration isolation more difficult. In addition, cables and rigid bolting short circuits the isolation paths, and the equipment vibrations are transmitted directly to the structure during normal operation.

Another source of vibrations to buildings is the wind. These random vibrations also excite structures to vibrate at their natural frequencies. The structure in the relative wind will generate vortices on the downwind side. These vortices can cause the structure itself to sway or to affect adjacent structures. Designing for wind resistance is part of the structural engineer's task. However, vibration measuring instruments with a low enough frequency range can measure building natural vibrations.

The third source of structural vibrations is machinery that is part of the building's utility systems. These are the structural vibrations that are the most common and we can do something about. By structural vibrations, I mean not only the columns and floor but also partitions, lights, and furniture. The following story illustrates furniture vibrations.

A colleague of mine named Walt had a desk at one end of a building that produced standing waves in his coffee cup. There was a "buzz" on his desktop. It could be felt in the floor through our shoes when standing near his desk. His secretary was convinced that the cause was the large scrubber fans installed on the mezzanine level immediately above.

I set up an accelerometer on his desk and viewed the output with a spectrum analyzer. Figure 5.52*a* is the vibration spectrum on Walt's desk. Notice that the amplitude scale is decibel volts. The 60-Hz vibration on his desk measures at −79 dBV. The tallest peak around was 60 Hz. Everything else was less than 20 percent of its amplitude. Figure 5.52*b* is the vibration on Walt's floor immediately below his desk. The 60-Hz vibration on the floor measures −91 dBV. The 60 Hz on his desktop was 12 dB, or four times higher than the 60-Hz amplitude on the floor.

The desk was not bolted to the floor, only resting on its four legs, but some amplification was taking place. That is, the desktop was responding to the 60 Hz incoming vibration from the floor with some amplification. The desktop was resonating. I knew I had to find the

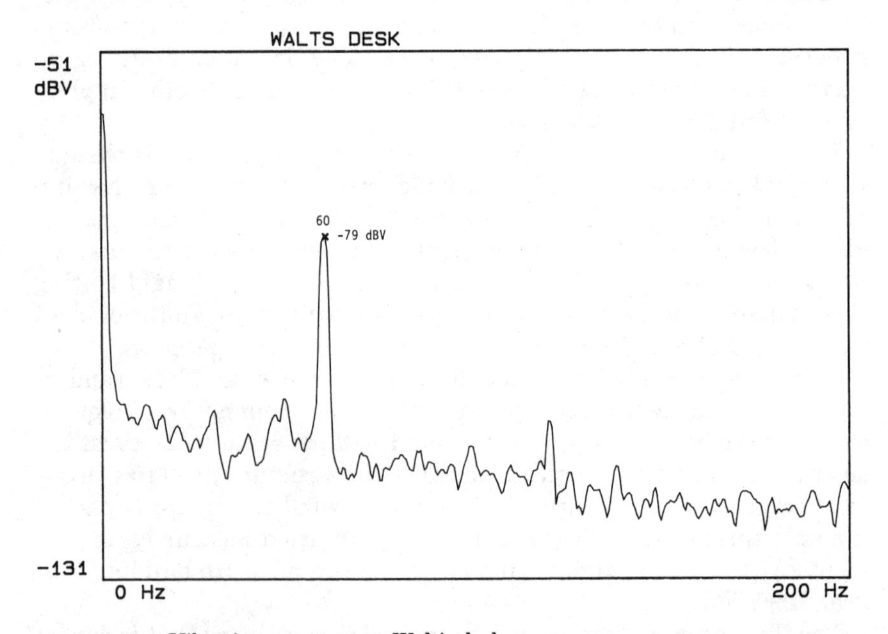

Figure 5.52(*a*) Vibration spectrum on Walt's desk.

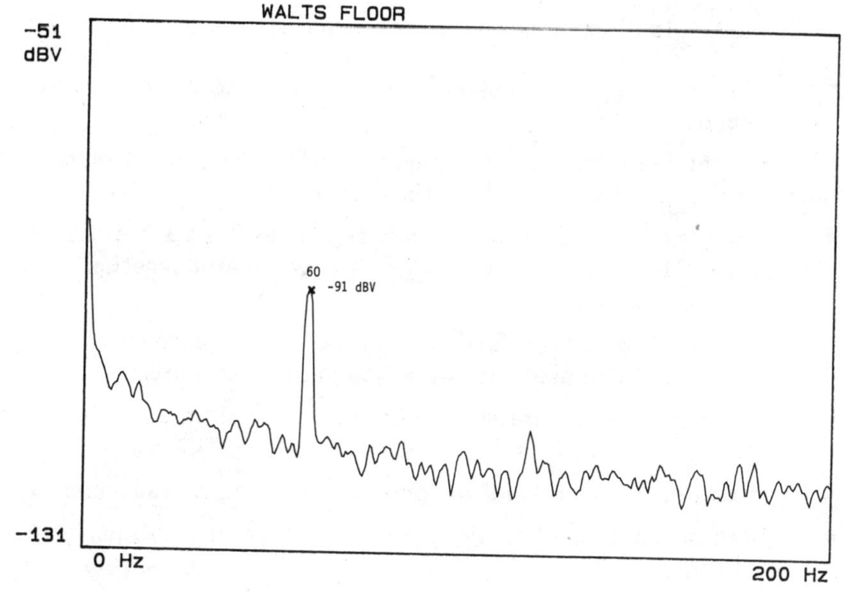

Figure 5.52(*b*) Vibration spectrum on Walt's floor.

source of the 60 Hz. The scrubber fans turned at 14.5 Hz (870 rpm), and they were known to be out of balance, but this frequency of 14.5 Hz was not arriving at Walt's desk. The scrubber fans were a separate, unrelated problem. To solve this problem on Walt's desk, I had to focus on the 60 Hz and nothing else.

I took amplitude measurements of the 60-Hz peak on the floor every 25 ft. First to the west, then north and south. The highest amplitude was near his desk. Next, I went up to the mezzanine level in the fan room over his desk. The 60 Hz was there, but smaller. I knew I was going in the wrong direction; I had to go down to the ground floor. There the 60 Hz was the largest yet. I tracked the 60 Hz to a vacuum pump below his desk. The piping to this pump was rigidly clamped to the structure.

A turn-off test was arranged for the next break in production at 2:00 P.M. When the pump was turned off, the 60 Hz on his desk dwindled away to nothing and promptly returned when the pump was restarted. "You found it," was what I heard over the telephone.

The corrective action was to replace the cork isolation pads under the motor and pump and align them. This made an improvement, but now that the occupants above were sensitized to this 60 Hz, it had to be completely removed. Flexible pipe connectors were installed, and nothing more was heard of this problem, and no more standing waves in coffee cups.

This example illustrates several points about structural vibrations:

- The source of many structural vibrations is mechanical rotating equipment.
- With improper isolation, the machine's vibrations are transmitted directly to the structure where they travel far.
- The frequency of vibration is unchanged as it goes through the structure. Measure the frequency at the complaint location as the first step in analysis.
- The amplitude decreases farther from the source. More importantly, the amplitude increases the closer you get to the source.
- Furniture can vibrate into resonance even though it is not bolted to the structure.
- Don't rely entirely on what anyone says. Rely on measurements.
- Turning off equipment while monitoring structural vibrations can identify the source.
- People are more sensitive to higher frequencies.

To solve existing structural vibration problems:

1. Measure the frequency of vibration.
2. Move throughout the structure tracking that frequency. The maximum amplitude will be on the offending machine itself.
3. Correct the machine vibration by balancing, alignment, or other fix. It is always best to fix a vibration problem at its source.
4. If that is insufficient, then isolate the machine better from the structure. Also isolate all its connections to the structure, i.e., pipe, supports, hangers, duct, and conduit.
5. Move the machine across the street. This may sound like displacing the problem into someone else's court, but it must be remembered that a problem of perception only exists with a sensitive receiver. If the machine is not experiencing short life, then the problem is only one of location. An appropriate fix is to put more distance between the source and receiver.
6. Replace it.
7. Move sensitive optical equipment closer to columns where floor deflections are smaller.

The last source of structural vibrations is activity going on within the building and outside nearby. This includes foot traffic, vehicular traffic, construction activity, and other transients. As transients, these vibrations are best viewed in the time domain, and accurate measurements of peak amplitude are possible in the time domain. For example, Fig. 5.53 shows the frequency and time domain views of the floor of a semiconductor manufacturing facility while jackhammering was going on 75 ft away. The frequency domain view (Fig. 5.53*a*) has much broadband vibration between 200 and 800 Hz, which is probably structural resonances being excited. This spectrum was a peak hold averaging over a 20-sec period. The peak acceleration was recorded at 0.0055 *g* at about 240 Hz. This is a gross underestimate. The time plot, Fig. 5.53*b*, captures one of these jackhammer blows, and it clearly measures 0.063 *g* peak to peak. The 240-Hz vibration has a period of 4 msec and is clearly visible in this time domain view. This example simply illustrates the superiority of time domain measurements for transients. These transients were imperceptible on the floor, but two precision optical instruments nearby tripped off during this jackhammering. These optical instruments had on-board sensors that discontinued the process at a preselected level of vibration.

In sensitive facilities, the production equipment itself and people activity are the largest source of vibration, especially transients. In microfabrication and ultraprecision manufacturing operations, these are the variables that presently limit the attainable precision and yields.

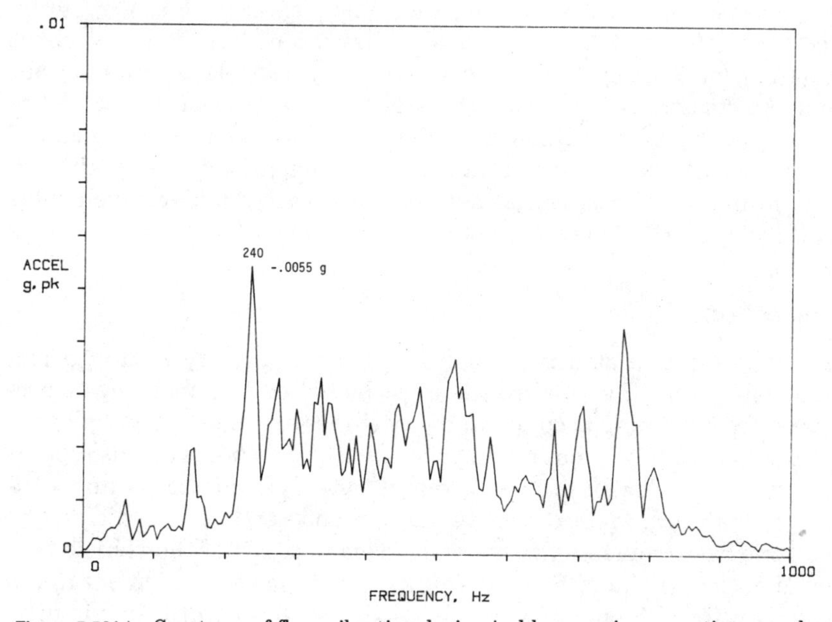

Figure 5.53(*a*) Spectrum of floor vibration during jackhammering operations nearby.

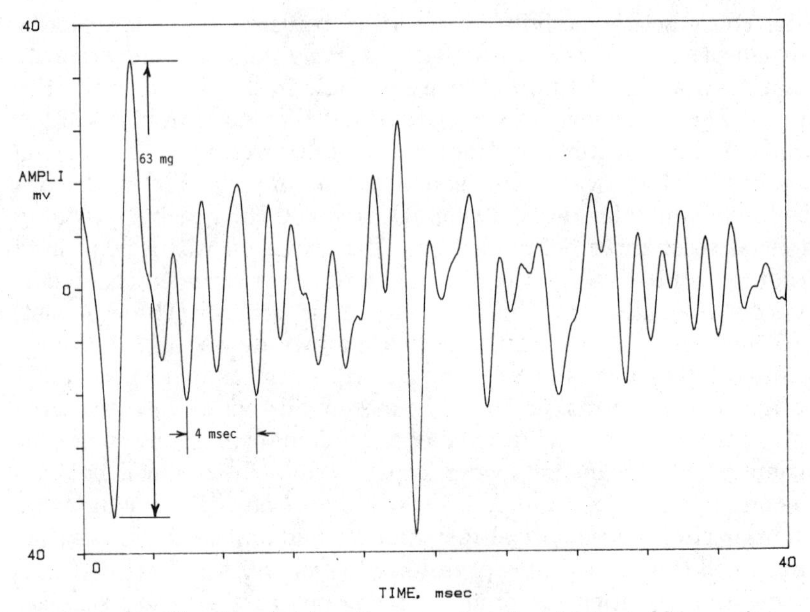

Figure 5.53(*b*) Time domain view of a severe transient during jackhammering.

The lowest floor vibrations that I have had the opportunity to measure at sensitive facilities are well below 0.001 *g* during normal operations. This is not an overall, but a horizontal limit across the frequency spectrum which no vibration peak exceeds. For very quiet floors built as a slab on grade, this limit can drop down to about 0.0001 *g*, or 0.1 mg. These levels are imperceptible to humans but could be disastrous to optical processes that are not well isolated. The highest peaks are always at the frequency of common rotating equipment, i.e., 20, 30, 60, and/or 120 Hz. For comparison, the ground vibration in a quiet residential neighborhood with no nearby vehicular traffic is typically below 0.000 01 *g*, or 0.01 mg.

Foundations

A foundation is a structure that supports the gravity load of a mechanical system. Every machine must be bolted to something to prevent it from walking around. Many machines also need a rigid plate to keep things in alignment during operation (Fig. 5.54). The base serves this function. The base also takes up any dynamic loads and ultimately transfers them to the foundation and earth. For vehicles, the machine base transfers the dynamic loads to a part of the vehicle that has significantly more inertia. Think of the base as the element in the load path from the machine to the earth that has significant rigidity.

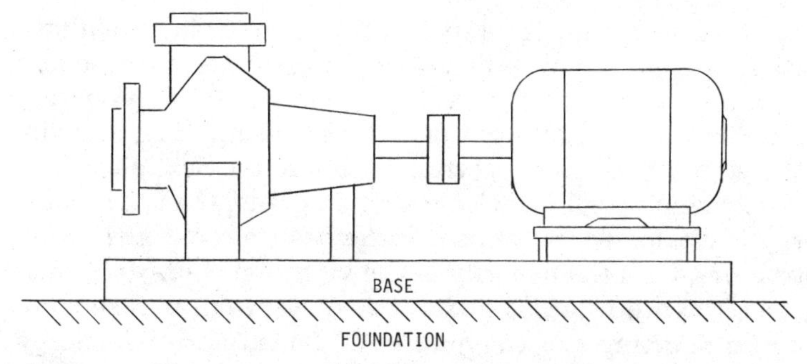

Figure 5.54 Motor-pump combination on a base and foundation system.

For hanging equipment, the load path is through the hangers. Hence, the base could be the ceiling beams. Since the machine is bolted to some structure that forms its base, the machine and the base need to be considered as one entity. Whatever vibration the machine produces, the base experiences the full impact of that. It, like any other member in the load path, transmits some of this energy, absorbs some, and reflects some. On larger machines, the base and foundation are paramount to successful dynamic behavior.

For purposes of vibration, there are two kinds of foundations: rigid and flexible (Fig. 5.55). The foundation is the structure below the springs, or downstream of the springs in the force transmission path. The base is above the springs, and its purpose is to supply rigidity for alignment. The base is usually always intended to be rigid. A rigid foundation is one that does not flex. A flexible foundation is one that takes up significant deflection just due to the weight of the machine. Upper stories and roofs in modern buildings can be considered to be flexible for machines weighing more than a hundred pounds. The deciding criterion is really deflection, and if the floor deflection is more than 0.1 in, then the floor should be considered flexible. All materials (even concrete) have some elasticity, and this elasticity along with the mass are the determining factors for its natural frequency.

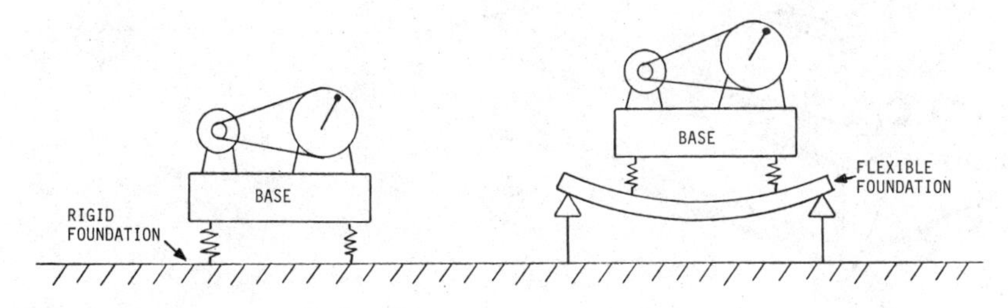

Figure 5.55 Two kinds of foundations: rigid and flexible.

As mentioned above, many machine parts need to be kept in alignment for proper operation. This is done by bolting the various machine components to a concrete floor or steel base. This type of foundation has such a high natural frequency that it forms a rigid link, and virtually all the vibrational energy is transmitted to the next element.

If the next element is soil (as with a slab on grade) then the vibrational energy is dissipated in the soil. If the base is connected to the building structure, then the majority of the vibration energy is transmitted to the structure where it is dissipated eventually. Depending on the material damping and the quality of joints, vibration energy can travel a long way before dissipating. Steel has very low damping. Vibration energy can literally travel for miles along steel structures and still be detectable.

A concrete inertia block (Fig. 5.56) is useful in two typical situations. First, if the machine assembly needs a rigid foundation and it also must be isolated from the building structure, a concrete inertia block can be used. The machine is bolted to the block, and the block is mounted on springs. In some applications, the machine structure is so flexible that it is impractical to mount it directly on springs, and the block is used as an intermediate link. Second, some machines generate such large forces during operation that the overall movement tends to become excessive. This movement is hazardous to utility service lines and piping. To limit this motion, the effective mass of the machine is increased by attaching more concrete and steel to it. Typical applications are reciprocating engines and compressors, vibration testing machines, and forging hammers. To be effective, the mass of the concrete block should be four to six times the supported mass.

Flexible foundations are usually not a good situation because with

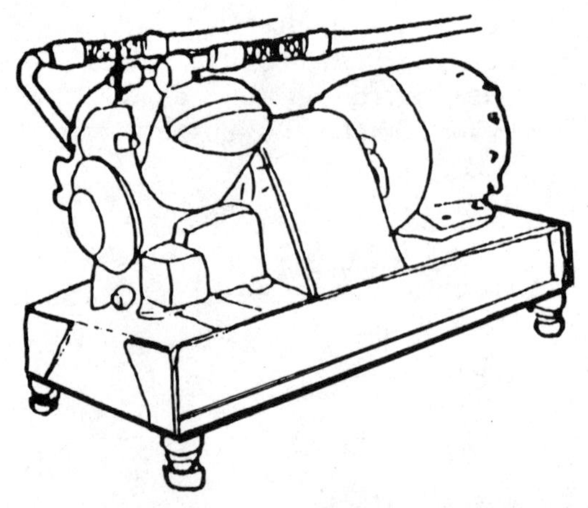

Figure 5.56 Reciprocating compressor on a concrete inertia block.

such foundations, there is no longer a single-degree-of-freedom system. There will be at least two springs and more than one mass involved. There will be an exchange of energy between the supported machine and its foundation which may have unexpected results.

Machinery mounted on the roof or on upper floors can generally be considered on a flexible foundation. The hazard with flexible foundations is that the single-degree-of-freedom, mass-spring model is no longer valid. The springs selected for isolation may not do their job properly. The floor will go into motion, and it may even amplify the vibration if the machine operates near its natural frequency. Floors are large panels and just very slight floor motion can become an objectionable acoustic problem.

The symptoms of foundation problems are usually cracks in the concrete. This means that some relative movement has occurred. Filling in the cracks is not a fix. The machinery will need to be realigned and possibly the old foundation chipped away and repoured. Finally, the source of the cracking will need to be corrected to prevent a recurrence.

There is no such thing as a rigid foundation for all time. The earth moves, expansion and contraction take place, and freeze-thaw cycles occur. There are cases of daily changes in vibration caused by thermal distortion of baseplates and foundations due to sun exposure. For machines exposed to the sun and weather, solar heating is a valid concern. I even remember hearing of a large machine on sandy soil in a tidal zone. The hydraulic soil pressure due to changing tides caused a corresponding change in alignment.

After a few years machinery faults may develop such as leaky seals, noisy bearings, and motor burnouts. These problems may have developed because the foundation moved and ruined the original alignment. Many more problems with machinery are due to foundation movement than we would like to know about. A good maintenance procedure is to check alignment and the foundation at every maintenance action. A simple periodic vibration monitoring program can easily detect such changes. For optimum performance, some companies do not wait for breakdown but have crews go around to balance and align every piece of equipment on a periodic basis. Whether it needs it or not, it is at least checked. Imbedded in the vibration signature is the balance and alignment condition. What is the purpose of balancing and alignment but to reduce vibration? It would be managing smarter to replace the periodic balancing and alignment crews with a vibration monitoring technician. The vibration measured at any point on a machine is the sum of the sources attenuated by the restraint system. The foundation is the restraint system, and the vibration signature contains information about its condition.

Unstable foundations cause variable amplitude and phase readings during balancing procedures. This makes it difficult if not impossible to balance. If you have a similar problem during balancing, check the foundation. The foundation affects the vibration readings on the machine's bearings. There are numerous case histories of rotating machines that could not be balanced because of unstable amplitude and/or phase readings. Once the faulty foundation was repaired, then the same machine displayed stable vibration readings and could be balanced.

I have a tabletop machine used for teaching purposes on which I can change the phase reading 45° by pressing down on one corner. This is analogous to tightening a hold-down bolt. I can also make a dramatic change in amplitude by lifting the machine off the table. A new vibration peak comes up and dominates the entire spectrum. This simple demonstration illustrates two related points:

1. Changing foundations grossly affects amplitude and phase measurements.
2. Vibration measurements can easily detect a changing foundation–hold-down system.

A final concept concerning foundations is their ability to transmit, absorb, and reflect dynamic energy. Maximum energy is transmitted when mechanical impedance is well matched. That is, the source vibration and the foundation have similar natural frequencies across the connection. The frequency of vibration that is being transmitted is not a factor, only the natural frequencies of the two connected systems. If the two natural frequencies, and hence mechanical impedances, are well matched, then maximum vibratory energy will be transmitted between them at all frequencies below their resonances. Changing the mass or stiffness of one of them will alter its natural frequency and disturb the impedance match. Less vibratory energy will be transmitted and more will be reflected or absorbed. An example will make this concept clearer.

A vibrating screen at a gypsum wallboard plant was experiencing clogging. This screen weighed 8000 lb. It was caused to vibrate by massive rotating imbalances. Its purpose was to vibrate and sift bulk material through a screen. Unfortunately, it clogged due to insufficient amplitude of motion. The 10-ft-high steel structure supporting it also vibrated excessively. To vibrate the screen with increasing amplitude and eliminate clogging, the steel structure underneath was stiffened. This reflected more vibratory energy back into the screen where it could achieve its full 0.2-in displacement motion. Stiffening the structure achieved the desired result without the need to increase the rotating imbalances. The previously flexible steel structure was absorbing too much energy.

The technology of measurement and analysis of machine-foundation systems is not well developed or well communicated. I suspect that no one understands it well enough to transfer the knowledge into a set of general rules. As mentioned, for large machines the foundation design is critical to satisfactory performance. The design of such foundations is based more on previous successful designs that are known to work, rather than on dynamic analysis from a purely analytical approach.

Sympathetic Vibrations

Two machines of identical geometry on the same base, or platform, is not a good design, especially if one is mostly in a standby mode (Fig. 5.57). The problem is that the working machine transmits vibration to the standby machine and causes damage. The standby machine sounds and feels as if it were running. The housing of the standby machine vibrates in sympathy with the operating machine particularly when resonances are involved. The housing thrusts against the inertia of the rotor, and these cyclical forces are transmitted through the bearings. The roller elements are not rolling but stationary in one spot against their races, and brinnelling damage accumulates. When the standby machine is put into operation, its bearings are already damaged and failure comes quickly.

The only effective protection that vibration monitoring can provide is to closely monitor the bearings in the time domain for shock pulses.

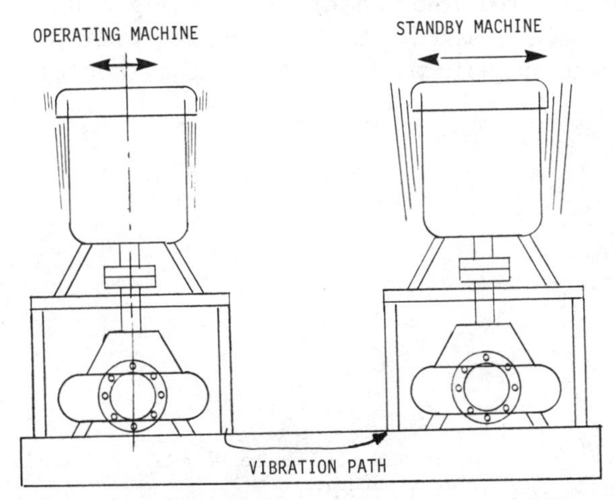

Figure 5.57 Sympathetic vibrations transmitted to a standby machine.

This situation has been brought up so that you can recognize the true cause of frequent bearing failures on tandem machines. These tandem machines will continue to be plagued with bearing changes unless action is taken to prevent this transmission of sympathetic vibrations. Two alternative steps can be taken to alleviate this problem. One is to stiffen the machines such that motion is restricted. This decreases the forces transmitted through the bearings as the housing oscillates against the inertia of the rotor. The other alternative is to break the transmission path by cutting the common support.

Since this subject of brinnelling of bearings has been introduced, a related problem with stored motors must be mentioned. The same type of brinnelling damage to bearings can occur to motors that are stored on a hard concrete floor near a vibrating machine. Shock pulses coming through the floor transmit directly through to the bearing. An effective protection against this is to store the motors on a neoprene pad. I know of one company that was aware of this damage and made it a standard procedure to install new bearings immediately before placing a stored motor into service. Over the years, with accumulated experience, they had developed a method of avoiding the consequences of this damage. With a rubber pad under stored motors, they are now avoiding the damage in the first place.

Machinery Soft Foot

Machinery soft foot is a phenomena whereby a machine's vibration level can be decreased dramatically by loosening one or more hold-down bolts. Every vibration engineer or technician has experienced it. Three points are all that are required to define a plane (Fig. 5.58). A fourth point (or foot) must necessarily distort something when it is clamped down, if it does not lie perfectly in the plane of the other three

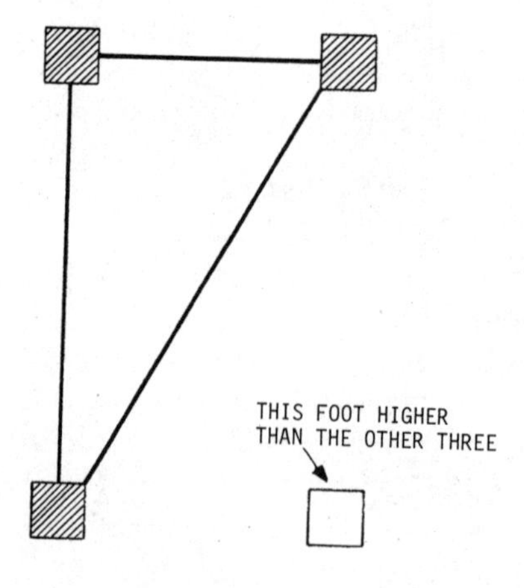

Figure 5.58 The cause of machinery soft foot.

points. When the fourth hold-down bolt is tightened, the machine casing becomes distorted. This changes its stiffness and could shift its response to a region of resonance. It also disturbs any alignment, either internal or external alignment.

Over a hundred years ago three-legged lathes were reported to produce better precision parts. Even today, high-speed compressors cannot operate on four feet. They are supported by two legs and a "wobble plate."

When a machine's vibration level cannot be reduced by balancing or alignment, the following procedure is recommended:

1. While monitoring the vibration level, loosen and retighten each bolt, one at a time.
2. Note which ones produce a dramatic decrease in vibration.
3. When finished with all the bolts, go back to the ones noted and see if the decrease is consistently repeated.
4. If so, shim that foot. Determine the amount of shimming with a dial indicator.
5. Repeat the above procedure again (loosen all the bolts one at a time while monitoring vibration) until there are no more dramatic vibration decreases.
6. Now proceed to balance or align the machine to get further vibration decreases.

Motors and generators are more susceptible to soft foot. This is believed to be because their 120-Hz hum and harmonics are matched closer to casing resonances.

Assembly Variables

During the assembly process, there are variables that can change the balance condition of a rotor. One is the bearings themselves. If a rotor is balanced on a balancing stand, it rides on the bearings of the balancing stand. When it is installed into its own bearings in the machine casing, it may respond differently.

Another variable is keys. The rotor may have been balanced with a full key, a half key, or no key. The key used at the balancing shop may have been a different length than the one you use.

Another variable is couplings, and coupling bolts. Couplings are not all identical in concentricity, and their bolts may be of different weights. When disassembling parts, mark which bolts were installed in which locations. This may seem like a small point, but it is critical for high-speed turbines. The final trim balance in-place compensates for all these assembly variables. To maintain this good balance, parts of variable weights must be reassembled in their original locations.

Balancing shops recognize that assembly variables can change the results of their work. They usually have the customer sign a form that includes a statement to the effect that they guarantee nothing because they have no control over the assembly process.

This concept of coupling bolts applies to any weight that you may add to or remove from a rotor. I know of a case of fan bolts that were threaded to different lengths, and thus caused balance problems. The bolts were even supplied by the fan manufacturer who claimed they were of equivalent weight. Actually there was a 0.4-gram difference in weight. Twenty of these bolts were around the fan hub, and two bolts of different weight on the same side will definitely cause an imbalance problem. Eight-tenths (0.8) of a gram would be enough to exceed the original factory balance specification.

How do you guard against assembly variables? The following two procedures have proven to be effective.

1. Trim balance the rotor in-place in its own bearings, with all of its own couplings, keys, bolts, and other accessories that have weight. There is no superior balance that can be achieved other than in-place at running revolutions per minute.
2. Measure the vibration amplitude at 1X running speed before and after any maintenance work. This will reveal if the balance condition has changed.

Beats

Beats are two separate vibration sources that interfere with each other. The interference causes addition and subtraction of the two waveforms and produces a new vibration that has a frequency which is the difference between the two source frequencies. The most common beating phenomena in machinery applications is two nearby machines that rotate at almost the same speed. The vibrations from both machines mix, and a beat vibration is produced that has a frequency which is the difference between the two:

$$\text{Beat frequency} = f_1 - f_2 \tag{5.11}$$

where f_1 and f_2 = two source frequencies

For instance, in Fig. 5.59 the wave pattern of two machines with a beat is shown. Let's assume that one machine rotates at 600 rpm and the other rotates at 540 rpm. The difference between the two speeds is

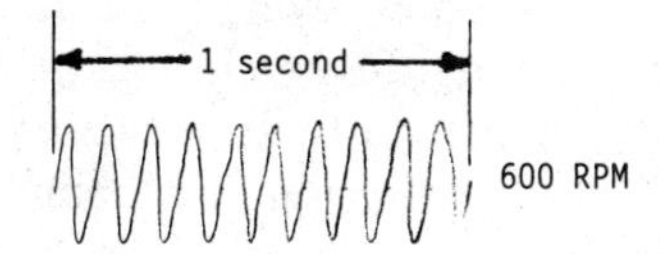

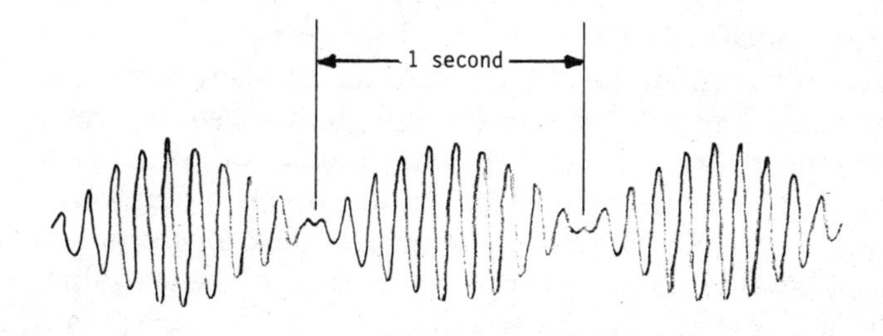

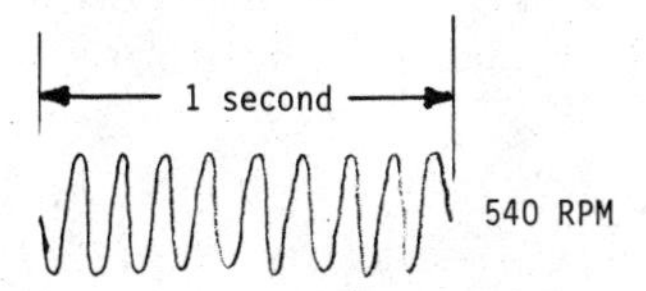

Figure 5.59 A beat of 1-sec period created by mixing two vibrations of 600 and 540 rpm.

60 rpm or 1 Hz. This corresponds to a beat frequency of 1 cycle/sec. That is, the addition of the two vibration sources will build up to a maximum cyclically every second.

The best way to illustrate this is with a twin-engine airplane. When the two propellers are not turning at exactly the same speed, there is an alternating rise and fall of the vibration (and sound) level. This is the beat frequency.

This phenomena of beats is relatively common around mechanical equipment rooms where there are many motors turning at almost the same speed. It not only causes a variation in amplitude but also a variation in phase. With these two quantities changing, it is almost impossible to balance.

I remember a story relayed to me by another vibration engineer about his first field balancing job. He spent all morning attempting to balance a centrifugal fan with poor results. He took a break for lunch to eat his ham and cheese sandwich. He sat down on the mechanical room floor and leaned his back against the wall to have his lunch. As he sat and ate, he could feel, through his back, the vibration come and go on the wall. It built up to a maximum, then retreated, and again to a maximum every 4 min. After lunch, he aimed his strobe light at the target on the shaft of the fan, and saw it go around 360° every 4 min. No wonder it was difficult to balance with a constantly changing phase angle. There was another fan nearby turning at almost the same speed. The two vibration sources interfered and created a beat with a 4-min period. In fact, the difference in speeds between the two sources was only ¼ rpm.

A single machine can also create a beat by exciting a panel or structural part into resonance. The stationary part may have a resonance near the machine running speed but not exactly coincident with it (Fig. 5.60). The machine speed, being close enough in speed to a resonance, will excite the stationary part into resonance. The stationary part will respond by resonating, then dampen down, only to be jolted back into resonance again by the nearby machine. The machine and the stationary part, at slightly different frequencies, then beat against each other. This is a rather common condition on HVAC equipment with sheet metal panels. The sheet metal resonance appears in the frequency domain as a peak near the running speed that alternately rises and falls. That is, it does not maintain a constant amplitude.

Turbulence

Fluid (air or water) flowing around an object will generate vortices downstream of the object (Fig. 5.61*a*). At faster flow velocities, these vortices can turn into turbulence. This turbulence is a random vibration and shows up in the frequency domain as broadband noise. Knowing this helps one to recognize cavitation in pumps, and around valves.

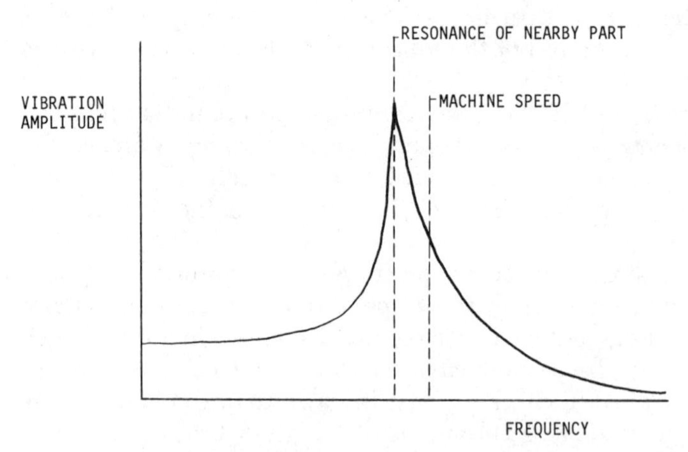

Figure 5.60 Single machine creating a beat by exciting a nearby resonance.

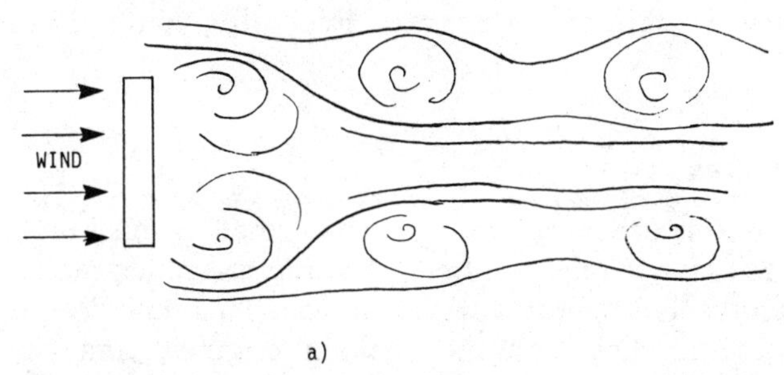

Figure 5.61(*a*) Relative wind over an object creates downstream vortices.

Random vibration has the capability to initiate and sustain resonance (Fig. 5.61*b*). The resonance appears as a sharp peak, or pure tone, in the presence of broadband noise. If you have an older car, you may recognize something "singing" in your car while traveling down the highway at higher speeds. Tire noise and wind are random vibrations that can initiate and sustain resonance.

Wind turbulence, and the resulting random vibrations, cause many problems on aircraft such as:

- Flutter
- Loosening of fasteners
- More energy consumption

The next time you are in a commercial aircraft, place your ear against the cabin wall. You will hear the broadband, random turbu-

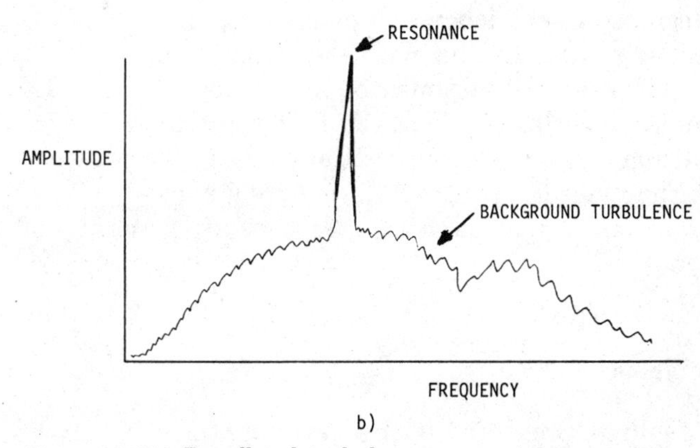

Figure 5.61(*b*) Broadband turbulence can sustain a resonance that it may encounter.

lence of air flow over the skin of the aircraft. Hopefully, you will not hear a pure tone.

Reciprocating Machines

Reciprocating machines cause more vibration than rotating machines, and they are more difficult to analyze. Reciprocating machines can be defined as any machine that operates with a piston. This includes all piston engines—gas and diesel engines, steam engines, and the Stirling engine. It also includes reciprocating compressors and pumps.

Reciprocating engines have inherently high vibrations at 1X rpm and harmonics. The vibrations are caused by gas pressure forces and imbalance. The gas pressure forces come from the exploding fuel-air mixture in the combustion chamber. The imbalance comes from the connecting rod and pistons that are continuously changing the radius of their mass centers. This can only be partially compensated for by counterweights. Vibrations of ½X rpm are common on four-cycle engines where the camshaft rotates at one-half crankshaft speed.

High vibration amplitudes at these frequencies (½ and 1X rpm) and harmonics do not necessarily indicate trouble. Reciprocating machines are made massive and rugged, and they can tolerate high vibration amplitudes for extended times. However, the presence of so many driving force frequencies has a high potential for exciting structural resonances, especially on a variable-speed engine. On accelerating and decelerating, as these harmonics traverse the frequency spectrum, almost every possible resonance can be excited.

Excessive vibration also occurs on reciprocating engines due to operational problems such as misfiring, piston slap, compression leaks, faulty fuel injection, and valve clash. On four-stroke engines this occurs at ½X rpm if one cylinder is affected. These operational problems will also cause a decrease in efficiency and power output.

Worn gears, bearings, and chains will also cause vibrations at higher frequencies. These amplitudes may be low compared to the 1X rpm and its harmonics, but this is where the FFT spectrum analyzer really shines—by displaying low-amplitude vibrations in the presence of other high-amplitude vibrations at a different frequency.

The vibration problems on reciprocating machines are generally of two types:

1. Resonance
2. Operational problems

Tests for resonance will be presented in the next chapter.

Operational problems (like misfiring) can be detected by trending the overall vibration amplitude. It is important to bring the engine to the same speed and load condition, and the measurement must be taken at the same identical point for accurate comparison.

Figure 5.62*a* is the vibration spectrum captured on the valve cover of a well-running V-8 piston engine. All cylinders were firing normally. Notice the 0.10 *g* overall vibration. Figure 5.62*b* is the same engine with one spark plug wire disconnected. Notice the increase in harmonics and ½ harmonics, and the overall increase to 0.14 *g*. Going one step further, Fig. 5.62*c* is with two spark plug wires disconnected. The harmonics and ½ harmonics show a further increase, and the overall is up to 0.20 *g*. The throttle position remained unchanged during these tests. The speed dropped with each additional misfiring cylinder, from 1170 rpm, to 1140 rpm, to 1080 rpm. Monitoring the overall vibration at known load conditions is an effective method of trending operational problems on reciprocating machines.

Excessive clearance can be detected by a noticeable increase in vibration while accelerating or decelerating.

Gas pressure forces cause torsional vibrations in the crankshaft. These torsional vibrations couple to the linear mode, and it is possible to get significant linear vibrations. These vibrations due to gas pressure forces occur at ½X rpm on a four-stroke engine, plus harmonics.

Finally, unbalanced reciprocating forces cause the typical vibration

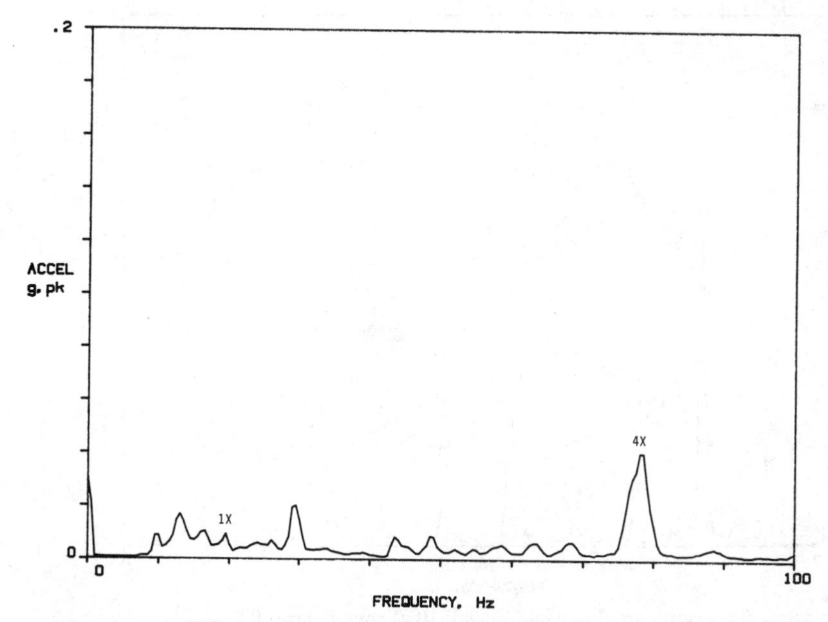

Figure 5.62(*a*) Vibration spectrum measured on the valve cover of a 5.0-L, V-8 piston gas engine. All cylinders were firing normally. Speed: 1170 rpm, overall 0.10 *g*.

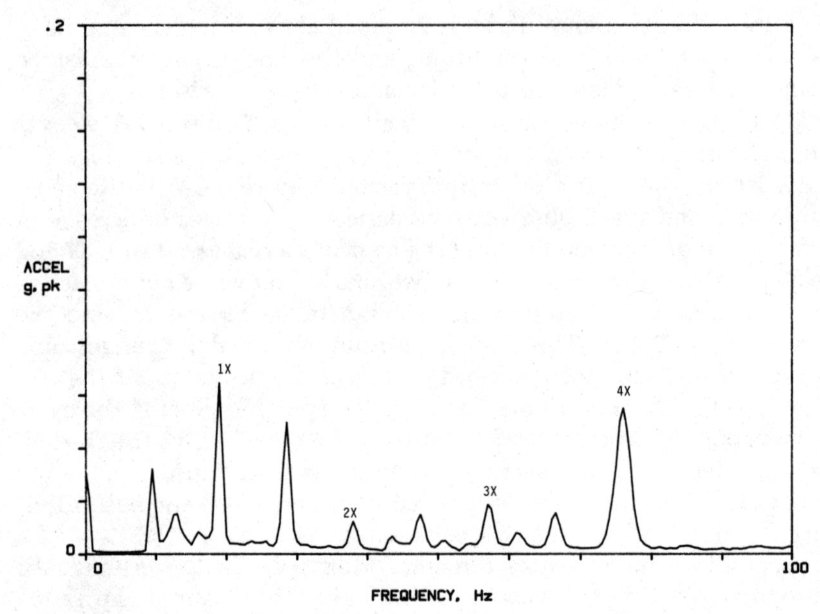

Figure 5.62(*b*) One cylinder misfiring. Speed: 1140 rpm, overall 0.14 *g*.

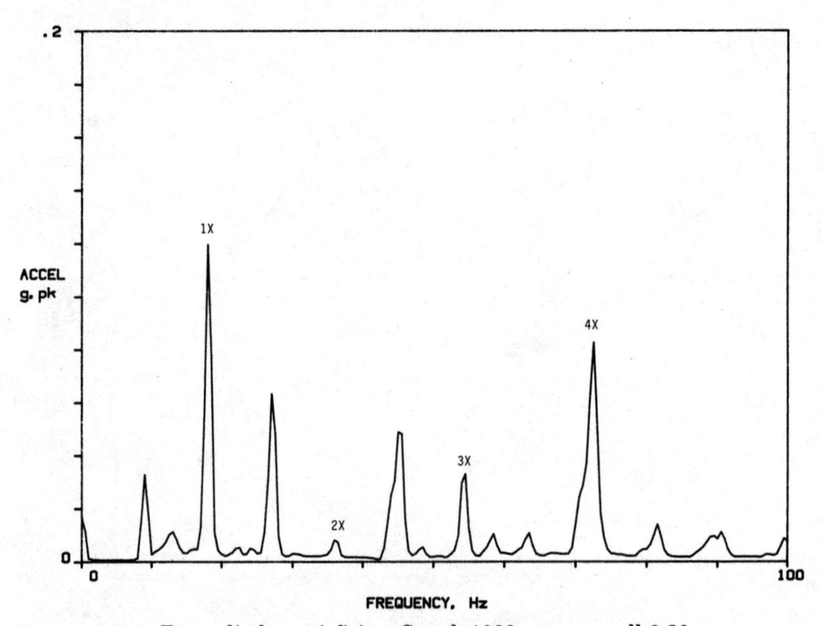

Figure 5.62(*c*) Two cylinders misfiring. Speed: 1080 rpm, overall 0.20 *g*.

at running speed. New engines are reasonably well balanced, but after an overhaul, if a piston or connecting rod is replaced with one weighting differently, then the vibration can be expected to increase. A good check is to measure the vibration before and after any major repair work, to judge the quality of the repair.

Once an increase in vibration has been detected by trending, then additional techniques can be used to analyze the problem. One technique is simply to listen with a stethoscope. The amplification with a mechanical stethoscope can be overwhelming on a piston engine. Use an electronic stethoscope with a volume control. Then listen to each cylinder or valve in turn without changing the volume control. In this way, you can hear a defective piston or valve. Also, be sure to use an accelerometer for your transducer. The higher-frequency response of an accelerometer will be necessary to hear gear or timing-chain problems.

Another technique is to observe the vibration in time waveform. With a long-term time waveform of an engine, the analyst can "see" the firing of each cylinder and the opening and closing of valves. Each one of these events is a crash and shows up in the time domain as a shock pulse.

Vibration analysis of reciprocating machines is inherently complex. One method that makes this analysis easier is to have a prior baseline vibration signature available, in both the time and frequency domains. Then, when a problem develops, another vibration signature is captured. By comparing this signature to the baseline when the engine was in good running condition, the differences can be identified.

Turbomachinery

Rotating equipment is generally easier to analyze than reciprocating machines. Rotating equipment runs smoother but at much higher speeds. Consequently, rotating equipment is made lighter and less rigid than reciprocating engines, but the effects of vibration can lead to a catastrophic failure faster due to the higher operating speed and lighter construction.

Turbomachinery includes steam and gas turbines and centrifugal compressors. The most common problems that cause vibration on turbomachinery are the same as on other rotating equipment, i.e., imbalance, misalignment, and resonances. Figure 5.63 is a spectrum from a velocity probe bolted to the casing of a stationary turbine engine. The turbine rotated at 15,000 rpm (250 Hz) and drove a 1800-rpm (30-Hz) generator through a gearbox. The imbalance in the turbine and generator are clearly visible.

Figure 5.64 was taken on the gearbox. Notice the prominent peak at 10,000 Hz. An accelerometer was stud mounted to the gearbox to ac-

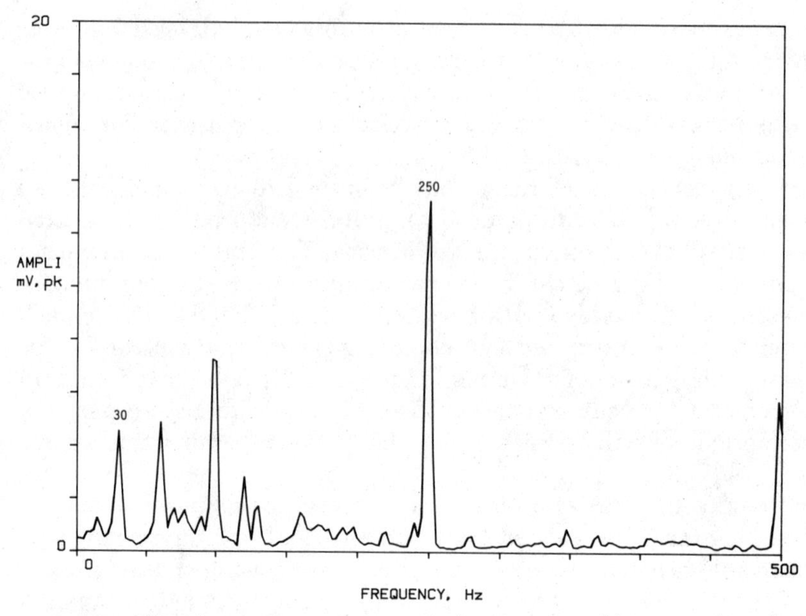

Figure 5.63 Vibration spectrum from a stationary gas turbine operating about 15,000 rpm (250 Hz).

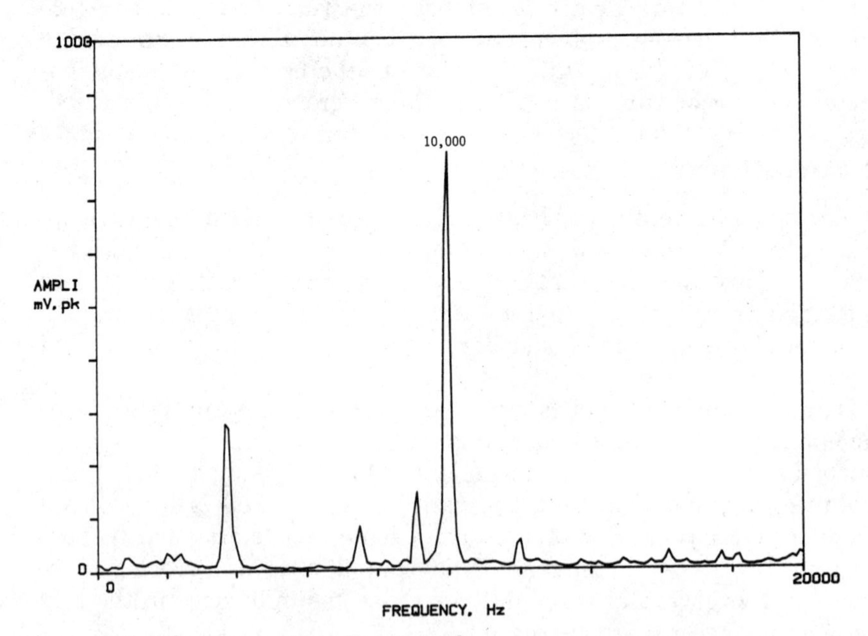

Figure 5.64 Vibration spectrum from a gearbox.

quire this data. This 10,000-Hz peak would be invisible to a velocity transducer because it would be outside its maximum frequency range.

Turbomachines frequently operate above their first critical speed and sometimes between the second and third criticals. At these speeds, the rotors are flexible. The bearings and their lubrication system are much more important at these higher speeds to provide damping and limit vibration. Some vibration problems on turbomachines can be corrected by changing the oil viscosity. Abrupt loss of damping in fluid film bearings has led to catastrophic turbine failures.

Because the rotors are flexible and light, proximity probes measuring shaft motion are better indicators of shaft vibration. However, be aware of the frequency limitations of proximity probes of about 2000 Hz. If you want to see higher frequencies (like on the gearbox), you must use accelerometers.

Two vibration problems unique to centrifugal compressors are surging and choking. Surging occurs when the discharge pressure is too high. Gas then flows in the reverse direction through the compressor. This creates a turbulent condition, and the vibration characteristics are random and broadband. If allowed to continue, surging will cause extensive damage. Surging sounds like a screeching elephant.

Choking is the opposite of surging. Choking occurs when discharge pressures are too low. When the discharge pressure is low, the velocity increases. When the velocity in the diffuser section approaches MACH 1 (about 700 mph), a turbulent or circulating flow between the blades will occur which has the effect of blocking the flow of gas. The vibration level increases because of the turbulent flow condition. The vibration is also random and broadband, just like surging. A check of the discharge pressure will tell whether the problem is surging or choking. The fix is to change the flow through the compressor.

A partial surge or choking can take place where the compressor is approaching the conditions for a full surge or choke but is not fully there. The turbulent flow starts in localized areas in the boundary layer. The vibration level increases and can signal this condition prior to a full surge or choke.

By far, turbomachinery runs smoother than other rotating equipment. The reasons for this are the extra care taken in manufacturing and balancing the rotors. Also, most turbomachinery is monitored for vibration, usually with permanent monitoring systems. The financial losses for a turbomachinery failure are usually much higher than any other kind of equipment. Hence, the additional expense for monitoring and analysis is easily justified. Usually, the avoidance of the first potential breakdown can pay for the vibration monitoring system for years.

Torsional Vibrations

All reciprocating machines (piston engines, compressors, and reciprocating pumps) have torsional vibrations in their crankshaft. These torsional vibrations caused many crankshafts to break until the problem was understood.

A torsional vibration is an oscillatory twisting of a shaft. Figure 5.65 is an aid to visualizing a torsional vibration. Between points *B* and *C* there is a torsional vibration in the shaft as the pendulum oscillates. There is no torsion between points *A* and *B*. The shaft end between points *A* and *B* rotates as a rigid body with the swinging pendulum, but there is no torsional stress in this section. Torsion occurs only when there is differential rotation between sections. The shaft of an electric motor experiences a torsional force on every start-up. The armature provides the twisting force, while the load connected to the shaft wants to remain stationary. This start-up torsion is one-half cycle of vibration as the twisting force reaches a peak, then relaxes to a steady-state value. Any machine that has a varying load during one complete revolution will also generate a torsional vibration as it rotates. These are forced torsional vibrations. There are also natural torsional vibrations depending on the rotational stiffness of the shaft and

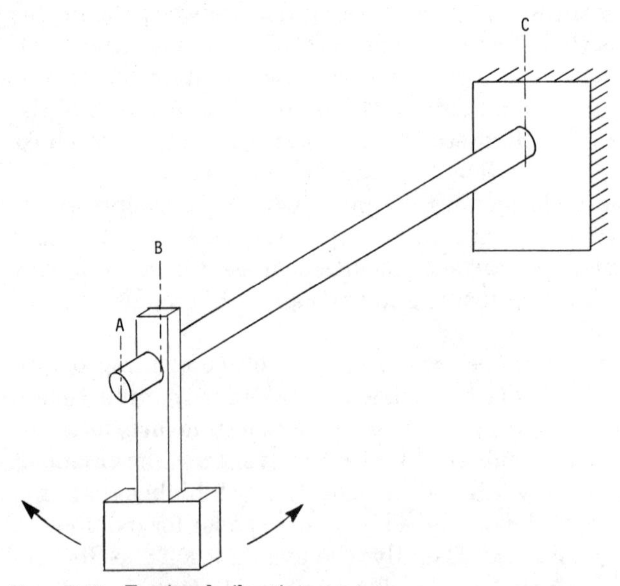

Figure 5.65 Torsional vibration.

the rotational inertia (mass) that the shaft supports. As with linear systems, the torsional natural frequency should not coincide with the forcing torsional forces if a condition of resonance is to be avoided.

There are two causes of torsional vibrations in reciprocating machinery:

1. Gas pressure forces
2. Unbalanced mass in the connecting rods

These torsional vibrations cause shear stresses on the surface of shafts and cracks develop at discontinuities. The continued cyclical torsional vibrations eventually cause the shaft to fail in fatigue. The crack surface is at 45° to the shaft axis. This 45° failure surface is a good indication that the part was exposed to excessive torsional stress. Torsional vibrations are difficult to measure. One must use:

- Strain gauges on the shaft and slip rings or telemetry to get the signal out
- Gear modulation
- Optical methods

Gear modulation involves attaching a gear to the shaft at the location where torsional vibrations are of interest. A lightweight material with minimum inertia, such as a thin aluminum gear, is preferred. A sensor, either proximity or light, detects the passage of each tooth and generates a high-frequency output modulated by the torsional vibrations. The signal must be demodulated and can then be observed on an analyzer.

A new technique is to attach a piece of tape to the shaft that has alternating light and dark bands. A fiber-optic sensor detects the passage of these bands, and a special demodulator removes the spike due to the tape end mismatch. This system is more economical to apply than the gear or strain gages.

In summary, Table 5.4 is a list of the typical vibration problems and statistical percentages of their occurrences. As can be seen, the majority of all vibration problems turned out to be imbalance, misalignment, and resonance. On belt-driven machines, belts and pulleys need to be added in as a big contributor. If history is a teacher, then we can expect the same probability of occurrence for future vibration problems. It is wise, in the analysis process, to investigate for the big four and eliminate them before delving into less likely causes.

TABLE 5.4 Typical Vibration Problems and Their Approximate Percentage of Occurrence

Imbalance	40%
Misalignment (on Coupled Machines)	30
Resonance	20
Belts and pulleys (on belt driven machines)	30
Bearings	10
Motor vibrations (120 Hz)	8
Cavitation in pumps	2
Fan and duct turbulence	5
Oil whirl	2
Sympathetic vibrations	3
Gears	2
False brinnelling	3
Piping	3
Bent shaft/bowed rotor	3
Looseness	5
Soft foot	5
Beats	2
Torsional vibrations	2
Vane passing	3

Note: Percentages do not add up to 100 percent because of compounding of problems. Several vibration problems may exist on the same machine, but one is usually dominant.

Case Histories

Out-of-balance impeller

A new boiler was installed in a central heating plant. This new boiler was the third in a lineup of three firetube boilers, 360 hp each. Immediately upon start-up, there was a serious vibration in the boiler fan. A maintenance man standing next to me remarked that this vibration was excessive. The mechanical contractor overheard this and remarked, "No sweat, you have a 1-year guarantee if anything goes wrong." The vibration was so bad that the control cabinet door would not stay closed. The latch vibrated open, and we had to tape the door closed. We knew we had two problems. One was vibration, the other was communication with the mechanical contractor. Without vibration measurements, it was a war of words.

Two months later we were well into the heating season, and this boiler ran 24 h a day, 7 days a week. The other two boilers were 22 years old but otherwise identical to the new one. By this time, I had acquired an FFT spectrum analyzer and put it to work on this problem. Figure 5.66 is a spectrum captured from Number 3 boiler fan. This clearly shows an imbalance at the fan running speed of 3510 rpm (58.5 Hz). This level of vibration, 0.8 *g*, was serious. The fan wheel was a 30-in-diameter cast aluminum impeller direct mounted on the motor. The maintenance department removed the impeller one day

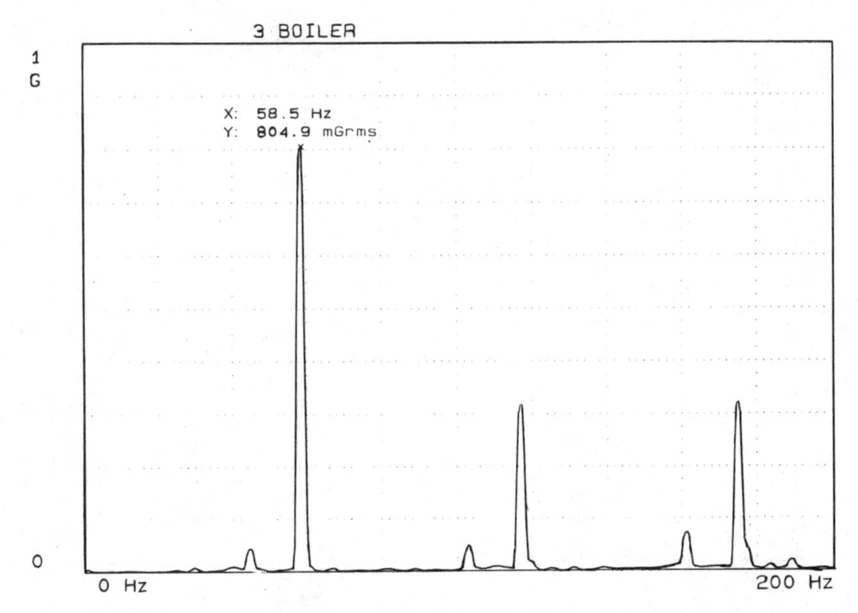

Figure 5.66 Spectrum from the new (Number 3) boiler fan.

and took it to a local hot rod shop to have it balanced. When it returned, another spectrum was taken which looked just like the one above. It appeared that no improvement was made.

A comparative spectrum was taken on the Number 2 boiler immediately next to the new one. This is shown in Fig. 5.67. This older boiler shows a much smoother running impeller. Its fan wheel was removed many years ago and rebalanced. It was also rough when new, and some rebalancing had to be done. It was so long ago that no one remembered where it was rebalanced, or by whom. But it had run smoothly for at least the past 15 years with no trouble.

One morning, 4 months after the installation of Number 3 boiler, a security person brought me some mechanical parts that the night guard found on the floor of the heating plant. It was gas linkage from the Number 3 boiler. We replaced it.

I captured a run-up plot of Number 3 boiler during start-up. This is shown as Fig. 5.68. This plot shows that at low speed, no serious vibration is present. The fan was balanced in the hot rod shop at 600 rpm. At 600 rpm, there is no significant vibration present, and it appears to be smooth. There is a resonance that comes and goes around 1000 rpm. The serious vibration does not begin until around 2000 rpm. Upon hearing this, the balancing shop owner suspected a crack in the rotor that opened up due to centrifugal force and absolved himself of any responsibility.

The following week, boiler Number 3 tripped off during the night

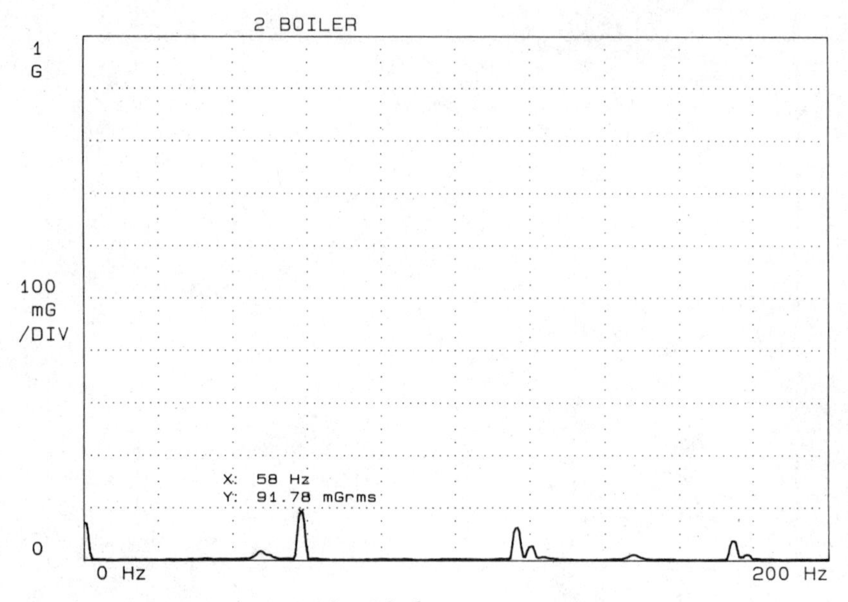

Figure 5.67 Spectrum from Number 2 boiler.

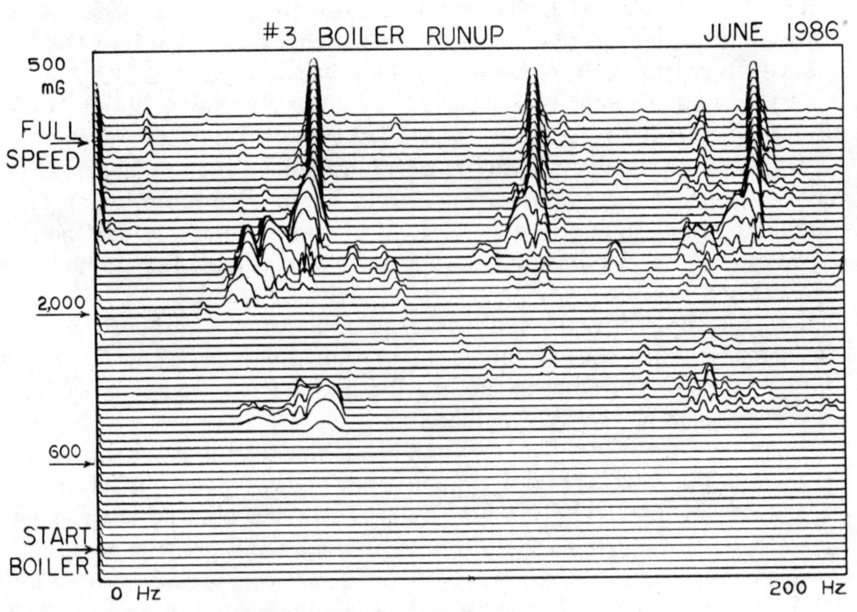

Figure 5.68 Waterfall plot during Number 3 boiler run-up.

with blown fuses. The fuses were replaced once and blew again. The next morning the electrical conduit that supplies 480 VAC to the motor was found shorted to the control cabinet. Where the conduit passed next to a metal edge, it had worn through the flexible metal conduit and the insulation and made contact to ground. This abrasion was clearly caused by excessive vibration. The mechanical contractor came out, replaced the wire and conduit, and restarted the boiler. A maintenance man asked him about the vibration, i.e., if he was going to fix it. The contractor responded, "No sweat, you have 6 months left on your warranty. If anything goes wrong, we'll fix it."

The vibration plots were gathered up and sent to an engineer at the boiler factory. Some phone conversations had previously taken place on this boiler vibration problem, but when the engineer saw the spectrums and other data, he quickly returned a call and said, "I will order a new impeller to be cast, dynamically balanced at 3600 rpm, shipped to your site, and installed at no charge." A vibration spectrum was taken on the new impeller after installation in boiler Number 3. Its plot is shown in Fig. 5.69.

As can be seen, its vibration amplitude at running speed was then comparable to Number 2 boiler. Notice that the harmonics of running speed have also decreased. It can be expected to run 15 years trouble-free.

This case history illustrates two points. First, balancing should be

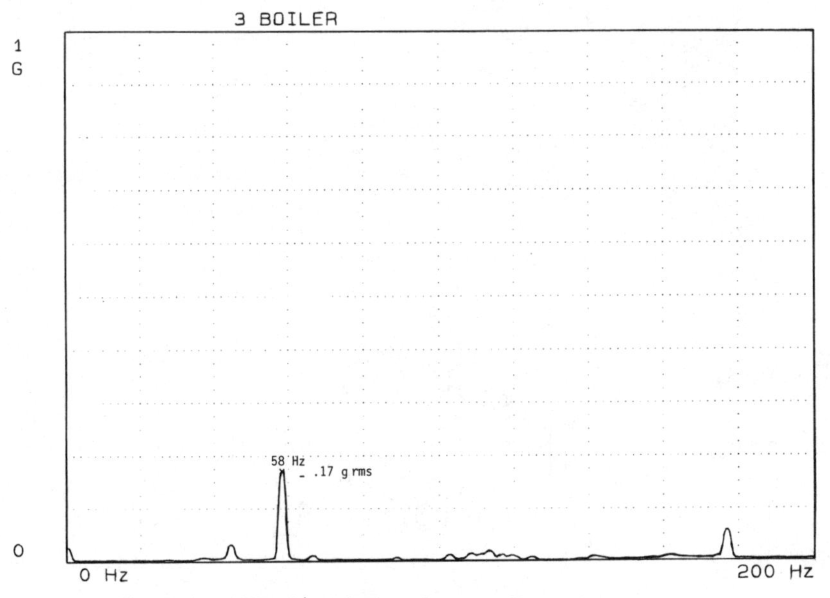

Figure 5.69 Spectrum of Number 3 boiler after impeller replacement.

done at the running speed on high-speed machines to be effective. Many balancing shops, without protective shrouds, will not balance above 600 rpm. It is unsafe for the balancing machine operators or anyone else in the shop. Much good balancing is done at low speed and results in a very satisfactory balance for 1800-rpm machines. Higher-speed machines sometimes need a trim balance at final running speed (and sometimes in situ) to perform satisfactorily. The second point is that without vibration measuring instruments and plots to communicate with, this problem could have remained a war of words. The key to the solution was to understand the nature of the problem and to communicate that to a responsible person. This was made possible with a vibration measuring instrument and hard-copy spectrum plots.

Scrubber fan

The vibration plot in Fig. 5.70 was taken on the outboard bearing of a large overhung rotor scrubber fan. The fan wheel speed was 870 rpm or 14.5 Hz. There is a peak at this speed with many harmonics. This fan was clearly out of balance. Notice the nearby peak at a lower speed of 780 rpm. This was coming from another nearby fan that was also out of balance and transmitted its vibration to this measurement point on the faster fan. The owner of this fan (C-2 scrubber) replaced the bearings in the second year of operation and kept a spare set of bearings nearby. This was a good idea, considering the effect this fan

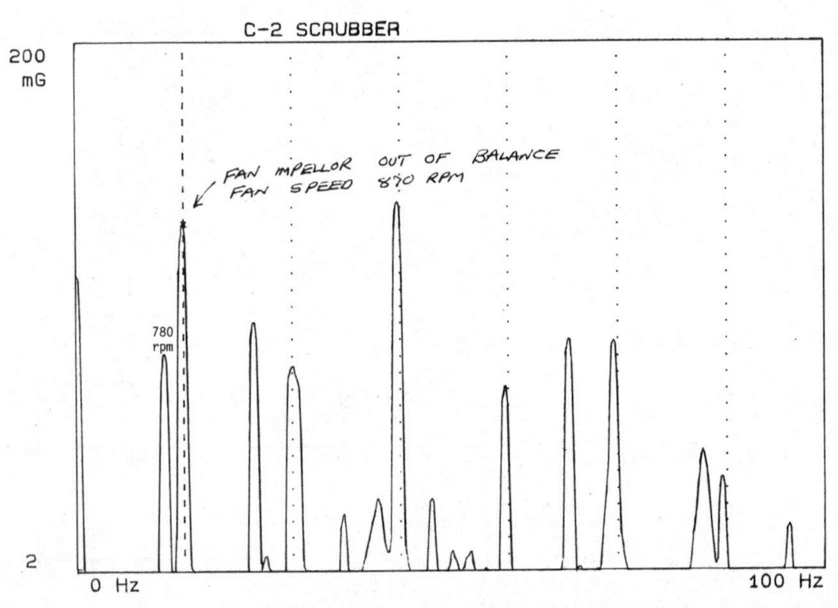

Figure 5.70 Spectrum from the bearing of an overhung rotor fan. Fan speed was 870 rpm (14.5 Hz).

had on the printed circuit operation. I could never understand why the owner could not afford a planned shutdown to balance, but he could afford an emergency shutdown to replace bearings.

The closeness of these two peaks illustrates the need to have good filtered amplitude measurements not only to identify sources but to balance without interference. Attempts were finally made to balance it, but the balance contractor was given too many constraints to do a good job. First, he was only given a 4-h window to complete his work. His tuneable filter balancer did not have a narrow enough window to cleanly separate the two fans. The vibration from the lower-speed fan confused his phase measurements. Finally, he could not apply any weight correction to this epoxy-coated fan. He was restricted to balance weight placements on the outboard pulley. This futile balancing effort was set up for failure before it was begun.

After only 3 years of operation, another larger scrubber fan was installed for a cost of approximately $50,000. This C-2 scrubber fan was retired to a standby mode, because it was a "troublemaker." It had a short life because of ignorance.

Vaneaxial fans

This next case history shows the variability in mechanical quality of equipment and how easy it is to detect with vibration measurements. Figure 5.71 is a vibration spectrum from a vaneaxial fan (AHU-20) turning at 1200 rpm (20 Hz). Notice the low peak at 20 Hz. This fan is well balanced, but it has a moderate motor hum at 120 Hz.

Figure 5.72 is a vibration spectrum from a supposedly identical fan (AHU-38). It came from the same manufacturer, was the same model number and horsepower, and had been operating continuously for 8 years, just like AHU-20 in Fig. 5.71. AHU-38's spectrum shows a significant imbalance, but a quiet motor. In fact, the imbalance on the AHU-38 exceeded the original factory balance specification. AHU-38 has also had a few bearing changes in the motor, which should not come as a surprise.

The difference in vibration amplitude at running speed between AHU-20 and AHU-38 represents a factor of 10. This is typical. In a random sample of any rotating equipment, fans, pumps, or motors, the roughest is 10 times the smoothest. In unusual cases, I have seen factors of 100 to 1. Ten-to-one variability in vibration amplitude of new equipment is typical. These fans had 10 blades on the impeller. The vane passing frequency is present on both fans at slightly less than 200 Hz.

Incidentally, this data was obtained as part of a project to speed up the fans from 1200 to 1800 rpm. This was accomplished by changing

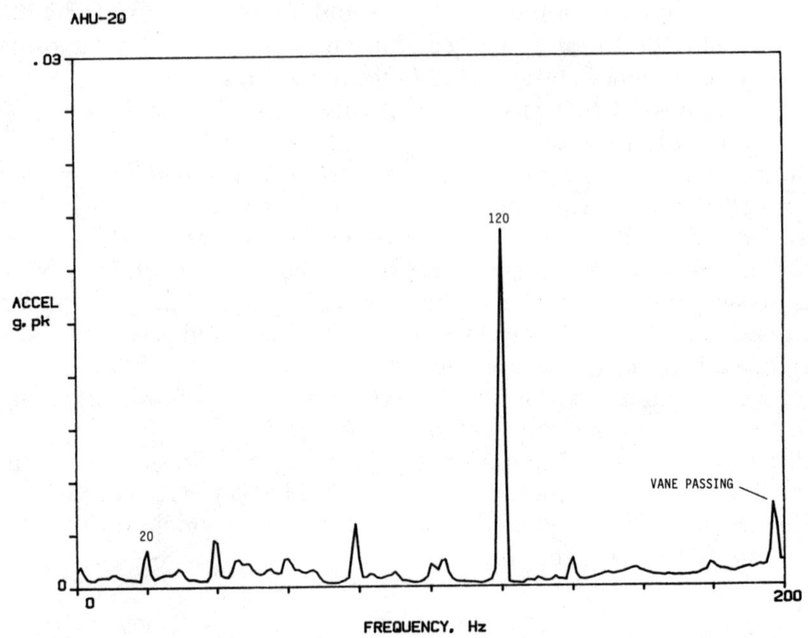

Figure 5.71 Spectrum of a vaneaxial fan that is well balanced but has a noisy motor.

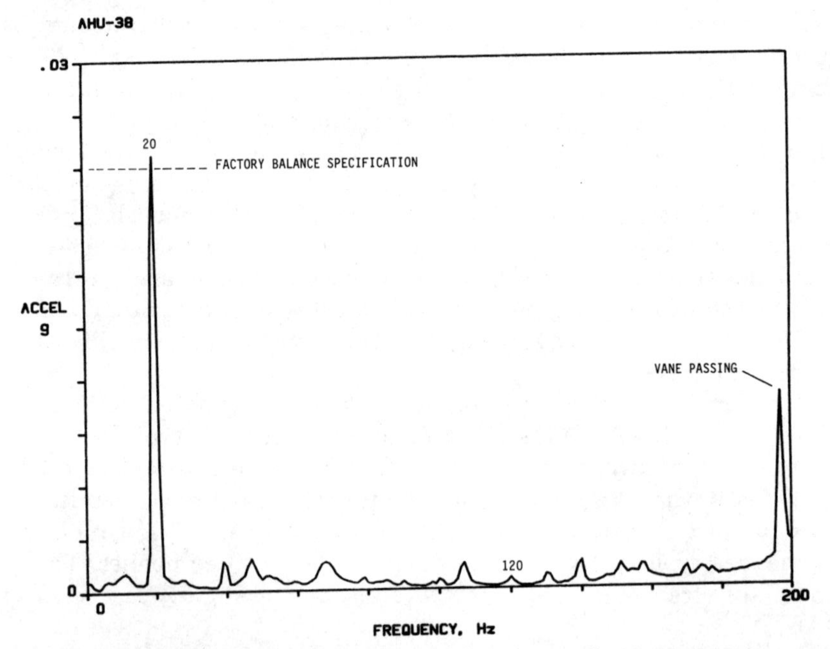

Figure 5.72 Spectrum of another vaneaxial fan (supposedly identical to the one of Fig. 5.71) that has a serious imbalance but a quiet motor.

the motor. The motor was changed on AHU-20 and the vibration spectrum at 1772 rpm is shown in Fig. 5.73.

Notice that the residual imbalance has increased some due to centrifugal force effects, but it is still a smooth-running fan. In fact, this amplitude increase due to a speed change is predictable. If linearity can be assumed, i.e., an increase in centrifugal force causes a proportional increase in vibration, then the vibration increase can be predicted from the centrifugal force formula:

$$F_1 = mr\omega_1^2$$

$$F_2 = mr\omega_2^2$$

$$\frac{F_2}{F_1} = \frac{\omega_2^2}{\omega_1^2} = \left(\frac{\omega_2}{\omega_1}\right)^2 = \left(\frac{1800}{1200}\right)^2 = 2.25$$

The observed change in vibration amplitude at running speed for AHU-20 was:

$$\frac{(\text{At 1800 rpm})\ 0.0065}{(\text{At 1200 rpm})\ 0.0022} \approx 2.9$$

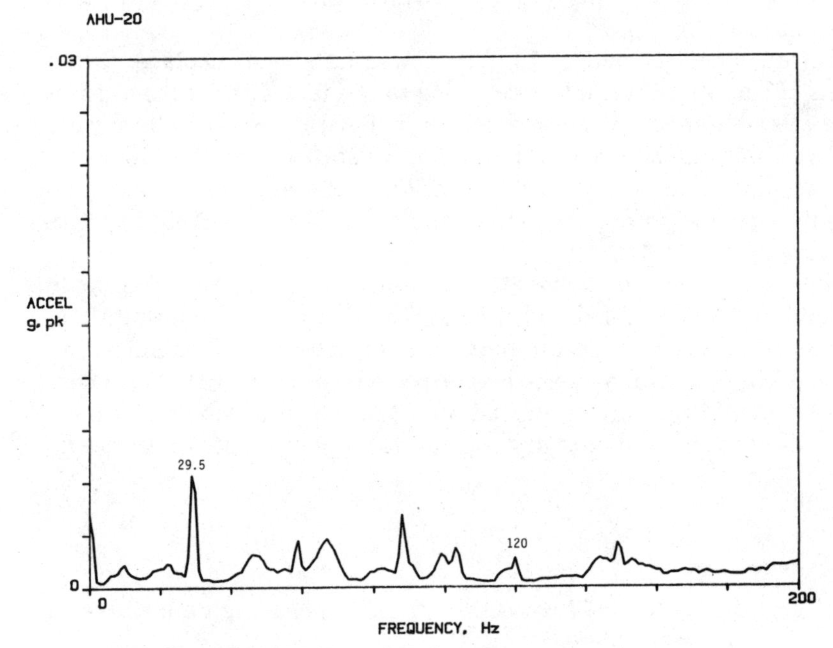

Figure 5.73 Spectrum of AHU-20 after changing motor.

Notice also that the 120 Hz peak is gone now since the motor was changed. The new motor is much better since the 120-Hz peak is hardly visible. This fan should operate trouble-free for many years.

Boiler fans

This case involves three boilers for a hospital. The boilers had small centrifugal fans, 12 in in diameter, that supplied combustion air to the boilers. The fans rotated at 3512 rpm (58.5 Hz). The fans in two of the boilers lasted about 6 months. The failure mode was that they would literally come apart, wedge inside their housings, and stall the motor. The fan wheels would need to be pried out from their housings, usually in pieces. Sometimes, the motor also needed to be rebuilt. A factory-new impeller would then be installed. The plot in Fig. 5.74 is of a brand-new impeller installed after a recent failure.

As can be seen, this impeller is seriously out of balance. The acceleration level at 3512 rpm is 0.4 *g* which is serious. Also, the large number of harmonics indicates that the dynamic imbalance is straining some material into the nonlinear region. The fact that this had been going on for some time told me that the factory had a problem with the balance quality on their fans.

By contrast boiler Number 2 had no such problem. Its fan was the original one supplied and had operated failure-free for 5 years. A vibration spectrum of boiler Number 2 fan is shown in Fig. 5.75. It is well balanced.

The fan wheel of boiler Number 1 was balanced in place. Seven grams of correction weight were added to the fan. The resultant spectrum after balancing is shown in Fig. 5.76. Six other impellers were balanced so that the hospital maintenance staff would have some extras on hand. They all required about the same amount of correction weight, 4 to 8 g, in approximately the same locations, 180° opposite the keyway.

This case history illustrates that new equipment is not necessarily as well balanced as it should be. Each fan failure cost the hospital approximately $800. The accumulated cost of these poorly balanced impellers to the hospital over the 5-year period was about $10,000. These failures should not have occurred, and the hospital should not have had to burden this unnecessary expense for replacements that were no better than the originals.

Grease-laden fan

The spectrum in Fig. 5.77 was taken from the bearing on a cafeteria

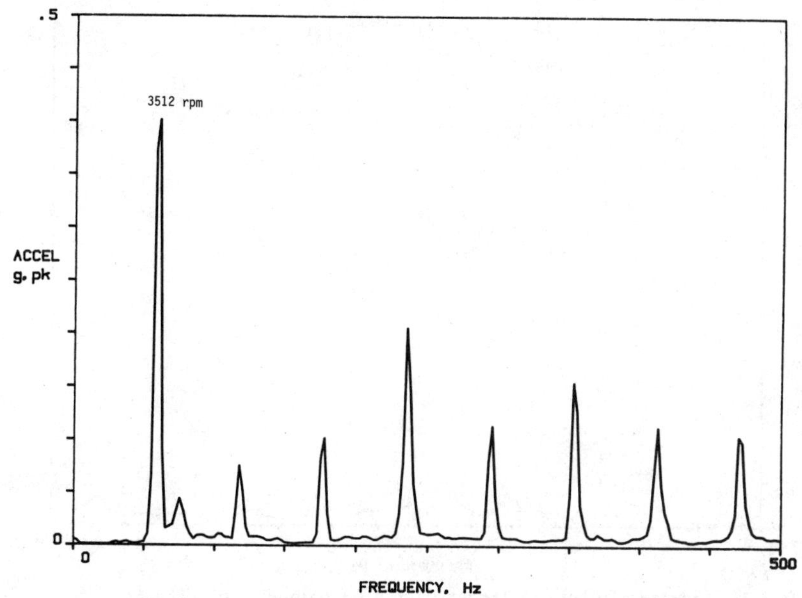

Figure 5.74 Spectrum of a new impeller installed on boiler Number 1. Fan speed: 3512 rpm (58.5 Hz).

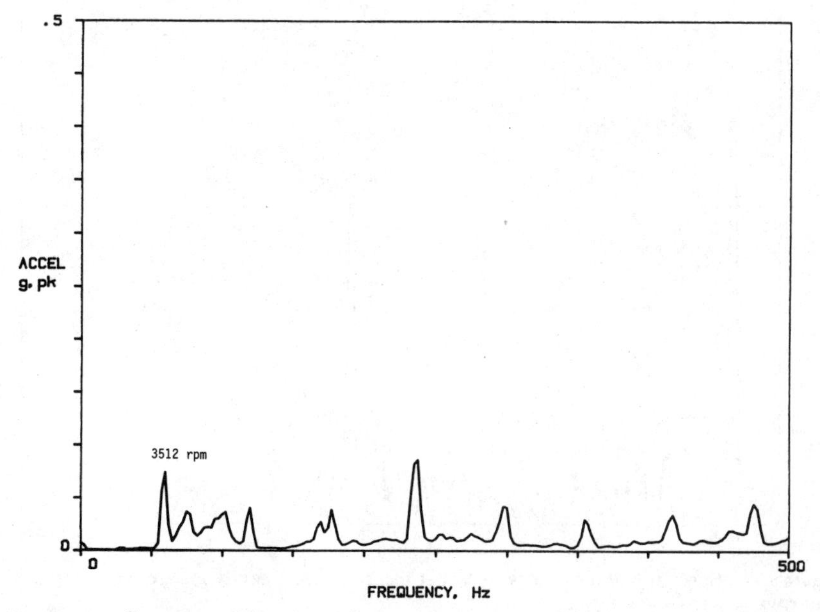

Figure 5.75 Spectrum of Number 2 boiler fan. It is better balanced than Number 1 boiler fan.

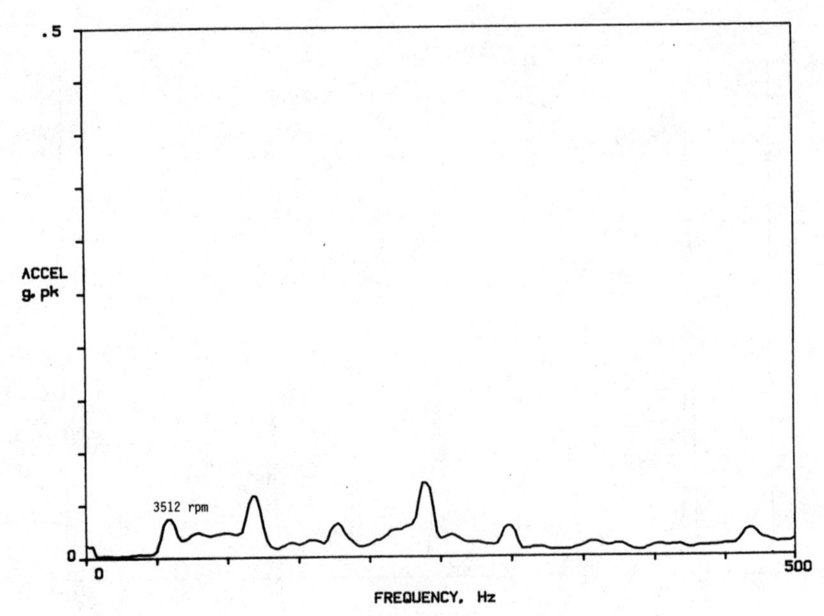

Figure 5.76 Spectrum of boiler Number 1 fan after balancing in-place.

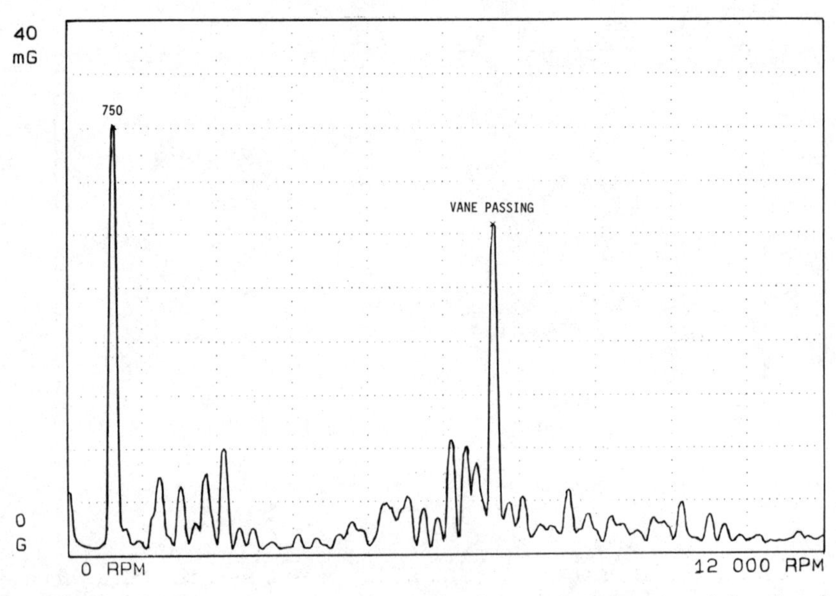

Figure 5.77 Spectrum of a cafeteria exhaust fan that was heavily grease laden. Speed: 750 rpm.

exhaust fan. This fan was heavily grease laden. It rotated at 750 rpm. While it shows a significant imbalance at 750 rpm, it should be cleaned before attempting to balance against grease and dirt. It was cleaned, after which balancing was unnecessary. This fan was monitored on a monthly basis, and the maintenance crews used my vibration data as a signal that it was time for cleaning.

Belt vibrations

The spectrum in Fig. 5.78 is typical of what is seen on belt-driven fans. The largest peak is at 1770 rpm and represents either an imbalance in the motor or an eccentric pulley. The first peak at 500 rpm represents the belt speed. Much of the other peaks across the spectrum are harmonics of the belt speed. The maintenance crew was actually trying to trap me with this problem. They knew that the belts were bad, and intended to install a matched set of belts. They wanted my opinion from vibration data to determine for themselves if this vibration technology was worth anything. The belts were changed and most of this vibration went away. What remained was the motor speed peak, the fan speed, which was relatively low, and a couple of minor resonances.

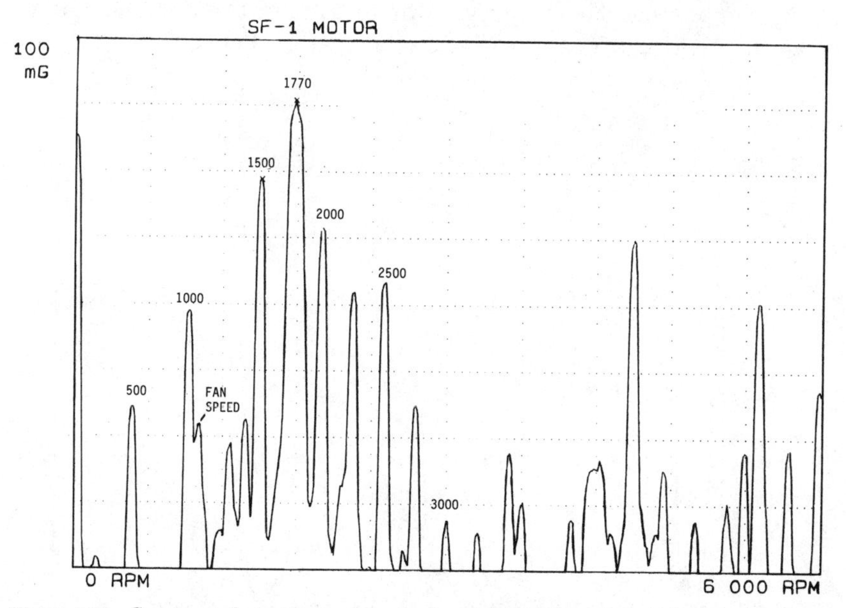

Figure 5.78 Spectrum from the motor of a large belt-driven fan. Belt speed was 500 rpm.

Compressor

The plot in Fig. 5.79 was taken on the exterior casing of an air conditioning chiller. The chiller operates by compressing freon. The compressor is basically a centrifugal fan mounted on a motor shaft. Most centrifugal chillers operate at 3600 rpm. This particular chiller had a two-shaft, 400-hp motor hermetically sealed deep inside. A centrifugal compressor wheel is on each end of the motor.

Chillers are very expensive machines, and many building occupants today feel like they cannot function without air conditioning. This is debatable, but some unscrupulous mechanical contractors are capitalizing on this situation. The owner of the machine in Fig. 5.79 was told that this chiller would not run another season and that it should be replaced immediately, for $500,000. In addition, a sense of urgency was created by estimating a mobile chiller cost of $2000 per day if the replacement was not completed by May 1. The only symptom was surging and some "strange vibrations." The surging, represented by the broadband noise above 400 Hz, could be easily fixed. There was the typical 120-Hz motor frequency present. The real problem was the imbalance in the compressor at 60 Hz. The subsynchronous peak at approximately 25 Hz was slight oil whirl due to excessive clearance in the bearings. This wear in the bearings was caused by the imbalance in the compressor. The diagnosis was to balance the impeller and replace the bearings. This work was scheduled in February.

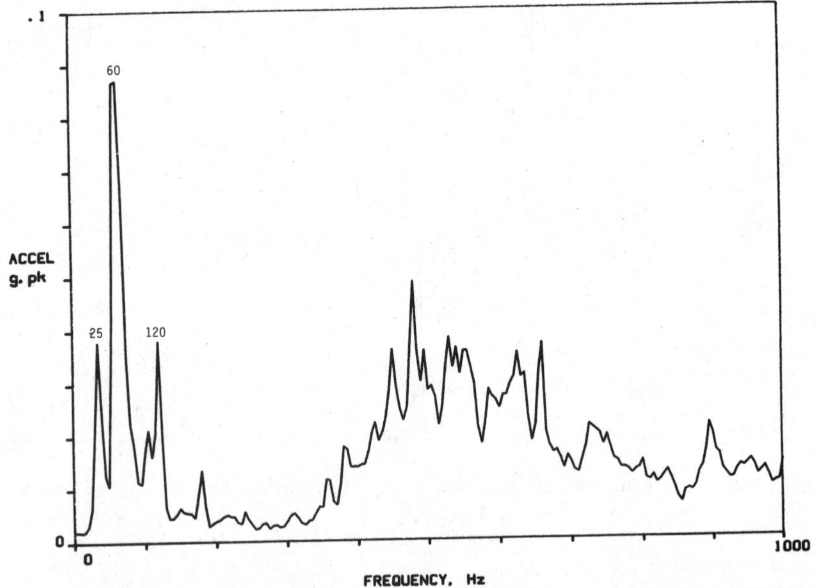

Figure 5.79 Spectrum of a 600-ton centrifugal chiller. Motor speed: 3600 rpm (60 Hz).

Upon disassembly, it was seen that a ⅛-in-thick layer of solids had deposited on the compressor wheel and had begun flaking off unevenly, causing the imbalance. The wheels were replaced, instead of just cleaned, because this was more expedient. The repair work was done, and the machine ran well afterward. The total repair bill was $12,000, and a mechanical contractor was wishing that I would go practice my diagnostic service elsewhere.

It is worthwhile to look at the return on investment of this diagnostic work. The cost of the vibration analysis was $400. The return on this $400 investment was the avoidance of spending $488,000 ($500,000 replacement estimate minus the $12,000 repair bill). The return on this investment is

$$\frac{\$488{,}000}{\$400} \times 100 = 122{,}000 \text{ percent}$$

This is not bad. It would be difficult to achieve this result at any financial institution. These are impressive numbers, and they are typical of savings that can sometimes be realized on expensive machines.

This example illustrates two problems in justifying a vibration monitoring program. The first problem is the infrequent nature of the benefit. Day after day, month after month, time is spent collecting and analyzing data which appears as a cash outflow to the manager with no apparent benefit. Occasionally, a problem arises as described in this case history that would easily justify the monitoring program if properly documented.

The second problem is the nature of the cash flow. It was mentioned that there is a hidden gold mine in applying a vibration monitoring program. The gold mine is not in the form of cash inflow, but the avoidance of cash outflow. The dollars are equivalent in value, but the human psyche does not view a savings the same as it views cash in hand.

A large motor

This case history illustrates how the diagnosis of machinery problems can be misleading. Interpretation of the vibration spectrum is usually easy and straightforward, but there are some difficult ones that keep our diagnostic skills sharp and provide that occasional humbling experience that we need.

The spectrum in Fig. 5.80 was obtained from a large 975-hp motor driving a compressor at 3588 rpm through a flexible coupling. The measurement was taken on the outboard end of the motor in the horizontal radial direction, as shown in Fig. 5.81. At first glance, this looks like a clear case of misalignment. That was the diagnosis, and an aligning session was scheduled. A small amount of misalignment

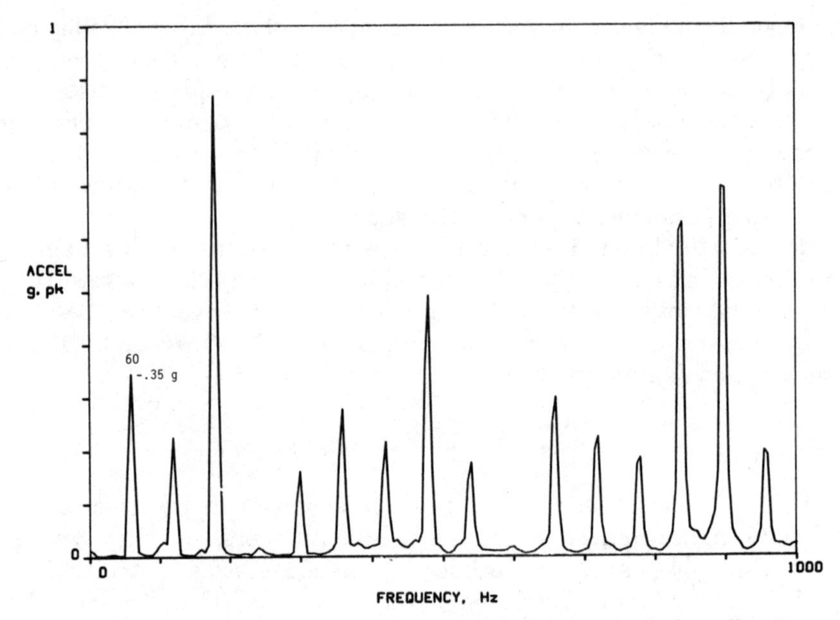

Figure 5.80 Spectrum from a large, 975-hp motor. Speed: 3588 rpm. Overall: 1.71 *g*.

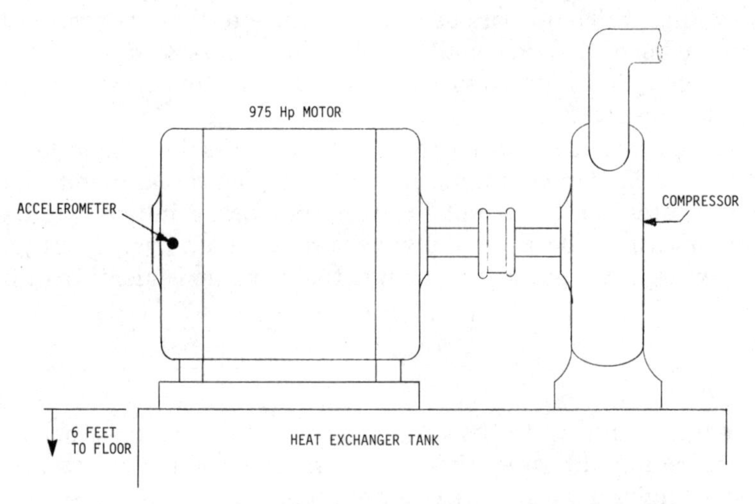

Figure 5.81 Diagram of chiller showing measurement location on 975-hp motor.

was measured at the first setup. A maintenance man who had been around this problem for some time suggested that we run the motor uncoupled. We did, and the resulting spectrum is Fig. 5.82.

The overall value dropped some, from 1.71 to 1.24 *g*. The interesting observation here is that the 1X-rpm peak increased significantly. The

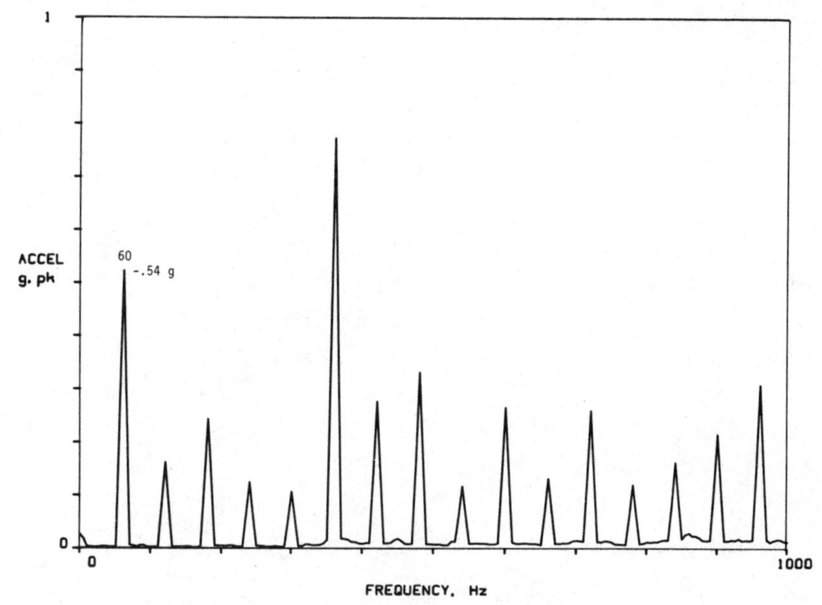

Figure 5.82 Spectrum of motor run solo. Overall: 1.24 *g*.

vibration effect due to the slight misalignment was reduced, but the motor now had more freedom to move due to the imbalance. The large number of harmonics still present when uncoupled was puzzling. Based on this data, a consensus was reached to first balance the motor, then align it.

Sixteen hours were spent attempting to balance this motor using the two-plane method. At first the motor responded to balancing, but each successive trim calculation was asking for greater weights at approximately the same locations. Something was wrong. After 2 days of balancing, the spectrum in Fig. 5.83 was obtained. There is a significant decrease from the previous spectrum, but not enough. There was still a serious vibration problem. It was not imbalance and it was not misalignment. What could it be?

Some previously obtained data was looked at more carefully. The machine vibration map in Fig. 5.84 was obtained with the motor coupled to the compressor and with the motor run solo. The motor uncoupled data has phase information. Notice that the radial vibration readings are similar in amplitude and in phase. The axial vibration readings are also similar in amplitude but opposite in phase. That is, the motor housing was rocking, or twisting. A bowed rotor was suspected. The vibration readings are missing from the shaft end because the author did not want to place his hands and head around a half coupling and shaft spinning at 3600 rpm.

The motor was coupled to the compressor. Soft foot was checked and

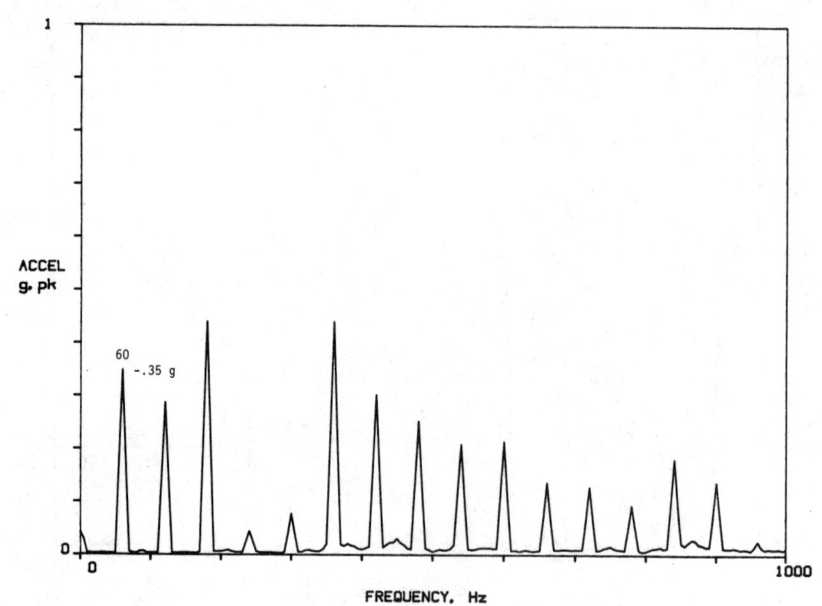

Figure 5.83 Spectrum of motor solo after balancing. Overall: 0.98 g.

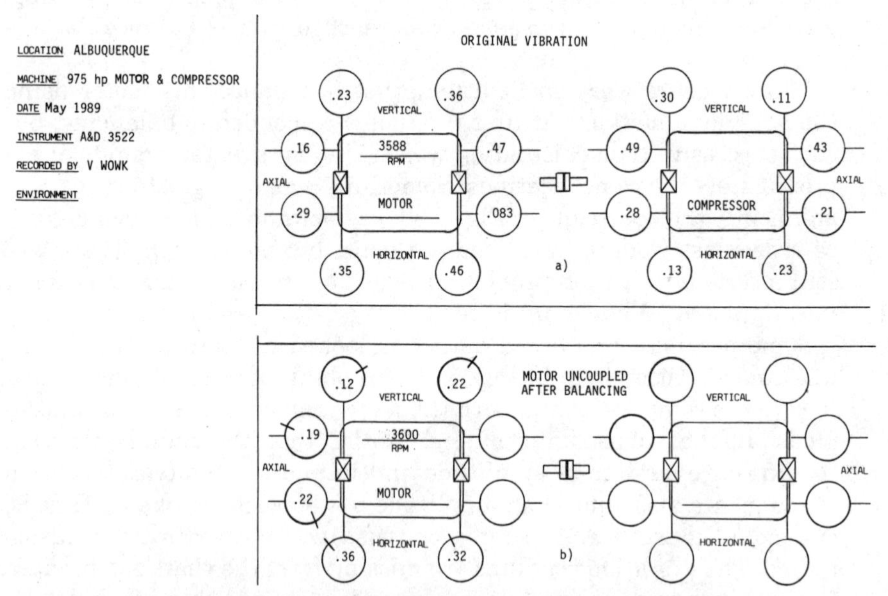

Figure 5.84 Amplitude and phase readings at 3600 rpm of (a) the motor-compressor coupled, and (b) the motor alone after balancing.

corrected for. A minor soft foot shimming of 0.005 under one foot was required. A final alignment was performed. Movements required were 0.024 horizontally at the rear feet, and small movements vertically of about 0.006 to correct for thermal growth.

Upon start-up, the vibration level was down, and it decreased as the machine warmed up. This was encouraging. After 10 min running, the chiller was put on-line and the motor was put under full load. This is when the vibration levels began to increase. Table 5.5 is a recording of the overall vibration level for the first hour after start-up. It was now clear that the motor was thermally sensitive and the rotor developed a bow when the temperature increased. This explains the difficulty in balancing and why the spectrum looks like a misalignment.

When this information was revealed, the owner stopped putting repair money into this machine and placed it on standby. Eventually, when funds became available, the motor was repaired. The bow was corrected by balancing, with the final trim balancing performed in-place at running speed with the motor thermally stabilized. Two years later the owner had a good motor, and the vibration analyst refrained from being premature in the diagnosis.

A bad pulley

This case history involves a rooftop fan that I was called out to balance. The fan was a 13-in-diameter centrifugal driven by a ½-hp motor. The fan was belt driven and turned at 1380 rpm. The motor was a nominal 1800-rpm type. This small fan was literally shaking the roof.

An accelerometer and spectrum analyzer were set up on the fan, and it could immediately be seen that there was very little vibration at 1380 rpm. The fan was already well balanced. The predominant vibration was at 1740 rpm, the speed of the motor. The accelerometer was moved to the motor, and the spectrum in Fig. 5.85*a* was recorded. The vibration at 1740 rpm in the horizontal direction was 1.38 in/sec. This is a serious vibration. In fact, hardware on the roof was coming loose because of this. It appeared that the motor was out of balance.

TABLE 5.5 Overall Vibration Level, *g*, for the First Hour after Start-up

	Inboard	Outboard
Start-up	0.79	1.16
+ 10 min	0.63	0.71
+ 20 min*	0.68	0.96
+ 30 min	0.71	1.13
+ 50 min	0.80	1.51
+ 60 min	0.85	1.37

*Chiller brought up to full load.

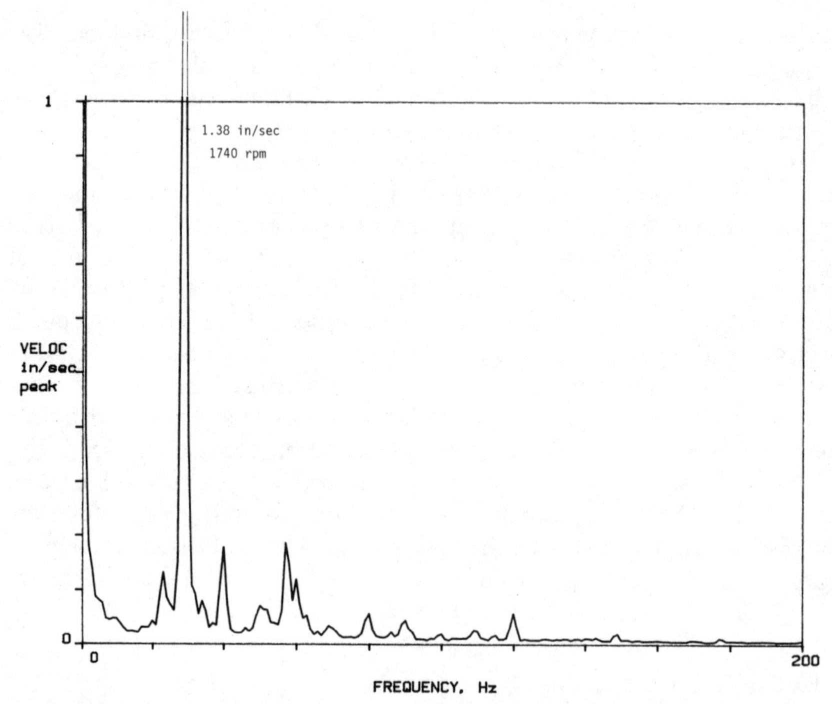

Figure 5.85(*a*) Vibration spectrum from a ½-hp motor, 1740 rpm. Pulley runout: 0.013 in.

However, before balancing any belt-driven equipment, I always check the pulleys. The motor pulley had a total runout of 0.013 in when measured with a dial indicator on the belt running surface. This was a new motor and pulley.

The belt tension was relaxed by rotating the adjustable pulley on the motor ½ turn. The vibration at 1740 rpm (29 Hz) dropped to 0.29 in/sec, a 79 percent improvement (Fig. 5.85*b*). This was almost acceptable to the owner. It was okay in terms of vibration, but the airflow reduction and loose belt were not. The motor pulley would have to go back to its original setting. Loosening the belts is an effective test for an eccentric pulley. If the belts cannot remain loose, then there are two other remedies—change the pulley in hopes of putting on a better one, or remachine the existing pulley.

Fortunately, there was a lathe below inside the building, and it was more convenient to remachine it. The remachining was in the form of clean-up cuts on the belt running surfaces. Unfortunately, this pulley had a rather loose fit on the shaft, and tightening the setscrew cocked it over to a runout of 0.005 in after machining. I would have liked to

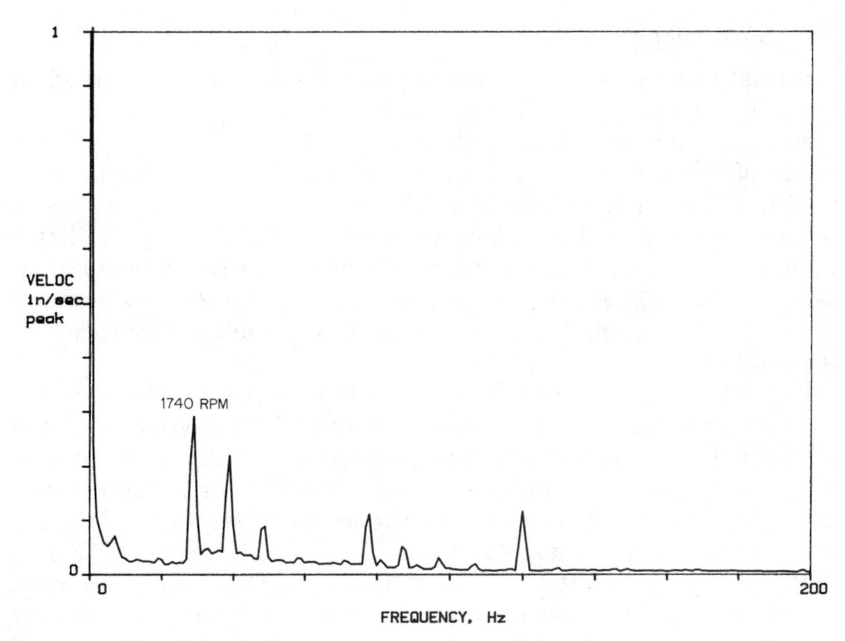

Figure 5.85(*b*) Relaxed belt tension.

have had less runout than this after remachining the pulley, but the resulting vibration was satisfactory. The final vibration with the remachined pulley and belt installed tight was virtually identical to Fig. 5.85*b*.

It is useful to review what happened here.

1. I was commissioned to balance the fan. This would have been a waste of time. A frequency analysis revealed the true cause.
2. There was a serious vibration at motor revolutions per minute that looked like imbalance in the motor. Any attempt to balance this motor would have been futile.
3. This problem was corrected by recognizing a pulley runout and correcting the root cause. This corrective procedure took about 2 h. It is standard practice for me to check pulleys for runout before doing any balancing on belt-driven equipment.

Perhaps my perception is biased because I see all the problems in my type of business, but I see too many bad pulleys than I would like to in a modern industrial society. Motor and fan manufacturers do a good job, generally, in balancing their equipment. Too often, this work at the factory is nullified by a lousy pulley.

Out-of-spec resonances

This final case history illustrates a problem with vibration specifications. Figure 5.86 is the axial vibration spectrum from a motor driving a centrifugal fan. This was a routine measurement during acceptance testing of new equipment. The specification called for a maximum allowable vibration amplitude of 0.2 in/sec rms at the rotation speeds. The fan rotated at 850 rpm, and the motor at 1750 rpm. By the letter of the specification, this fan passes. However, the two resonances at 1540 and 1940 rpm are above the limit. This spectrum was a summation average of 32 spectrums, so the peak amplitude is probably considerably higher.

With a tuneable filter analyzer tuned to only the two peaks at 850 and 1750 rpm, this fan would easily pass, and the analyst would have no knowledge of the two resonances. This points out the need to do frequency analysis across the entire spectrum of interest. More importantly, the specifications should require measurements across the entire spectrum, not just discrete points. If that were required, then it would save me embarrassment in situations like this because mechanical contractors frown on measurements of acceptance that are not specified.

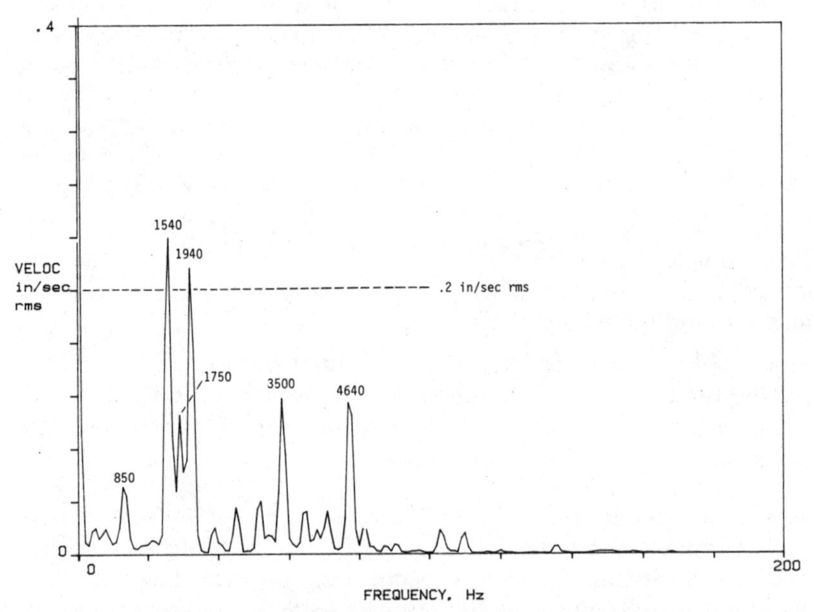

Figure 5.86 Spectrum from a centrifugal exhaust fan. Fan speed: 850 rpm, motor, speed: 1750 rpm. This measurement was on the motor in the axial direction.

Chapter 6

Techniques

Vibration analysis is broadly composed of two separate functions—acquiring data and then analyzing the data. The analyzed data is the input to a decision-making process that will ultimately lead to a quiet, smooth-running machine. This is illustrated in Fig. 6.1. The important point to be made in this flow diagram is that many dollars and long-term effects hinge on that data. The very first step, acquiring the data, is the most critical factor in this whole process. All subsequent efforts are based on this data. These subsequent efforts may require many dollars, time and labor, and the repair project could take as long as a year to complete. In light of this, it makes sense to spend sufficient time to acquire good data and not to hurry this process. The acquisition of good vibration data is the most important function that the analyst can perform. At the front end of this data acquisition is the transducer and how it couples to the surface. These transducer mounting methods will be covered in this chapter.

It is best for the vibration analyst to concentrate on the acquisition of good data when at the vibrating machine. The analysis can take place later in a quiet office. The noisy machine room, where communication usually takes place at a shout volume, is not conducive to

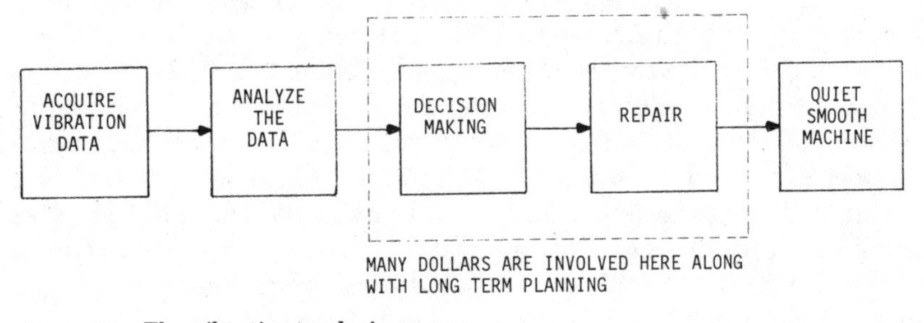

Figure 6.1 The vibration analysis process.

thinking and analyzing. A time factor may also be involved, and with time constraints, the priority at the machine should be getting good data and recording it. The analysis function works best with several heads together. The analysis, therefore, should be a process of reaching a consensus in an environment which allows good communication.

There are two good techniques that lend greater credibility to the data. The first is to repeat the measurements at a different time; an hour later, and then 24 h later. Each time, disconnect your instruments and set them up again fresh. This will give you confidence in your instrument setup. The time delays of 1 h and 24 h give you confidence that the machine really has a problem that won't go away. The second validation method is to use a totally separate instrumentation system, including a different transducer. If the data repeats, then you can have confidence that the problem does not lie with your instruments. As an alternative, your instrumentation system can be calibrated before and after the critical measurements.

These exercises in data validation are sometimes necessary because someone later may challenge the data. When the analysis and decision making begin to place a person in an unfavorable position, then his or her first defensive measure is to question the data. It is good to be prepared beforehand. Besides, whenever large repair dollars are at stake, so is your reputation as a vibration expert.

The two elements of good data are accuracy and repeatability. The separate instruments and calibration lend accuracy, and the time delay ensures repeatability.

Amplitude Mapping to Locate the Source

Several case histories throughout this book emphasize the most important two points of vibration analysis.

1. The frequency is transmitted unchanged.
2. The amplitude gets larger closer to the source.

In any vibration problem, the first step is to identify the offending frequency. Once the frequency is pinned down, then the amplitude is used to track it to the source. This is applicable to buildings as well as to machines. For buildings, mapping out the vibration amplitude on the floor will lead you to the offending machine. This amplitude mapping will also produce some useful information about the building transmission characteristics, and possibly its modal behavior. For machines, its exterior can also be mapped to pinpoint the probable cause. The traditional horizontal, vertical, and axial vibration measuring is a form of amplitude mapping.

In summary, the two rules of vibration analysis are:

1. The frequency is characteristic of the source.
2. The amplitude is characteristic of the severity plus the path.

The next chapter will elaborate more on this.

Loose-Parts Monitoring[1]

The nuclear power industry has made use of amplitude mapping to locate loose parts within the reactor vessel. This is done with a series of accelerometers attached to the outside of the reactor vessel. In this case, the transducers are stationary, and the loose mechanical parts within can be stationary or mobile. With an array of sensors, it is possible to estimate the locations within by triangulation, by the relative amplitude sensed by different sensors, and by time of arrival.

Figure 6.2 is an illustration of a typical installation. Accelerometers are mounted around the reactor vessel, steam generators, and piping. A typical system may contain approximately 20 accelerometers.

The signals detected are shock pulses in the time domain. They are actual metal impacts predicted by Hertz contact theory. These shock pulses look just like any other metal-to-metal impact in the time domains. Typical frequencies are 1000 to 10,000 Hz and amplitudes are 0.05 to 30 *g*.

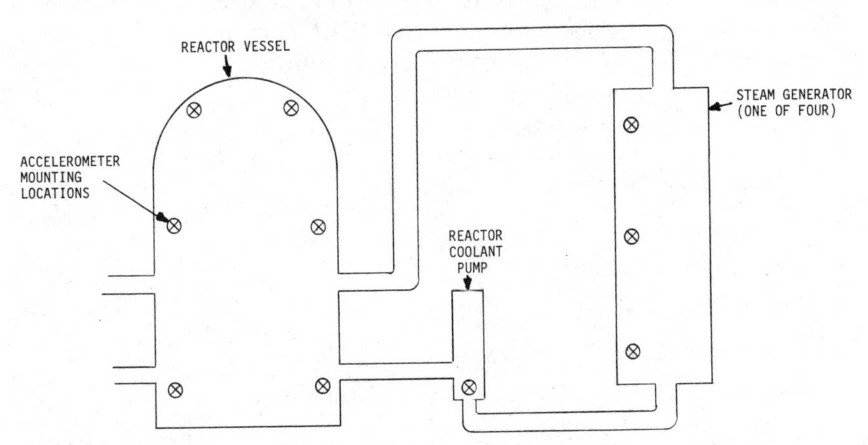

Figure 6.2 Loose-parts monitoring system for nuclear power plants.

[1]For more information on loose-parts monitoring systems, refer to "Loose-Parts Monitoring Systems Improvements," *Electric Power Research Institute (EPRI) Report No. NP-5743,* March 1988.

Loose parts within the reactor vessel can do a tremendous amount of damage. The loose parts are created by the thermal stresses within the reactor, or originate elsewhere and are carried to the reactor vessel by the coolant flow. The loose parts tend to collect at the reactor vessel where the fluid velocity slows down.

There are some unique requirements for the instruments used in loose-parts monitoring systems. Since the accelerometers are on the outside of the reactor vessel under the insulation, they are exposed to approximately 625°F. The use of charge-mode accelerometers is necessary with radiation-resistant cables. Physical access to sensors and preamplifiers inside the reactor building in a contaminated environment present some maintenance difficulties. Transducer reliability is a necessity. The accelerometers are permanently installed with a stud mounting.

This loose-parts monitoring system is mentioned for two reasons. First, this technique has applications in other large vessels and piping systems. Second, for those that may be unaware, it is common to find construction materials in closed piping systems, like tools, hardware, sawdust, and lunch bag contents.

Transducer Mounting Methods

Transducers can be mounted to the vibrating machine surface a number of different ways. The reasons for using one method over another are convenience and desired frequency response. It is desirable to couple the vibration transducer to the surface well enough so that it moves with the surface without separation. This discussion assumes that an accelerometer is the vibration transducer. Some of this discussion is applicable to velocity probes also.

The accelerometer can be:

- Permanently bonded
- Stud mounted
- Attached with a magnet
- Attached with beeswax
- Attached with double sticky tape
- Hand held, with or without a probe extension

As can be seen in Fig. 6.3, the frequency response varies with the mounting method. The accelerometer has a resonant frequency, typically about 25 kHz. It is always desirable to measure well below this resonant frequency to obtain a linear response and a valid measurement utilizing the transducer's sensitivity constant. Measurements

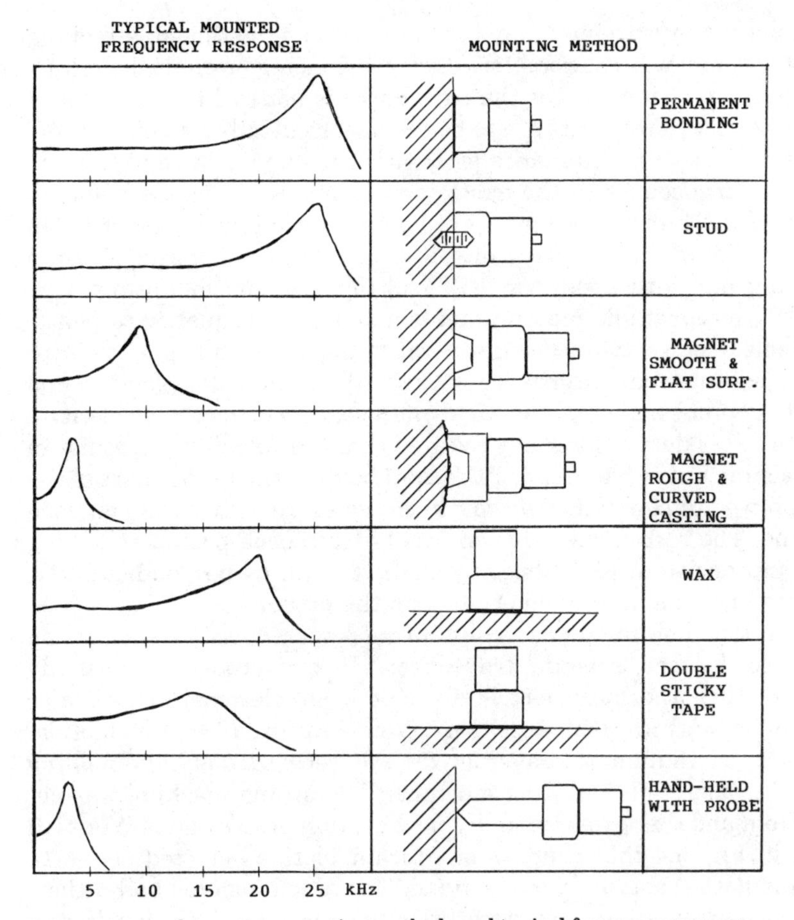

Figure 6.3 Accelerometer mounting methods and typical frequency response.

are possible at and above the resonant frequency; however, the published sensitivity or calibration constant is not valid, and you will need to come up with your own. The mounting method introduces another spring in series with the measurement path and lowers the useable resonant frequency of the accelerometer. This lower resonant frequency is called the *mounted resonant frequency*. For high-frequency measurements, keep the interface as short and stiff as possible.

Stud mounting is the best for high-frequency measurements. Hand holding an accelerometer seriously degrades the mounted resonant frequency. Hand-held probes are good for a quick "look-see" when probing around a machine surface, or if the frequencies of interest are less than 1000 Hz.

Magnet mounts are probably the most common method of attaching velocity probes or accelerometers during balancing or analysis. Adjust the magnet position so that the transducer is not rocking on the surface. When measurements are to be taken periodically for comparison, as in a predictive maintenance program, it is important to always return your transducer to the exact same point. It is also desirable to eliminate the human variability of holding techniques. Therefore, in predictive maintenance programs, it is best to permanently attach a mounting nut with epoxy, or drill and tap, for stud mounting. This will ensure repeatable measurements and a high-frequency response.

A nuclear power plant with several hundred machines to monitor began a vibration monitoring program with portable data loggers and hand-held probes. They gathered about a year's worth of nonrepetitive and worthless data before they realized that the variability was not in the machine but in the hand of the man holding the probe. Part of the solution was to conspicuously mark the measurement points on each machine. The rest of the solution was to train each person collecting data in a consistent probe-holding method, defining perpendicularity, pressure, movement, and where to grip the probe.

When hand holding a probe, significant low-frequency hand motion is sensed by the seismic transducer. If not properly thermally shielded, then the piezoelectric element in accelerometers will also sense hand warmth. It is sensitive to temperature as well as motion. Figure 6.4 is a summation average for a 30-sec period of hand holding an accelerometer on a milling machine. The milling machine was not operating, and there was virtually no vibration present on it. You will notice in Fig. 6.4 that there is significant motion detected below 15 Hz. This is the motion from the hand. The accelerometer had a thermal boot surrounding it so very little temperature variation is presumed to be detected. The overall vibration level is 0.033 *g*.

Figure 6.5 is an averaged spectrum of the same accelerometer on the same nonoperating mill, but now mounted with a magnet. Most of the low-frequency motion is absent. The small peak at 0.5 Hz is the accelerometer cable swaying. If the cable did not sway, then even this small detectable motion would be absent. The overall level is 0.002 *g*.

The message here is that if small amplitude vibrations at low frequencies are to be measured accurately, then human hands must not hold the seismic transducer, or the data below 15 Hz should be discarded. It is good practice to never rely on hand-held measurements, but sometimes it is necessary due to expediency. When higher-amplitude vibrations are present, such as on a normal vibrating machine, then these low-level, hand-motion signals are down in the noise level and insignificant. Even so, when someone wants to challenge your data, hand-held vibration data is open for attack. My business as

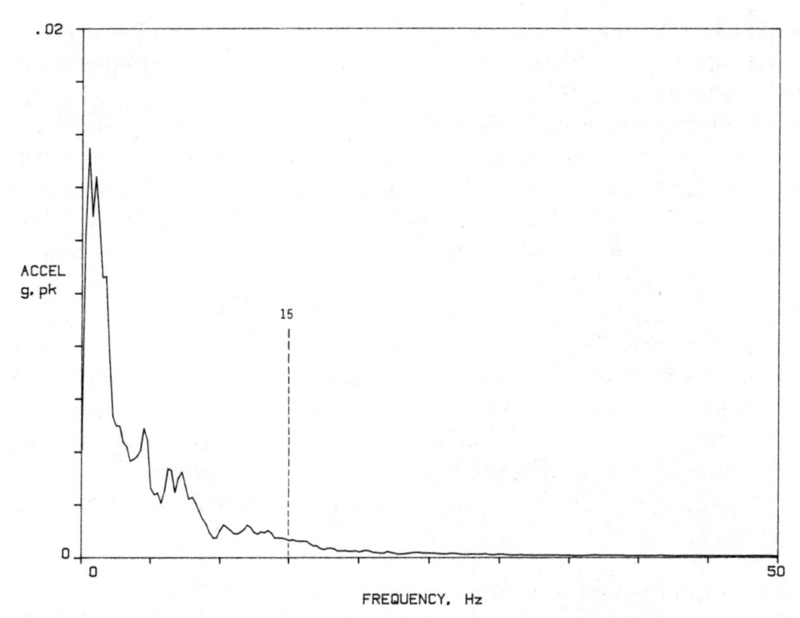

Figure 6.4 Hand holding an accelerometer on a nonoperating machine. Overall is 0.033 *g*.

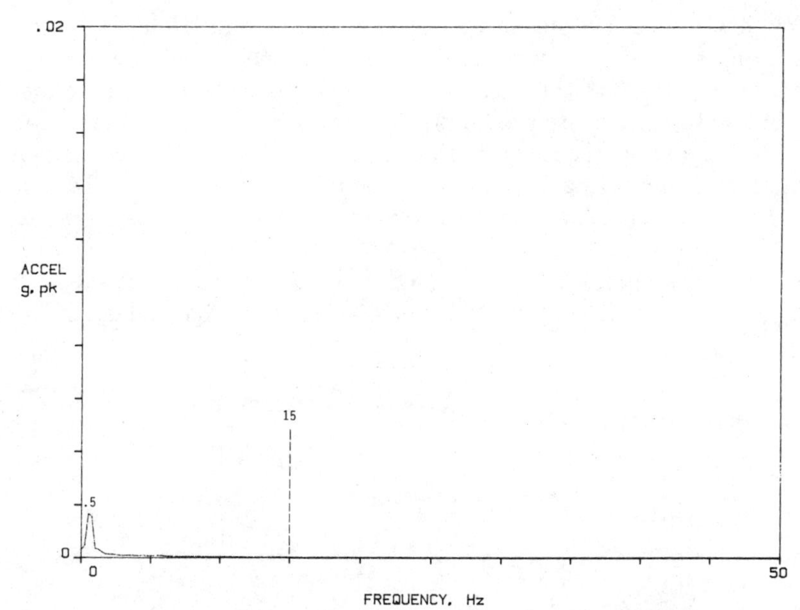

Figure 6.5 Magnet-mounted accelerometer on the same nonoperating machine as in Fig. 6.4, in the same location. Overall is 0.002 *g*.

a consultant has put me into conflict situations occasionally. Be advised that hand-held and magnet-mounted vibration data are inadmissible in court.

A magnet will faithfully track low-frequency vibrations, without introducing hand motions below 15 Hz. The magnet mounting is, therefore, a big improvement over hand holding in the low-frequency area. They both have limitations at high frequencies, above 2000 Hz, that can only be overcome with better coupling to the vibrating surface, such as stud mounting or epoxy.

If hand-held probes are to be used, such as are supplied with some portable instruments, then the exact measurement spot should be center drilled or punched and conspicuously marked. It is not possible to hold a probe perfectly still. Some hand motion below 15 Hz will be sensed and added to machinery vibrations detected by the probe, but if the machine vibration amplitude is high, the hand motion is insignificant.

If hand-held probes are to be used, these precautions should be observed:

1. With hand-held transducers, keep the probe length short and stiff.
2. Use a thermal boot on piezoelectric accelerometer.
3. Use a uniform pressure.
4. Keep the probe perpendicular to the surface.

For low-frequency measurement (i.e., less than 1000 Hz), any mounting method will work okay as long as the accelerometer remains in contact with the surface and the precautions about hand motion below 15 Hz are observed. For high-frequency measurements, the mounted resonant frequency is the limiting factor to accuracy. Even normal stud mounting becomes questionable above 10,000 Hz. For very high frequencies a film of oil or grease under the accelerometer will improve coupling.

A rarely used hand-holding method is to use a fishtail-shaped stick (Fig. 6.6). The stick is made of hardwood, or it can be a plastic. The

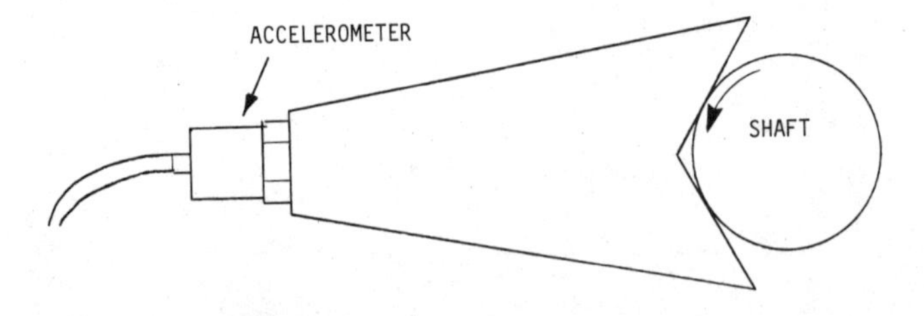

Figure 6.6 Fishtail-shaped stick used to measure absolute shaft motion with a seismic transducer.

seismic transducer is threaded into the stick, and the V tail is held against the rotating shaft. Plenty of lubricant must be used on the surfaces in contact. This method allows the actual shaft motion to be measured with a seismic transducer without the expense of mounting a proximity probe. The friction will generate a significant amount of broadband noise and should be filtered out. This mounting technique is seldom used, but there are some applications where it could be useful, such as modal balancing of long flexible rolls.

A question that frequently occurs at training workshops is, where is the best location on a machine to mount transducers? There is no correct answer, and the appropriate response is another question: What is the purpose for the measurement? For balancing, the horizontal radial direction is usually best. For bearing monitoring, the vertical direction is best. The axial position is good for detecting misalignment and thrust-bearing problems. For diagnostic work, measurements should be taken in many locations, with phase information. For periodic monitoring, almost anywhere on the machine will do. Some transducer makers would like you to permanently bond one of their transducers at each bearing in the horizontal, vertical, and axial directions. This would mean at least 12 transducers on each machine combination. Few companies can afford this. At the other extreme is no transducer at all, and hand feeling is the only measurement. Somewhere in between is appropriate, and some guidelines will be offered.

Figure 6.7 shows a typical machine and the typical locations for monitoring vibrations. The reasons for taking vibration measurements are generally of the following:

- Periodic monitoring
- Balancing
- Analysis

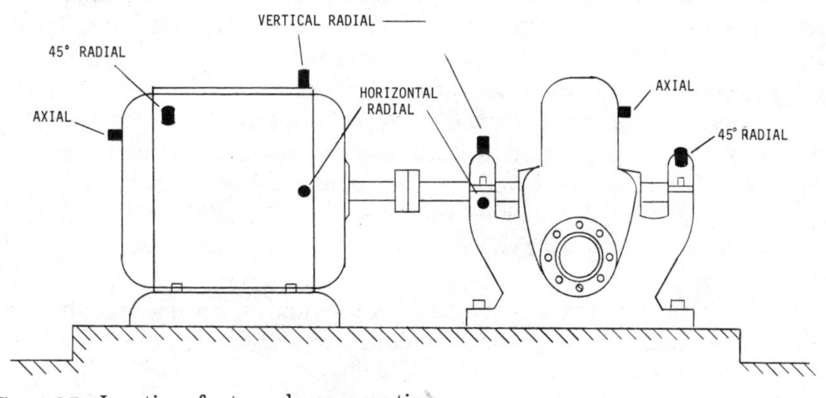

Figure 6.7 Locations for transducer mounting.

The purpose of periodic monitoring is to detect changes, and these changes will be detectable almost anywhere on the machine, or its support platform. It is only crucial to always return to the same point, to use the same instruments, set up in the same manner, and measure with the machine in the same load condition. One measurement point per machine is sufficient. This point should be as close as possible to one of the bearings, preferably in the direction of the load zone. The radial direction is preferred unless a thrust bearing is to be monitored, then the axial direction is better. For two coupled machines, it is beneficial to install two monitoring points, one on the driver and the other on the driven machine. A permanently installed mounting adapter is recommended into which you can couple the transducer. This ensures that the same point is used and that your hands are off the transducer during measurements. With these guidelines observed, then the best location to mount a transducer will be one that is easy to reach and not hazardous for the person who must take these measurements month after month. Permanently installed mounting adapters with a roving transducer are strongly recommended not only for cost reasons but also for calibration purposes. The transducer needs to be calibrated at least every 6 months on a shaker table. This is difficult to do with it permanently bonded to a machine. Some difficult-to-get-to locations may need a permanently installed transducer with a cable extended to a convenient measurement point. This discussion suggests that three measurements at each bearing, in the three directions, is unnecessary for trending. Good success has been achieved with monitoring only one point per machine.

Transducers for balancing are mounted at the bearings, usually in the horizontal, radial direction. This is because the machine is usually more flexible in the horizontal direction and the centrifugal force due to imbalance is radial. However, a true imbalance can be corrected by taking measurements in the vertical direction just as well. For machines mounted on flexible springs, the transducer can also be mounted on the platform for balancing. Actually, almost anywhere will do for balancing. I balanced a fan once with measurements in the axial direction. The axial vibration on this fan was stronger than the radial vibration, and it responded to balancing corrections. I once met a person who balanced a disk by measuring the vibration on a separate platform 6 ft away. The separate platform was the location where it was desired to reduce the vibration, and I might add that this technique worked very well. The radial direction as close as practical to the bearing is the conventional location to mount transducers for balancing, but actually almost anywhere that a good 1X-rpm signal is detected will do.

For analysis purposes, the judgment of the analyst becomes paramount. Analysis is composed of two parts—frequency analysis first and then later some measure of amplitude, and possibly phase. For frequency analysis, almost anywhere on the machine will do. The same frequencies will generally be present everywhere only differing in amplitude. The point where the analyst must apply crucial judgment is when he wants a measure of amplitude. If there are specifications for the measurement purpose, then they usually dictate locations. Architectural specifications for new construction typically specify the bearings, or the platform in the vertical direction if the bearings are inaccessible. Bearing frequencies will appear stronger in specific directions. Resonances of machine parts and nearby piping or ductwork can add to the amplitude. Background vibrations from other sources can transmit to the transducer. The analyst must keep in mind that the vibration detected at the transducer location is the sum of all vibrations arriving at that point. His or her transducer cannot separate them. He or she must use the spectrum analyzer to separate them along with his or her mind and the judicious selection of measurement location. This is the time to take vibration measurements at each of the bearings in the horizontal, vertical, and axial directions and record the amplitudes and phase. More data is better than less at this point. It can be sorted out later. It is also very useful, while there, to collect comparative data from identical nearby machines at the same locations. Another guideline can be offered for transducer location for analysis, if the bearings are inaccessible. It would be to measure, initially, where the motion on the machine is least as detected by hand feeling. This avoids flexible panels and starts at the stiffest points where mechanical coupling is the best. The purpose here would be to analyze internal sources. If, however, the purpose of analysis is to find the flexible parts and fix them, then measurements need to be made all over the machine, perhaps every few inches.

If you absolutely cannot get near the machine because of rotating parts, or temperature, then connect to some structure attached to the machine, like piping, conduit, beams, or struts. For frequency analysis this is valid. As a last resort, use a microphone and analyze the sound field. There are situations where you cannot get close to a machine because of hazards to life and limb. Examples are helicopter tail rotors, steam boxes, or airborne vehicles. In such cases, the acoustic energy emitted contains the same frequency information and can effectively be analyzed with a microphone.

Phase Relationships

For rigid machines the phase relationship of one end of the machine compared to the other end is useful in diagnostics. It is possible to visualize how the machine is shaking. In-phase vibrations are vibrations that are moving in the same direction at the same instant of time. That is, two separate points on a machine that measure the same phase angle are translating through space without rotation (Fig. 6.8*a*).

Out-of-phase vibration means that pure translational motion does not exist. Some component of rotation, or flexing, is taking place.

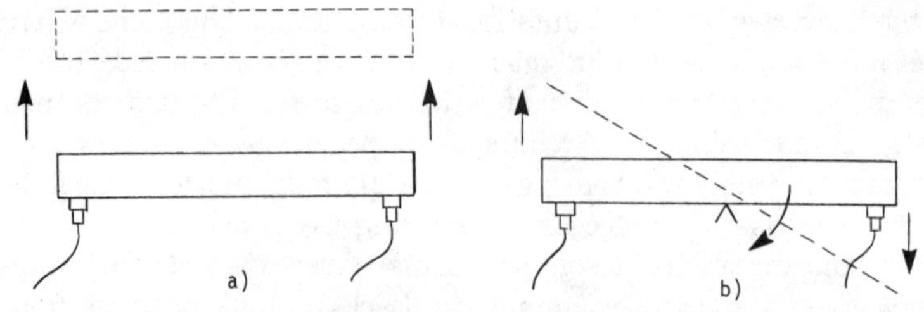

Figure 6.8 (*a*) In-phase motion. (*b*) 180° out-of-phase motion.

Figure 6.8*b* shows a bar pivoting on a fulcrum, like a see-saw. The two ends of the bar are moving in opposite directions at the same instant in time. Two transducers, one on each end, would measure phase angles that are 180° different.

Measuring the phase on the outside casing of a machine gives an idea of the forces within. This helps us visualize the motion of the internal parts. For example, a bowed rotor will cause a twisting of the support bearings (Fig. 6.9). Actually, the bearing and support frame will trace out a conical motion. Transducers placed on opposite sides of the shaft, either top to bottom or side to side, will measure phase angles 180° different.

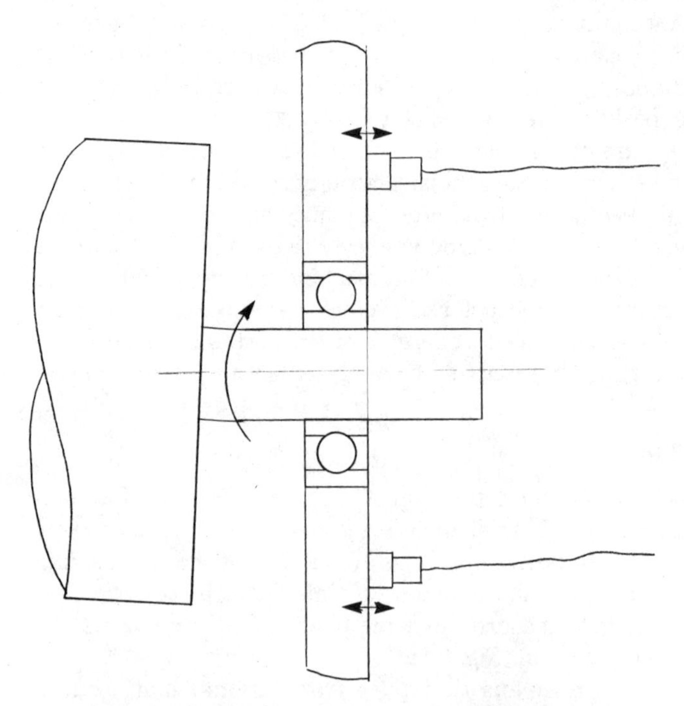

Figure 6.9 Axial vibration measurements on a bowed rotor.

A map of the machine vibration, shown as Fig. 6.10, can be made to visualize how it is shaking. Take vertical, horizontal, and axial readings at the rotational speed. Record the amplitude and phase at each point. Amplitude is recorded within the circle. Phase is recorded as a tick mark on the circle circumference.

A precaution is in order for the pickup direction. If the pickup is turned 180°, then the phase reading must also be turned 180° (Fig. 6.11). The pickup orientations should be consistent in the axial directions for the phase comparisons to be valid. However, this is not possible on opposing machine surfaces. Since the pickup orientation must be reversed, the phase reading must also be reversed.

Bodé, Polar, and Waterfall Plots

The Bodé and polar plots are just different ways of looking at the same vibration data. The Bodé plot is a graph of amplitude versus shaft revolutions per minute and phase versus shaft revolutions per minute. Figure 6.12 is a hypothetical Bodé plot, showing how the phase angle

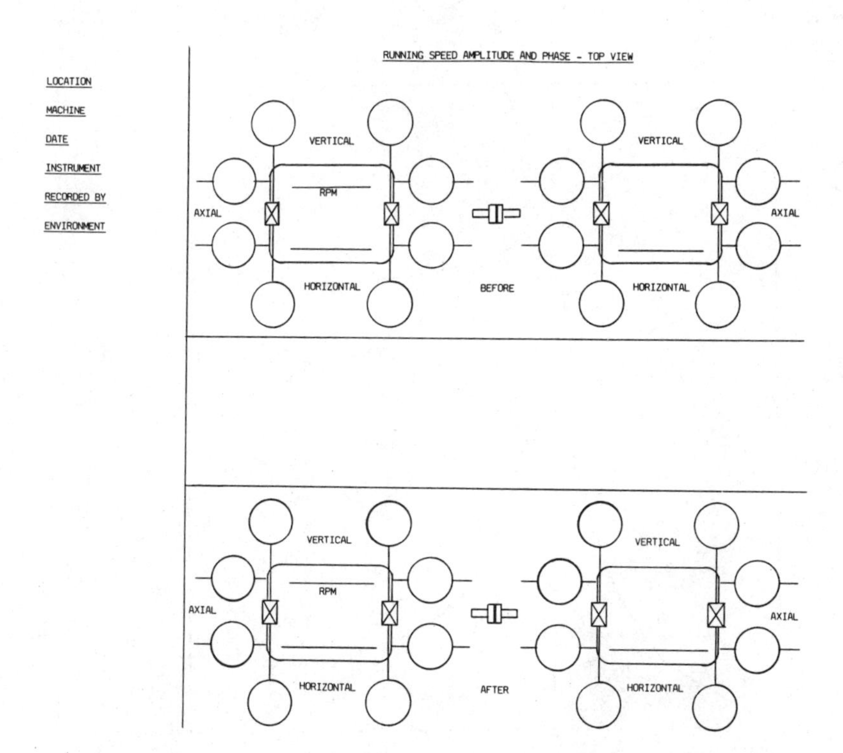

Figure 6.10 Machine map for recording amplitude and phase during analysis.

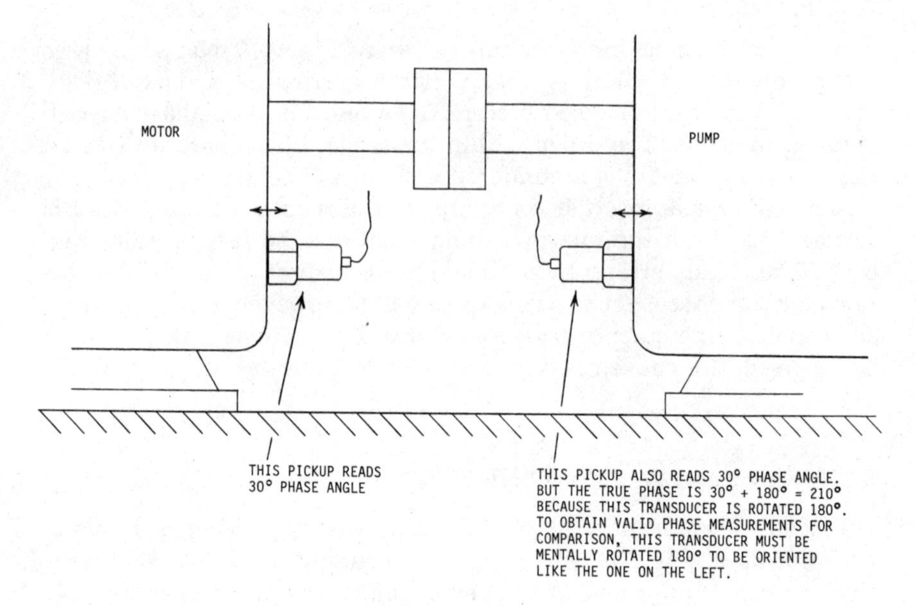

Figure 6.11 Precautions in the transducer orientation for phase measurements.

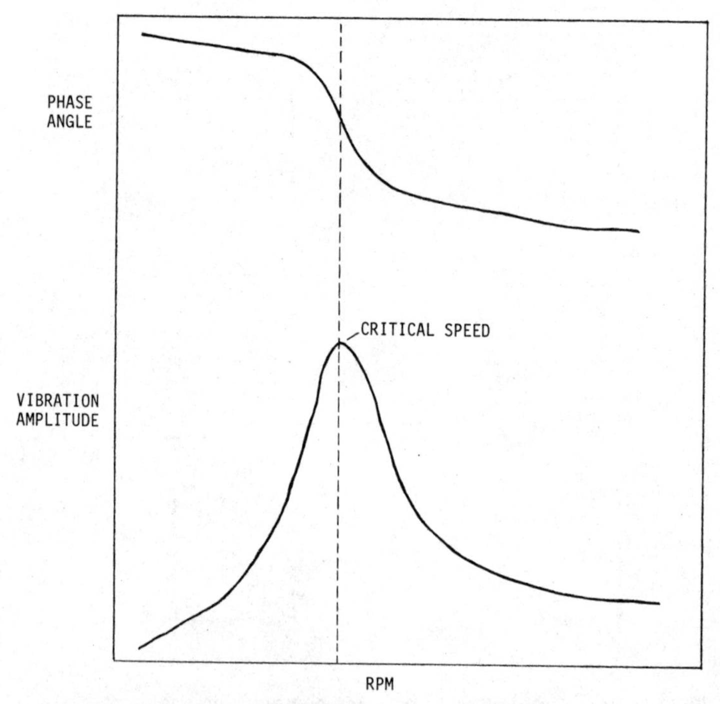

Figure 6.12 Hypothetical Bodé plot.

data is presented above the vibration amplitude with speed as the horizontal axis. A tracking filter is needed to obtain this plot. The tracking filter locks onto the shaft revolutions per minute and stays with it as it changes. In this way it records the amplitude and phase at different speeds. This is very useful in finding shaft resonances. At resonances, the amplitude peaks and the phase changes. The shaft revolutions per minute can then be read from the horizontal axis, and the critical speed or resonances are identified. Figure 6.13 is a Bodé plot acquired from a small machine during its run-up. Resonances are identified as peaks at 646 and 1418 rpm, with corresponding phase changes. There appears to be another, highly damped, resonance at 2921 rpm.

The polar plot of Fig. 6.14 contains the same machine run-up data

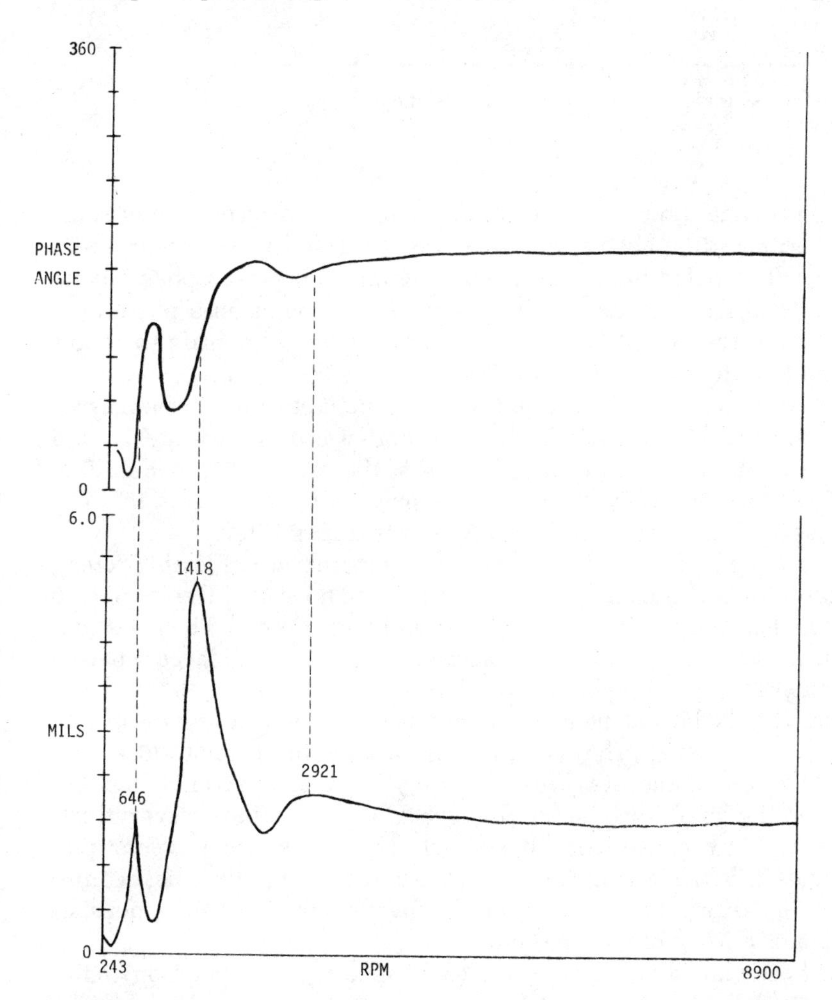

Figure 6.13 Bodé plot from the run-up of a small machine. (*Courtesy Cook & Associates.*)

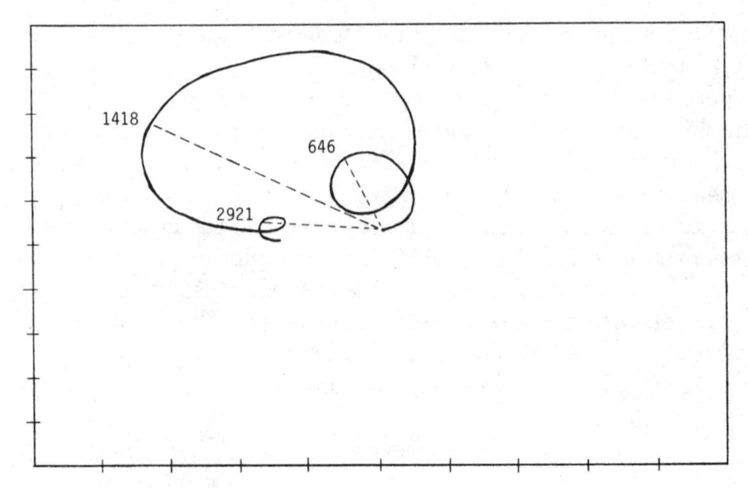

Figure 6.14 Polar plot during run-up from the same machine as Fig. 6.13. (*Courtesy Cook & Associates.*)

as the previous Bodé plot. Each loop represents a resonance of some kind because the phase angle underwent a 180° reversal before continuing. The polar plot is a graph of amplitude versus phase using polar coordinates. Polar coordinates identify a point on a plane by a vector from the origin. The two numbers needed to identify the point are the length and angle of the vector.

For example, point *A*, in Fig. 6.15*a*, can be identified in the normal cartesian coordinate system by its x and y coordinates: $7.07x$ and $7.07y$. In polar coordinates, in Fig. 6.15*b*, the same point is identified as $10\angle 45°$—a distance of 10 from the origin at an angle of 45°. The polar data must also be gathered with a tracking filter.

The polar plot is used frequently for balancing large flexible rotors; typically for turbines in the power utilities. The size of the large loop is a direct measure of the amount of imbalance present. The polar plot allows balancing to be done at, or close to, the critical speed where a reduction in amplitude is the objective.

Both the Bodé and polar plots are used for machine run-up and coast-down. As such, they both require a tracking filter to lock onto the shaft speed and stay with it. They both look at only one frequency—the shaft speed as it changes. They are blind to everything else going on because it is filtered out. There is another type of plot that is useful for viewing run-up and coast-down that does not require a tracking filter. This is a waterfall diagram (Fig. 6.16), sometimes called a *cascade plot* or a *3-D plot.*

The waterfall diagram is a series of spectrums stacked on top of each other. Each subsequent spectrum is stepped along in time. The

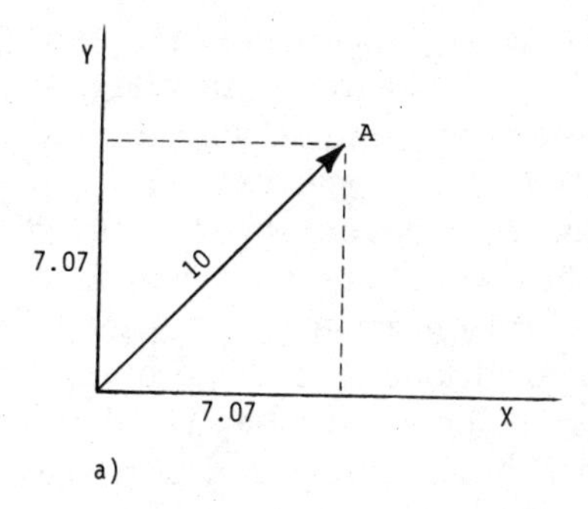

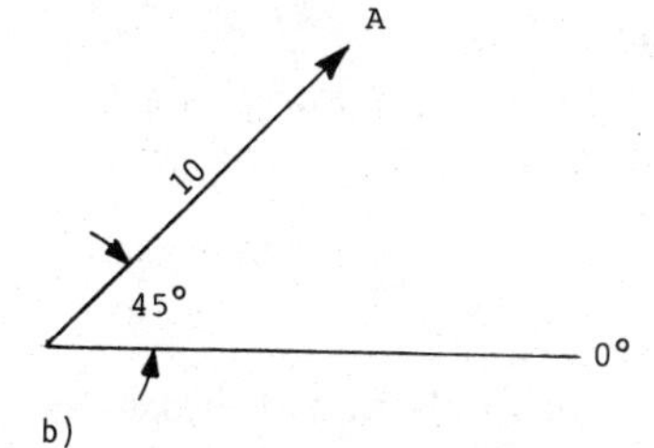

Figure 6.15 (*a*) Cartesian coordinate representation of a vector. (*b*) Polar coordinate representation of the same vector.

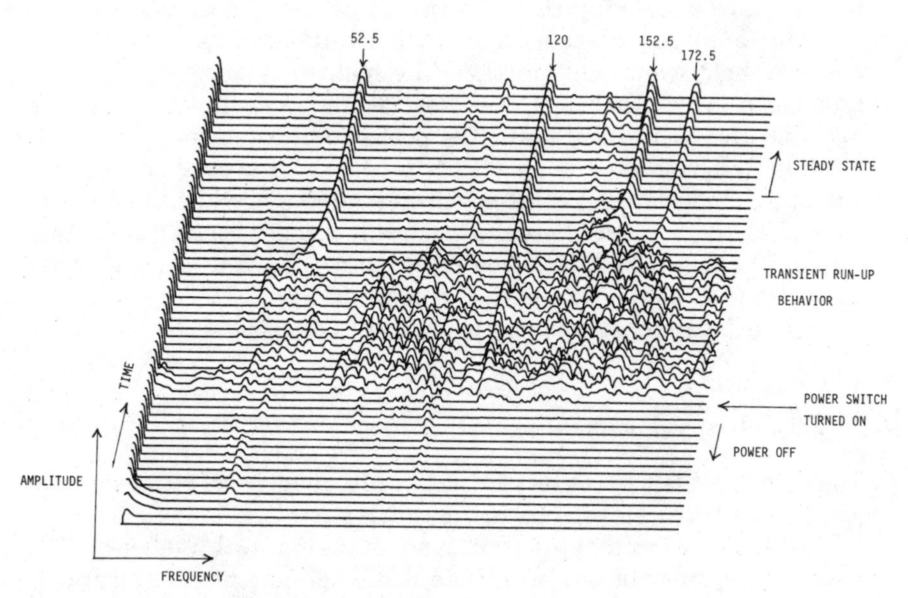

Figure 6.16 Waterfall diagram of the start-up of a laser printer.

waterfall diagram, unlike the Bodé and polar plots, looks at all frequencies simultaneously, so no information is lost. The waterfall diagram does not require a tracking filter, only a spectrum analyzer with this capability. It is three dimensional in amplitude, frequency, and time.

The waterfall diagram in Fig. 6.16 was taken during the start-up of a laser printer. The oldest data is at the bottom, with most recent data at the top of the diagram. Very little vibration was present prior to turning on the power switch. The small vibrations present were probably coming from the adjacent computer on the same table. Upon energizing the printer, the 120 Hz appears immediately and does not change much in amplitude or frequency. There is a significant amount of vibration during start-up, which mostly settles out. The peak at 52.5 Hz is probably a motor at 3150 rpm. The 152.5- and 172.5-Hz vibrations are possible bearing frequencies since they clearly change with the rotational speed. The remaining small peaks in the top 15 spectrums are probably resonances, because they come and go.

Tests for Resonance

Since resonance is such a common vibration mode and it causes serious damage, it is useful to have some tests for resonance. The Bodé and polar plots are two methods to check for resonance. However, they both require a tracking filter. In the absence of a tracking filter, a spectrum analyzer can be used to obtain similar information by using the peak-hold averaging function. The method is to set up the analyzer for a large number of peak-hold averages, then deenergize the machine and let it coast down. The peaks recorded during this coast-down are resonances. Figure 6.17 is a peak-hold averaging plot of a pedestal grinder during coast-down. Resonances are identified at 25, 29, and 41 Hz. Full speed, from whence this grinder began descending, was 3570 rpm, or 59.5 Hz. This is the same pedestal grinder whose run-up data is plotted in the time domain in Fig. 6.31.

There are two other common tests for resonance:

1. Bump test
2. Variable-speed shaker

The bump test is used to test for natural frequencies of stationary objects, i.e., structures, piping, and machine parts.

The resonant frequency is the same as the natural frequency. The natural frequency of motion is obtained when an object is struck. It rings at its natural frequency. So the way to check for resonance is to strike the object under test with a hammer while measuring the natural frequency. The hammer is usually a wooden mallet, but it could also be a 2- by 4-ft plank, your hand, or any convenient object to tap it

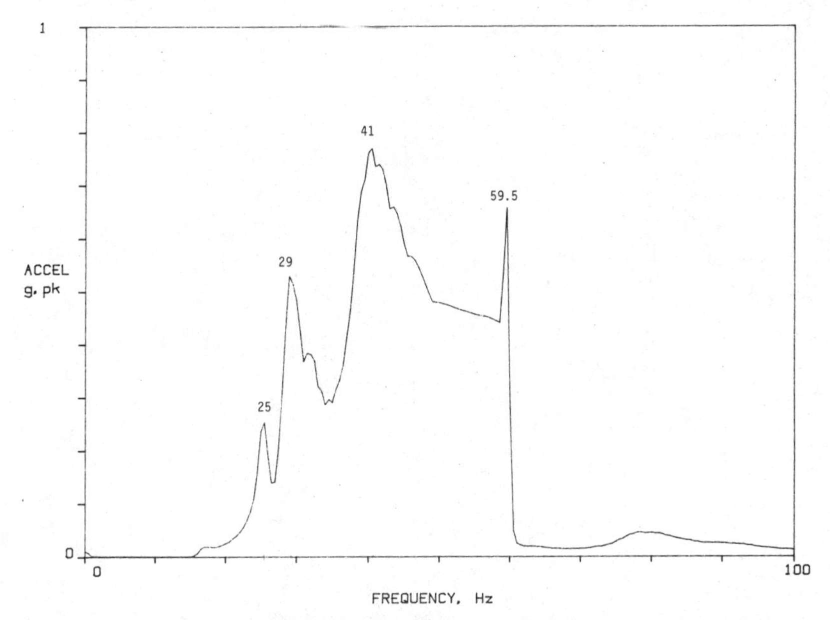

Figure 6.17 Peak-hold averaging during the coast-down of a pedestal grinder.

with. You could also stretch something and let it go. It will also ring at its natural frequency when released. This technique is called *step-relaxation.*

For very large structures, it is sometimes better to strain them with a rope, then cut the rope. Large structures require a significant amount of energy input to excite them into resonance, and even the straining with a rope or cable can be insufficient. The wind can provide enough energy for large structures, and it may be best to conduct the testing during strong wind conditions. Be advised that large structures, like buildings, have very low natural frequencies, approximately 0.1 to 10 Hz. Consequently, a low-frequency transducer will be needed.

Figure 6.18 shows a bump test being performed on a vertical pump. These machines have horizontal resonances very close to running speed in the 10- to 40-Hz range. It is also normal to have more than one resonance present.

Once the structure is excited, some way is needed to measure and record the frequency of vibration. This can be done with a tuneable filter analyzer, but a more convenient method is to use the peak-hold averaging function of a spectrum analyzer. The technique for checking for resonance is to attach your pickup to the piece of metal you want to check, start the peak-hold averaging, then rap the test piece with a mallet. Repeated light blows will sustain the resonance with-

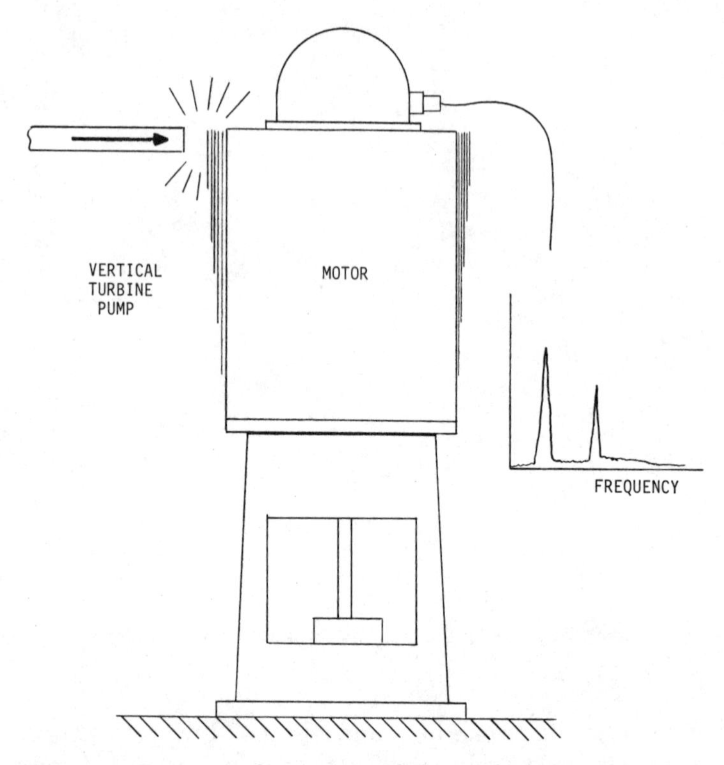

Figure 6.18 Bump test for resonance being performed on a vertical pump.

out overloading your instrument. The resonances will be recorded as peaks. Go to another piece, attach your pickup to it, restart the peak-hold averaging, then rap the new test piece with a mallet, and check its natural frequency. Go on like this. This technique of bump testing is usually done with the machine turned off.

The other technique for checking for resonance is testing with a variable-speed shaker. This is also usually done with the machine off. Figure 6.19 is an illustration of a commercially available variable-speed shaker. In addition to this shaker, a power amplifier is required to move it, and a signal source to generate a sine wave. The technique is to excite the part under test with the shaker while monitoring the vibration. The frequency of the shaker is changed with the purpose of finding where the vibration is the worst. A sweep of frequency from one extreme to the other will usually find the resonances. The resonant frequency can be read from the signal generator or the vibration measuring instruments when a resonance is found. Commercial shakers are electrodynamic in nature with a "stinger rod" attached between the movable element in the shaker and the surface under test.

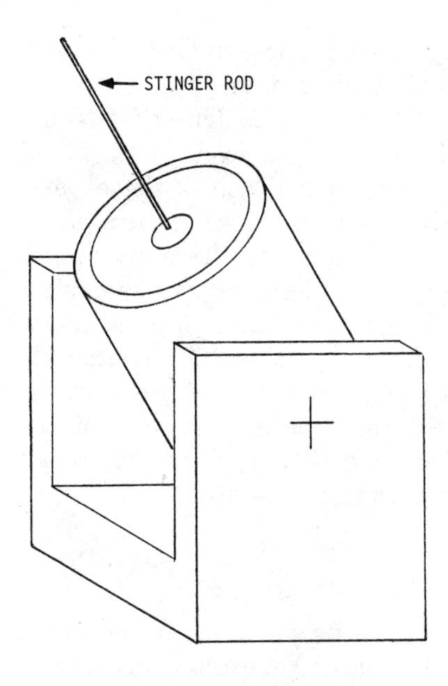

Figure 6.19 Electrodynamic shaker.

This rod transfers the excitation force without adding mass or significant transverse or torsional forces to the test object.

With either of these methods, bump test or variable-speed shaker, the resonant frequencies are identified. These frequencies must then be correlated to the problem when the machine is operating. That is, with the machine operating normally, these resonant frequencies could show up in the frequency spectrum. If a match is found, it can be tested by bracing. This stiffens the part and shifts its natural frequency higher. Run the machine with the braces or stiffeners and see if the problem vibration decreases dramatically. There should be a dramatic decrease if resonance was the problem.

Stiffening a part will usually produce a decrease in vibration. The key observation to identifying a resonance is a dramatic decrease in vibration. Stiffening a flexible part may produce a 30 percent decrease in vibration amplitude, but stiffening a resonant part will produce a 90 percent reduction. Typical temporary braces are made from available materials like angle iron, clamps, lumber, hydraulic jacks, wire, cables, or wedges.

Sand Patterns

The previous two tests for resonance, bump test and variable-speed shaker, require that the machine be stopped. The following two tests, sand patterns and deflection-mode shapes, require that the machine remain running and vibrating normally. The purpose of these following tests is to do simple modal analysis and find the best place to stiffen a part, but they can also confirm a resonance. The first technique, sand patterns, is applicable to horizontal plates.

Plates can vibrate in many different patterns. Each of these patterns is a different mode corresponding to a different natural frequency of vibration. With natural modes of vibration there are areas that vibrate with large amplitudes and nodes where very little movement takes place. It would be nice to know where the high-movement areas are. That is where stiffeners should be attached. On horizontal plates, you can sprinkle some sand on the plate as it vibrates (filings or powder also work well). The sand will not remain where the vibration amplitude is greater than 1 *g*. The sand will settle to the nodes (Fig. 6.20). In simple terms, brace it where the sand ain't.

Deflection-Mode Shapes

The purpose of modal analysis is to visualize how a structure is shaking, i.e., determine its mode shapes. This usually requires many transducers, a very big computer, an engineer, and a lot of time. Sand patterning is a simple method to visualize mode shapes on horizontal plates. Another method will now be presented that will allow you to do

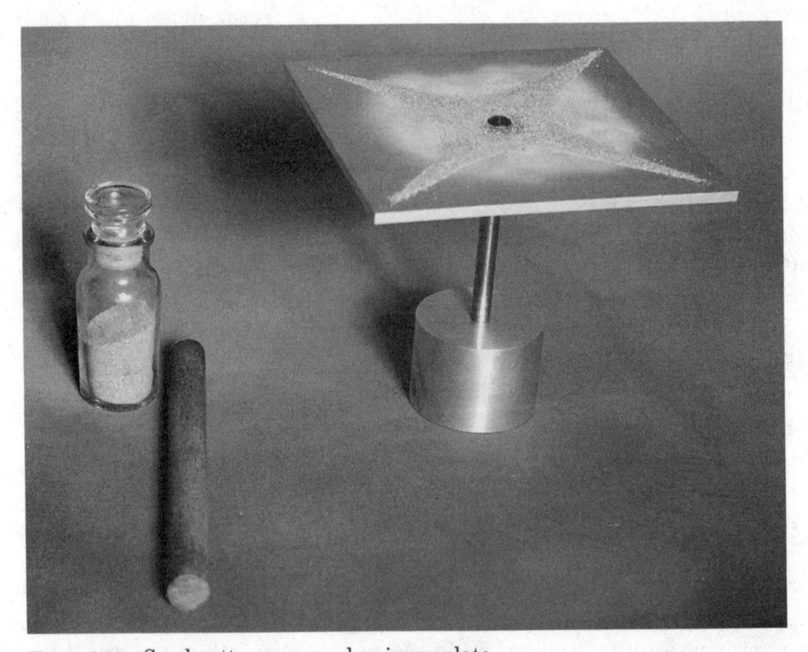

Figure 6.20 Sand pattern on an aluminum plate.

simple modal analysis on any shape structure in any orientation. This method only requires a simple vibration meter.

The technique (Fig. 6.21) is to mark off equal divisions (about 10) along the part and take readings at each point. The vibration meter records the maximum amplitude at each point. Frequency analysis is not necessary, since usually only one frequency is dominant. Plot the amplitudes on a sheet of graph paper. Figure 6.22 shows some typical plots that could be produced.

This method is very useful because the machinery is not shut down. The mode shape can be determined for both resonant and nonresonant parts. This method has been used to prove or disprove suspected resonance, thereby saving the time and embarrassment to futilely alter a structure that is nonresonant. Resonance problems will have a characteristic mode shape with nodes and antinodes. In Fig. 6.22*a,* the part should be braced at its unrestrained end, point *E*. In Fig. 6.22*b,* the part should be clamped in the middle, point *A*. Clamping the middle of a second mode resonant part will have little effect. This is an antinode, point *N* in Fig. 6.22*c* and moves little. This resonant-mode

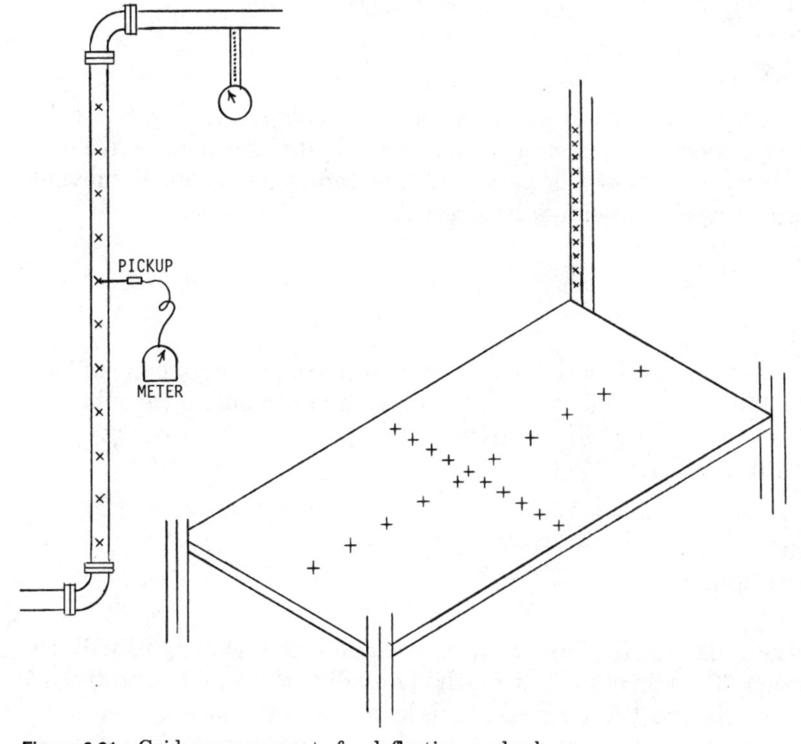

Figure 6.21 Grid measurements for deflection-mode shapes.

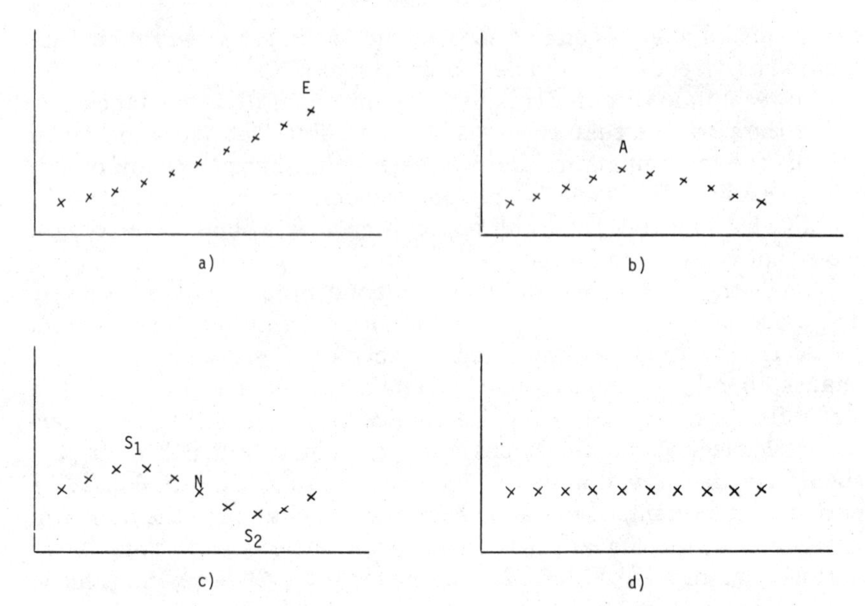

Figure 6.22 Typical plots of deflection-mode shapes. (*a*) Unrestrained end: brace at point *E*. (*b*) First mode resonance: brace at point *A*. (*c*) Second mode resonance: brace at points S_1 and S_2. (*d*) No resonance. Rigid body motion.

shape is best clamped at points S_1, and S_2. The plot in Fig. 6.22*d* displays rigid body motion, and stiffening along its length will have little effect. Any point on this part could be connected to an immovable structural part with equivalent results.

Uncoupled Driver

A simple technique used frequently is to uncouple the motor and run it solo. Measure the vibration level on the motor alone. If the vibration level is significantly lower, then the problem is probably not in the motor. It is either:

1. Unbalance in the driven machine
2. Misalignment

However, be aware that with no load on the motor, it will run smoother. To determine how much of an effect the load has on the motor, measure the 120-Hz peak, both loaded and unloaded.

Also, listen to the motor on start-up and shutdown (look at the 120-

Hz peak also). This will tell you if the motor vibrations are electrical in nature, or something else. Upon initial start-up, the rotational speed is zero and builds slowly. The 120-Hz magnetostriction comes on instantly when 60-Hz power is applied. Upon shutdown, a similar situation occurs. The 120-Hz motor vibrations disappear immediately when power is disconnected. The motor is still at full speed and begins to slow down, so imbalance or other problems associated with rotation (i.e., bearings) will slowly decrease.

In summary, running the driver (motor) uncoupled can narrow the vibration problem to the motor or attached equipment. Looking at the 120-Hz peak can further narrow motor vibrations to electrical or mechanical causes.

Orbits

Machines with journal bearings transmit less vibration to the external housing because of attenuation of the vibratory forces going through the oil film. If significant vibration is felt on the outside of a machine with journal bearings, then there is a serious problem because the shaft is probably hitting the inside of the journal. Usually the housing vibration is low, and significant vibratory motion is resident on the shaft. It is, therefore, superior to measure shaft vibratory motion directly.

A single proximity probe can be installed and its output viewed on a meter or an analyzer of some kind. The more typical setup is to install two proximity probes at 90° apart (Fig. 6.23). The output from these two probes is input to the x and y channels of an oscilloscope. In this way the actual displacement of the shaft could be seen, greatly magnified, on the screen (Fig. 6.24). The proximity probe outputs are filtered, usually to only look at the 1X running speed. Otherwise, a "dirty" signal would be seen (erratic). The output of a Keyphasor could be input to the oscilloscope trigger input to cause an intensification and/or blanking, to mark the timing. This method of orbits is still used today, particularly in the power utilities industry. Since the electron beam in the oscilloscope traces the motion of the shaft (assuming a smooth, round shaft with no defects), any irregularities in shaft motion will be displayed. Hitting or rubbing can be directly detected on the oscilloscope display.

Synchronous Time Averaging

In addition to frequency spectrum averaging, some analyzers have the capability to average in the time domain. Without some kind of tim-

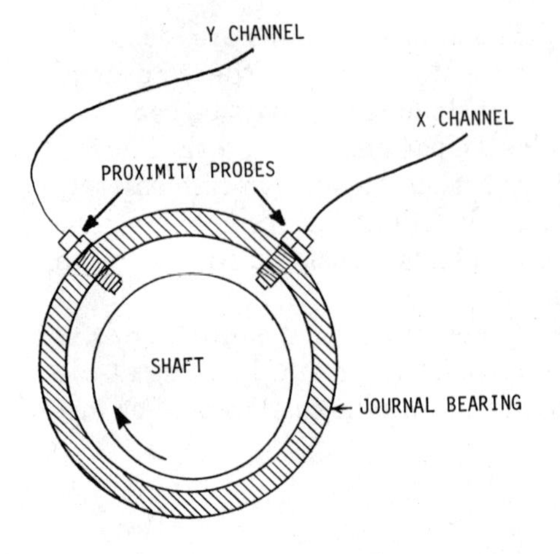

Figure 6.23 Proximity probes installed 90° apart.

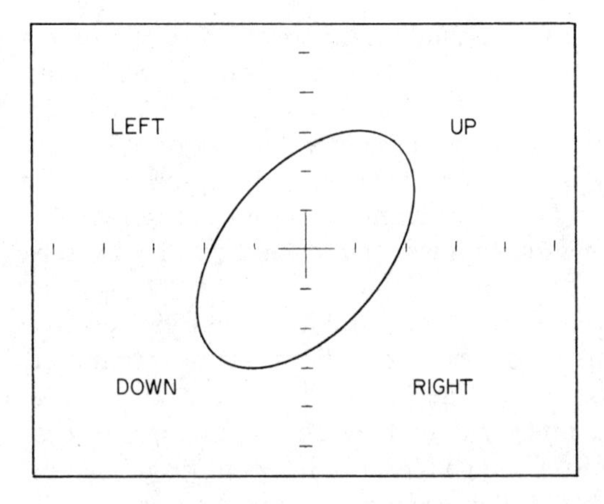

Figure 6.24 Typical orbit display on an oscilloscope.

ing mark, time domain averaging will average out to a straight line and zero. This is because the start of the next block of data captured by the analyzer is initiated when it is finished processing the previous block and has no correlation with the rotation of the machine. When an external trigger is added, then the analyzer will wait for the next passage of the trigger mark to begin processing the next block of data. The external trigger can be photoelectric tape, a Keyphasor, or something else attached to the moving part. In this way, the analyzer begins processing each block of data with a common starting point on the moving part. Every force event that repeats with each rotation will be averaged in, and every force event not associated with rotation will be averaged out. Random noise, such as turbulence, is averaged to zero during synchronous time averaging.

Let's see how this works. Figure 6.25 is of a complex vibration signal in its raw form in the time domain from a lathe rotating at 822 rpm. It appears like there may be a sine wave imbedded in there, but it is not easy to discern, much less analyze.

If this signal is averaged in the time domain without a trigger over a long period of time, then the result is as shown in Fig. 6.26. The vibration averages to zero and a straight line. This trace is the result of 1024 averages.

If the raw vibration signal is averaged with a trigger, then the result is very different. Figure 6.27 is a plot of the complex vibration using an external trigger. The external trigger was a piece of photoelectric tape on the chuck that sent a pulse to the analyzer once per revolution. This is called *synchronous time averaging* because the start of sampling is synchronized to some timing mark. Figure 6.27 shows some interesting results. A very definite sine wave now emerges after 1024 averages. The speed of the lathe was 822 rpm, which corresponds to a period of 73 msec. The vibration on the headstock of this lathe does not display any 1X-rpm vibration, but there's

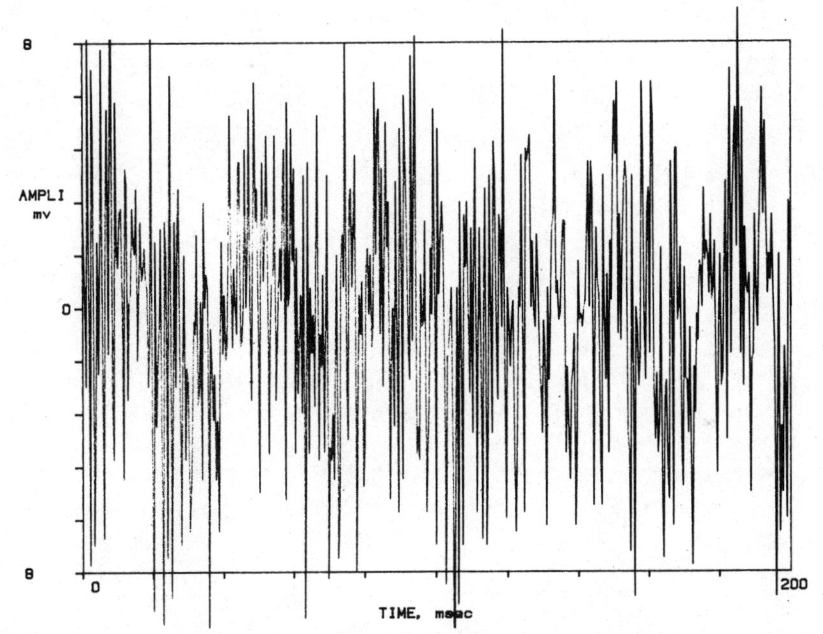

Figure 6.25 Vibration from the headstock of a lathe. The chuck rotational speed is 822 rpm.

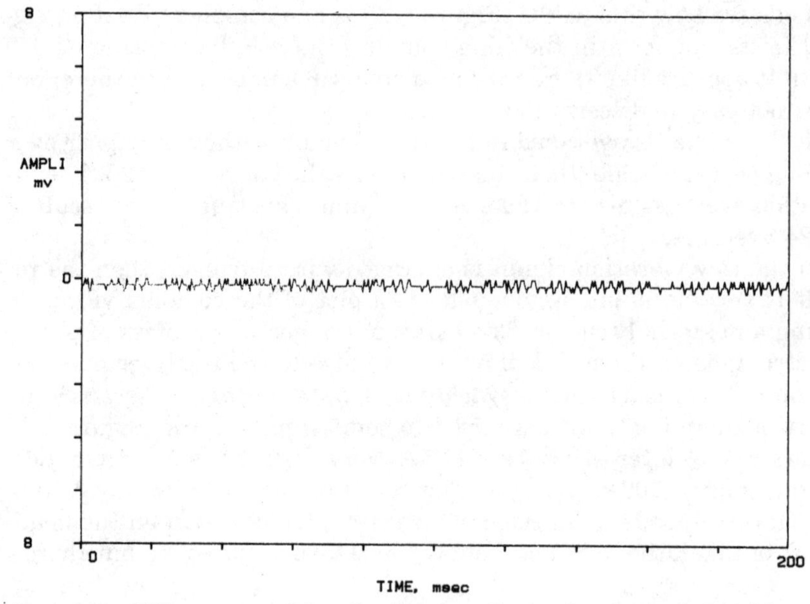

Figure 6.26 1024 averages of the vibration from the headstock of a lathe with no trigger.

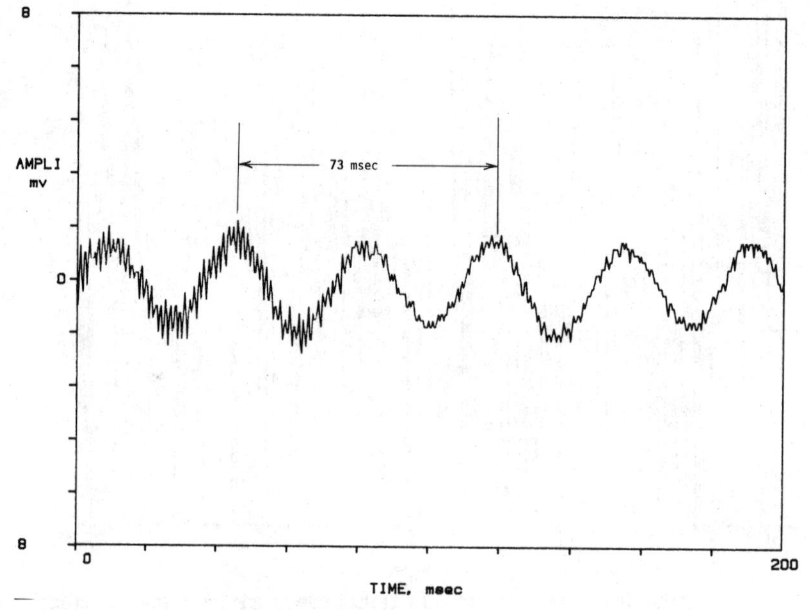

Figure 6.27 1024 averages of the vibration from the headstock of a lathe using an external trigger.

something definitely going on at 2X rpm. For comparison, Fig. 6.28 is the frequency domain view of this lathe vibration. This spectrum confirms the 2X-rpm vibration and the lack of any significant vibration at 822 rpm. The vibration at 7140 rpm is evidently unrelated to the rotation of the chuck since it does not appear in the time domain averaged view of Fig. 6.27. It averaged away.

Of what use is this? It is useful for detecting defects in gears. If the timing mark is placed on a shaft which has a cracked gear, then over a number of averages, the vibration signal will not resemble a perfect sine wave. All the other vibrations (bearings, motor, etc.) will be averaged out. As the crack comes into the area of gear mating, the vibration signal will be different than in another orientation because of the difference in elasticity as the crack comes into the load path. Therefore, the distortion in the sine wave. For gears, it is possible to "see" a crack develop and to predict which tooth is cracking. The result, after a large number of averages, is a vibration signal of every gear tooth as it comes into the load path. A cracked gear will show up as a discontinuity in the normal sine wave pattern. This is a very powerful technique, because it is possible to detect cracked gear teeth from the outside vibration measurement on a machine without stopping it. The conventional methods of detecting cracks is to stop the machine, disassemble it, clean up the parts, and inspect each part individually with a nondestructive technique like magnetic-particle

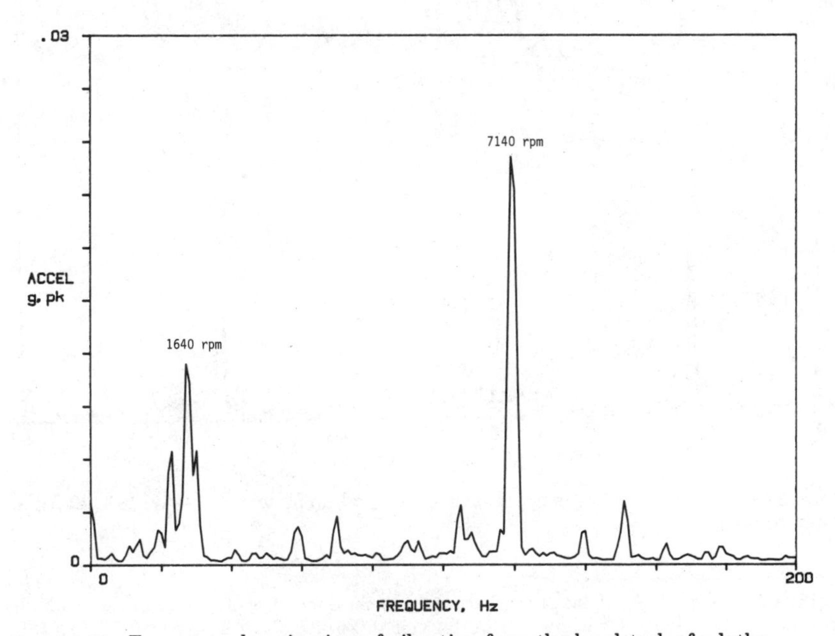

Figure 6.28 Frequency domain view of vibration from the headstock of a lathe.

inspection, dye penetrant, x ray, or eddy current. The savings in inspection time and labor are obvious. Dynamic signal analysis techniques, such as this, are also superior because they lend themselves to computer and statistical processing methods like kurtosis, frequency shifting, and others. The conventional methods rely heavily on the experience and skill of the inspector.

Let's look at a case history example of where synchronous time averaging could have detected a cracked gear in a helicopter. Figure 6.29 is a spectrum of the gearbox of an Australian helicopter that crashed in December 1983, off the coast of Australia killing both crew members.[2] The 940 Hz is a tooth-meshing frequency, and the 1880 Hz is a harmonic of tooth mesh. These lines also appear in a healthy gearbox. Nothing unusual was indicated by this spectrum. This gearbox was periodically monitored by capturing vibration data on magnetic tape. This taped data was analyzed by processing into the frequency domain and looking for changes. This system of analysis failed in this

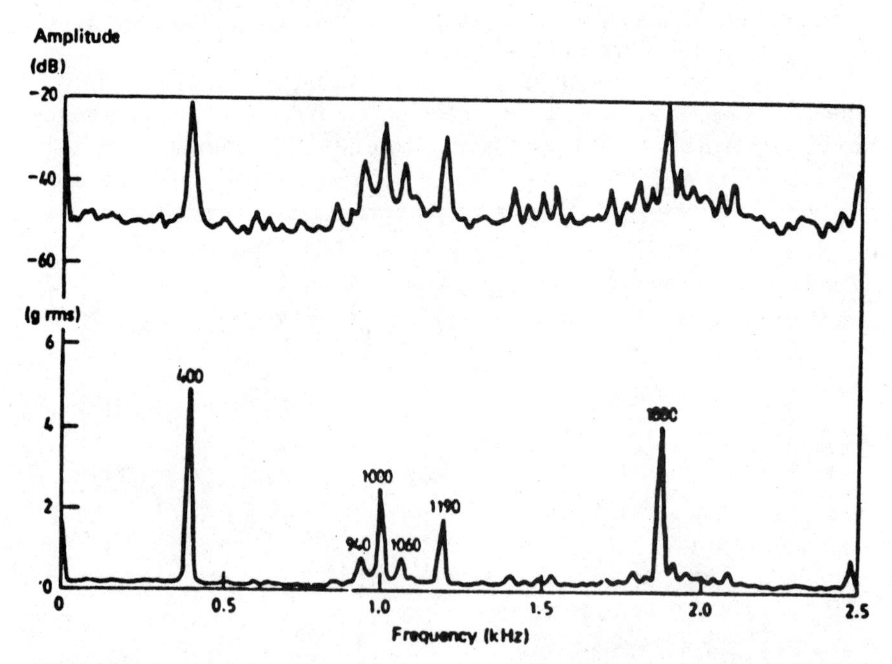

Figure 6.29 Spectrum from the gearbox of an Australian helicopter (WAK 143) 42 flight hours before failure.

[2]P. D. McFadden, "Analysis of the Vibration of the Input Bevel Pinion in Ran Wessex Helicopter Main Rotor Gearbox WAK143 Prior to Failure," *ARL-AERO-PROP-R-169*, Aeronautical Research Laboratories, Melbourne, Victoria, September 1985.

case and did not detect a serious crack in the input bevel gear to the main rotor. A postmortem evaluation examined this gear and found signs of corrosion in the crack and other indicators that this crack had been developing for some time. So why did spectrum analysis fail? The answer is that spectrum analysis is not a sensitive enough indicator in this case.

The taped data was replayed many times and averaged in the time domain. A trigger signal from the output shaft was available on the tape, and synchronous time averaging was applied. After about 10,000 averages, the signal in Fig. 6.30 emerged. The crack shows up as a discontinuity in the normal sine wave pattern. This plot correlated exactly with the location of the crack from the timing mark. Each sine wave is a new tooth that engaged and carried some load. The discontinuity was caused by a change in local stiffness due to the crack. With this defect present, the helicopter continued to fly for 42 h more. If the vibration data had been interpreted differently at the time, this tragedy could have been avoided.

Crack Detection

The synchronous time average technique is applicable to detecting cracks in rotating parts where the cracked area alternately comes into the load path and retreats. For most rotating parts, the power transmitting members stay loaded continuously. For these parts, and stationary nonrotating parts, two other techniques are available for detecting cracks. These are changes in frequency and changes in phase.

It should be evident from Eq. (3.8) that the natural frequency is dependent only on the stiffness and mass.

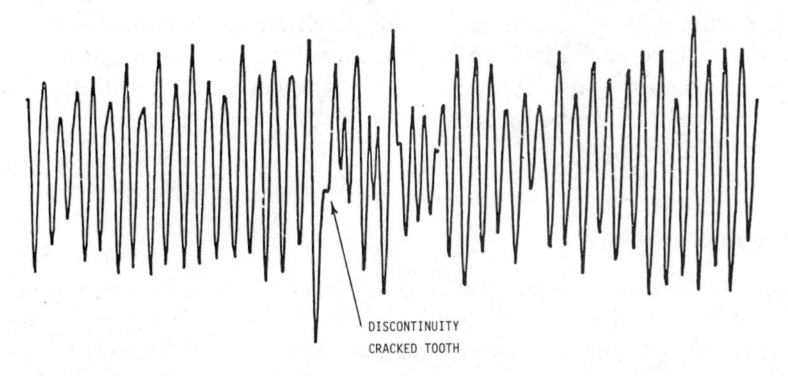

Figure 6.30 Time domain average view of WAK 143, 42 flight hours before failure.

$$\omega_n = \sqrt{\frac{k}{m}} \tag{3.8}$$

A crack has no effect on the mass, but it does change the stiffness, and therefore, the natural frequency. Not only is the first mode single-degree-of-freedom natural frequency altered, but the natural frequencies of all vibration modes are changed. The change is usually a decrease in frequency.

In reality, rotating systems have dynamic stiffness and dynamic mass. These are analogous to static stiffness and mass, only modified by the centrifugal force effects. For a constant-speed rotating system, these quantities should not change, nor will the natural frequencies, unless a crack causes a change in dynamic stiffness. This technique works equally well with stationary vibrating parts. Their natural frequencies are also excited by the general machinery motion and will show up in the spectrum.

The method is to record a vibration spectrum when the machine is in known good running condition. It is helpful to separate out the frequencies which are potential resonances. Thereafter, these frequencies are monitored for changes. The higher-mode frequencies are more sensitive to cracking and will show larger decreases in frequency. The amplitudes are irrelevant. They may show no change or possibly even a decrease. The key indicator for potential cracking is an unexpected decrease in natural frequencies. The appropriate action to take when this occurs is to first monitor more frequently until the next shutdown and then to inspect the suspect parts with one of the conventional nondestructive evaluation methods for confirmation.

In a similar manner, an abrupt change in phase can signal cracking. A phase change indicates a change in the force transmission time between the two transducers; specifically, a change in timing between force initiation and arrival at the sensing location. This change can be caused by many factors that affect the mechanical impedance, like looseness, balance condition, and foundation integrity in addition to cracking. However, phase changes are very sensitive to cracking because near resonance very small frequency changes cause large changes in phase angle.

Time Waveform Analysis

Prior to tuneable filters and spectrum analyzers, the oscilloscope was used for vibration analysis. The advantage of the oscilloscope is its virtual undamped, instantaneous response. The electron beam is very fast, and it can follow very fast input signals faithfully. This is useful for analysis of transient events, such as shock pulses, metal-to-metal impacts, and spikes in bearings. These events generate vibration, but the actual impact is better seen on an oscilloscope. Transients do not display very well in the frequency domain because they are not periodic for very long. Burst and pulses are best seen in the time domain with either a waveform capture or directly on an oscilloscope.

The time waveform can also be used to look at very long duration events, such as resonance during run-up or coast-down. Figure 6.31 is a 4-sec-long time trace of the run-up of a pedestal grinder. Notice the two clear resonances. Also notice the beat frequency during steady-state operation after it attains full speed of 3570 rpm.

So, in the time waveform, very long duration as well as short-duration events can be studied. The time waveform can also be used to measure phase directly. A trigger signal is still needed as a reference to measure phase, and a filtered vibration signal is needed to obtain a clean waveform.

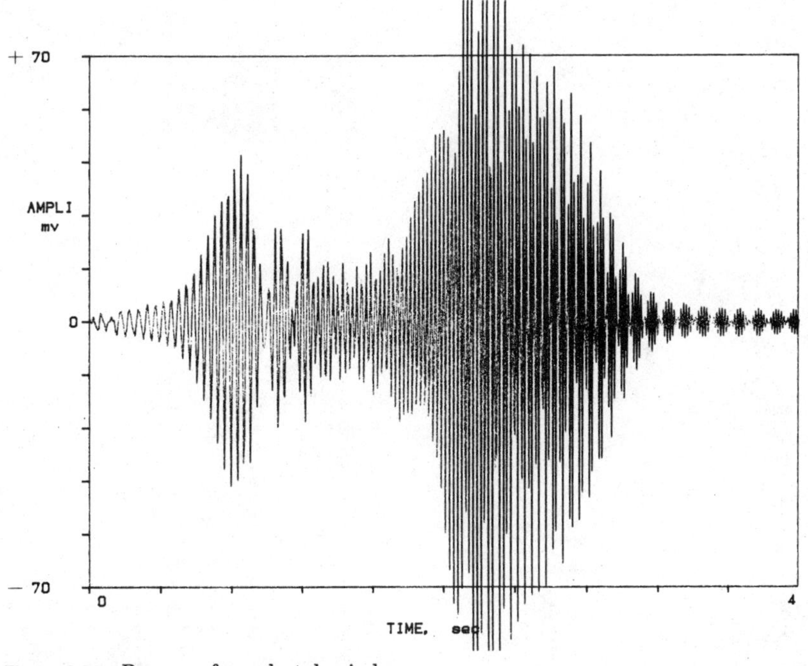

Figure 6.31 Run-up of a pedestal grinder.

Chapter

7

Machine Diagnostics

I remember a story told about how new information was obtained prior to the scientific method. The setting was in medieval Europe several hundred years ago. A group of adult men, intellectuals presumably, were discussing a question posed to them: How many teeth does a horse have? One philosopher stood up and quoted from the Bible. Another argued from Aristotle's writings. This discussion carried on for some time, when an errand boy came in and heard all this. He could not understand why grown men had idle time to waste on this subject, and during a pause he just burst out with, "Sirs, excuse my interruption, but why don't you just go and count the teeth in a horse's mouth?" The philosophers laughed and thought that to be below their dignity. They ignored the boy's suggestion, continued their lofty discussion, and did not find the answer to their question.

We have come a long way in scientific investigation. Rather than ask this authority, or consult that reference, we investigate. This is the age of scientific enlightenment where every effect has a cause. The scientific method is to observe and collect information in search of the cause. Once the cause has been identified, then the corrective action is rather straightforward.

Sometimes it seems that an inordinate amount of time is spent up front investigating the cause. More effort is spent in the investigation than the remedy. This is good if it avoids wasted effort on the wrong fix.

Analogous to the boy's suggestion for counting the teeth of a horse, why don't we go examine the machine? The machine can talk, but it speaks in its own language. It communicates in the form of vibration. Imbedded in that vibration signal are all the internal defects of the machine. It is telling us all about itself. All we have to do is listen. Sometimes we need a translator. That is what the FFT spectrum analyzer does. It translates the complex vibration signal coming from the machine into something that we humans can understand.

During diagnostics, there never seems to be a shortage of outside observers who are long on opinions but short on relevant knowledge. It is best to keep them away, if possible, to avoid their distractions.

Figure 7.1 is a cartoon of a sick fan on a doctor's examining table. In reality, the machine doctor must make house calls and go to the sick machine with instruments. This method of going to a machine and listening from the outside is rather similar to what a doctor does. A doctor will observe on the outside, ask some questions, listen to the heart, etc. The machine doctor's task is more difficult. A body is self-healing. Given time and rest, the human body has the capability of repairing itself. Rarely will a machine repair itself. The vibration usually gets progressively worse.

In machine diagnostics, we are playing the role of a machine doctor. Would you go to your doctor with a backache and tell him or her to take your back apart and fix it? More damage than good would probably occur in the process. Similarly, more damage than good frequently occurs in overhaul maintenance. It is far superior to hook some instruments to the machine and let the machine tell us what's in need of repair.

Vibration signals are a major source of information available from the machine itself for fault detection and diagnosis. There are other sources, i.e.:

- Oil analysis
- Temperature
- Electrical parameters
- Performance measures

Figure 7.1 A sick fan.

All these factors should be considered, along with the machine history, in any good diagnostic effort.

A note on analysis: This is still a highly technical and analytical function best done by the human mind. There is very little repetitiveness which characterizes the tasks that the computer is best at. Each problem is new and different. The reasoning power of the human mind to connect new data with past experience has been the algorithm that has proven to be most successful. In fact, the legwork is the real drudgery of the analysis function. Getting instruments to the job site and operating them to get the new data is most of the work. The cause is usually clear once the data is viewed in the frequency domain. If it is not, then more data must be acquired with more tests, i.e., run-up, bump tests. The human analyst is still best at applying this type of reasoning. Vibration analysis is a decision-making process with new questions at each step, not a repetitive, arithmetic task.

Vibration is an effect. In diagnostic work, we are always looking at the effect to try to find the cause. The cause is coded in the frequency information. Diagnostics is always a two-step process:

1. Acquire the necessary data in a systematic way.
2. Interpret the data to identify specific problems.

We have covered data acquisition in-depth in previous chapters. This chapter will focus on the interpretation of the data, but first let's take another look at data acquisition from the perspective of coarse overall data versus detailed data.

Narrowband Frequency Analysis

Amplitudes change according to the impedance in the force transmission path. The frequency information, however, is preserved. In diagnostic work, the amplitude is of least importance.

Figure 7.2 is an example to illustrate this point. In this illustration there is an overall meter reading of 0.058 *g*. There is no previous vibration data to establish any kind of trend, only a snapshot in time. If I now ask you to diagnose the problem, could you? Probably not. The meter gives an amplitude indication but no frequency or phase information.

Now suppose that a vibration spectrum were available as in Fig. 7.3. This spectrum is from the hard disk in my desktop computer. It is noisy and concerns me. I also know that the disk rotates at 3600 rpm. Now can you tell me what the problem is? You can clearly see that the problem is imbalance in the rotating parts, and you can recommend a

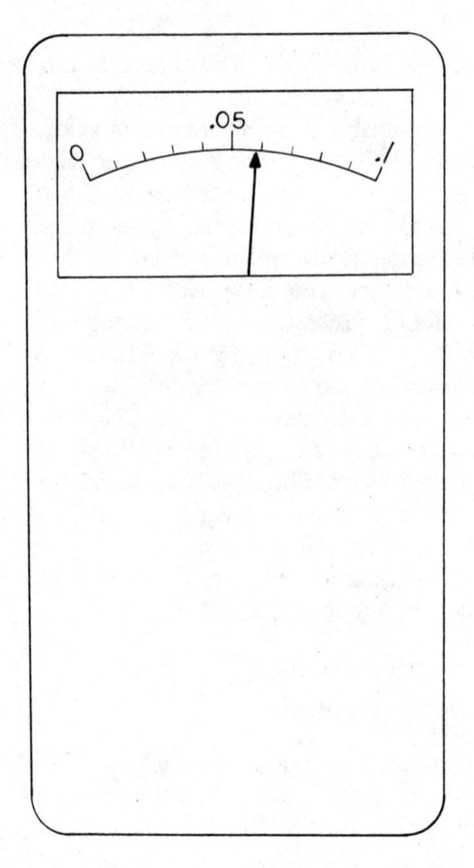

Figure 7.2 Overall meter reading of 0.058 g.

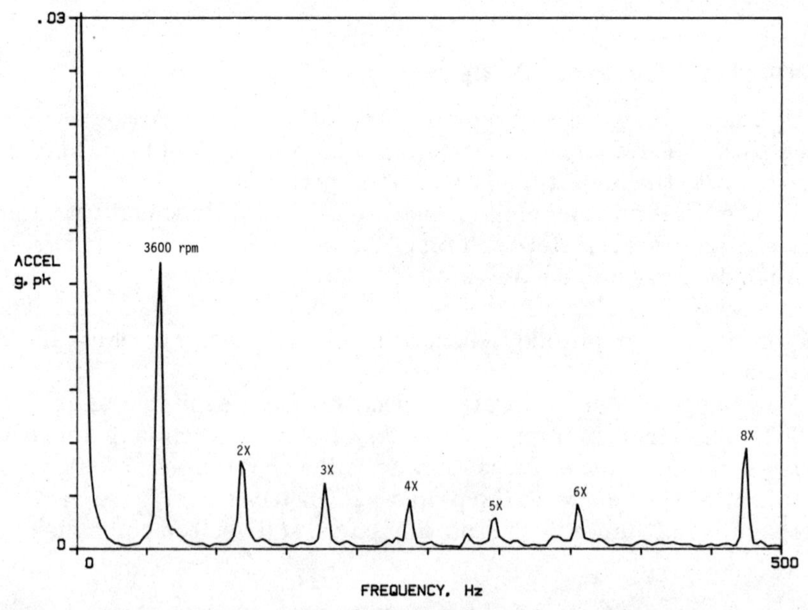

Figure 7.3 Vibration spectrum from the hard disk in a desktop computer.

fix; to balance the rotor. In addition, after it is balanced, you can judge the quality of the balancing by the relative drop in this peak at 3600 rpm. This gives you a means of quality control on maintenance activities.

Let's review this process:

1. We had a problem.
2. We put an instrument on the machine to measure the symptom. The first instrument (with an overall reading) didn't tell us much. It provided only coarse amplitude, no frequency information.
3. We captured a narrowband vibration signature from the machine.
4. The spectrum showed us that the problem was imbalance.
5. The amplitude of the peak at 1X running speed is a measure of how badly the machine was out of balance.
6. After balancing, another spectrum can judge the quality of repair.

The fundamental aspect to this whole process is obtaining a narrowband vibration spectrum that contains the key frequency information. The frequency allowed us to diagnose the problem and pinpoint the cause. The amplitude was a measure of its severity.

Table 7.1 is a chart of common machinery faults and how they appear in the frequency spectrum. There is very little new information in this chart that has not been previously covered in this book. Table 7.1 mostly summarizes previous knowledge. This table is an aid for making the connection between frequency and causes of vibration.

The analysis effort is measuring the effect of oscillatory motion at remote exterior points on a machine to try to understand the internal forces that cause this motion. When looking at a spectrum, try to visualize what force is repeating at that frequency within the machine. Study cross-section drawings, looking closely at mechanical contact points like bearings and gears. What forces, either mechanical or electromagnetic, are being generated during operation, and in which direction?

When diagnosing machinery ills, always look for and correct root causes. Imbalance and misalignment are root causes. They will, in time, cause bearings to wear. The bearings will become noisy. Replacing the bearings treats the symptoms, not the root cause. The new set of bearings is doomed to premature failure, just like the original bearings, if the root cause is not corrected. Likewise, welding a cracked motor foot does not address the root cause. It is a bandage fix. What caused the cracking in the first place? If the same problems continue

TABLE 7.1 Common Machinery Faults

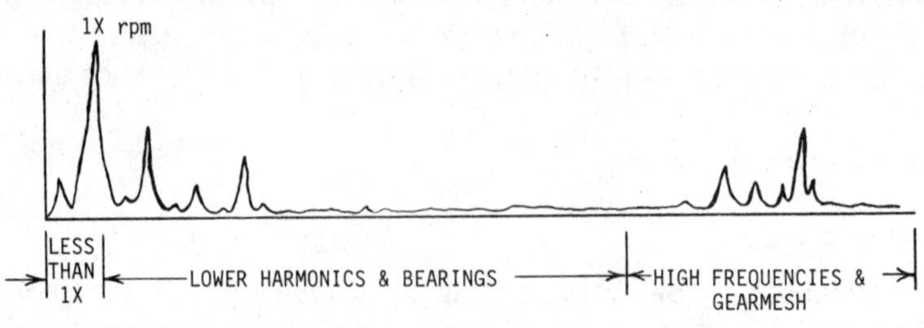

Cause	Frequency	Amplitude
	Less than 1X rpm	
Beats	Difference Frequency	Comes and goes; caused by two machines running at close rpm.
Loose journal bearing	Subharmonics ½, ⅓, etc.	Decreases with load.
Oil whirl	Approx. 45% of rpm	Applicable to high-speed machines with journal bearings.
Looseness	½, 1½, 2½, etc.	Decreases with load.
Belts	Belt rpm plus harmonics	Strobe light helps to see defect. Slipping belts can cause a periodic buildup and decline.
	1X rpm	
Unbalance	1X	Mostly radial; a common fault.
Misalignment	1X plus harmonics	High axial, high 2X and 3X; a common fault.
Eccentricity	1X	Looks just like unbalance; can be corrected by loosening belts.
Bent shaft/bowed rotor	1X	Looks just like unbalance; can be corrected with massive balancing weights.
Soft foot	Variable	Dramatically decreases by loosening one hold-down bolt.
Reciprocating	1X plus harmonics	Cannot be easily corrected; isolate machine better or provide inertia base.
	Medium frequencies 1X to 10X rpm	
Misalignment	2X, 3X, plus harmonics	High axial; a common fault.
Motor (electrical)	120 Hz; also 60 and 240 Hz	Stops immediately upon disconnecting power; also causes 120-Hz sidebands at higher frequencies. Not usually destructive; an indication of the quality of construction. Present to some degree in all motors and transformers.

TABLE 7.1 Common Machinery Faults (*Continued*)

Cause	Frequency	Amplitude
	Medium frequencies 1X to 10X rpm	
Looseness	½, 1½, 2½, etc.	Decreases with load.
Bearings	4–10X	FTF ≈ 0.4 × rps OR ≈ 0.4 × rps × N IR ≈ 0.6 × rps × N N = number of balls rps = rev/sec High-frequency shocks in time domain.
	High frequencies	
Gears	Gear mesh = rpm × no. of teeth	Sidebands at tooth mesh 2X gear mesh typically larger.
Blades	Blade passing = rpm × no. of blades	Not usually destructive.
Resonance	Discrete peaks	A serious condition; usually high amplitudes; slight speed changes cause a dramatic decrease.
Cavitation	Broadband 3–5 kHz	Changing operating conditions (i.e., pressures) alleviates problem.
Cracking	Unexplained drop in frequencies	Amplitude changes are insignificant. Phase changes are significant.

to recur, then you can be guaranteed that the root cause has not yet been found and corrected.

Standards

Once the cause has been determined with a frequency correlation, then it is appropriate to make some judgment based on the amplitude. As mentioned earlier, the frequency allows us to diagnose the fault, and the amplitude is a measure of its severity. There are industry standards for various classes of machines. The standards define the acceptable level of vibration.

Most standards address only the amplitude at running speed and ignore everything else. Where the frequency is unclear, it can be assumed to be at the running speed. Chapter 19 of the *Shock and Vibra-*

tion Handbook[1] contains graphs and tables from some of these standards. The pertinent standards have also been assembled by Ronald L. Eshleman[2] in *Machinery Vibration Standards* which is available from the Vibration Institute, and these standards will not be repeated here. Most standards suffer from the weakness of focusing only on the running speed, and from the complexity of different measures for amplitude; mils of displacement, inches per second of velocity, and g's of acceleration. The unit in the standard of interest is probably different from your measurement. Conversion is possible with precautions by using the following formulas.

$$A = 0.1022f^2D \qquad A = \text{peak acceleration, } g \tag{7.1}$$

$$V = 61.7\frac{A}{f} \qquad V = \text{peak velocity, in/sec} \tag{7.2}$$

$$D = 9.78\frac{A}{f^2} \qquad D = \text{peak displacement, in} \tag{7.3}$$

$$f = \text{frequency, Hz}$$

These formulas come from the basic solution of a single-degree-of-freedom system as presented in Eqs. (3.3) to (3.5). These values are simply the maximum amplitude with conversion factors for the units. The formulas are valid for a pure sine wave vibration at a single frequency that is repetitive. Since one vibration frequency is usually dominant, these formulas are generally applicable. For a complex vibration signal, the instantaneous velocity or displacement can exceed these calculated values when different frequencies become additive. For nonrepetitive events, like transients, the true amplitude must be read from a time domain capture. Any other measurement method will grossly underestimate the amplitude.

The displacement formula gives the peak displacement in inches. Many of the standards define displacement limits in peak to peak. Be aware that the formula-calculated number may need to be multiplied by 2 to get the correct peak-to-peak value.

The standards are useful as a foundation to negotiate from when dealing with an equipment supplier. They represent an industry standard which is a consensus from within the industry itself. It makes sense that a motor should meet the industry standards of the National

[1]Paul H. Maedel, Jr., and Ronald L. Eshleman, "Vibration Standards," *Shock and Vibration Handbook,* 3rd ed., McGraw-Hill, New York, 1988, Chap. 19.

[2]Ronald R. Eshleman, Director, Vibration Institute, Willowbook, Ill.

Electrical Manufacturers Association (NEMA). Most do, and these standards obligate the suppliers to take responsibility for their own quality control.

The majority of vibration standards, as mentioned, focus primarily on the running speed of a machine and ignore other frequencies. Other standards specify an overall measurement within some frequency band. These are serious shortcomings because they discourage frequency analysis. A much finer discriminator for machine health is resident within FFT spectrum analyzers; however, I have yet to see any standard favor this measurement instrument.

There is another shortcoming in vibration standards. This is the amplitude levels that are acceptable. Most of these standards were originally composed many years ago. The acceptable levels are crude by modern expectations for machine longevity and smoothness. I am regularly called upon to reduce vibration levels on machinery that meets industry standards. In addition, I see equipment from the best manufacturers that is consistently far below these levels. Vibration levels lower than industry standards are achievable and expected. This makes me doubt the usefulness of these standards. The most aggressive and quality-conscious organizations are striving for their own levels of excellence which are lower than the norm.

In an operating environment, where the equipment was purchased and installed a long time ago, the production manager wants to know the machine's condition in terms he or she can understand. He or she may not have a feel for mils or inches per second, but he or she understands terms like "good" and "rough." The industry standards can be useful in this judgment, but then you have to dig up the appropriate one. For these reasons, I have produced the chart in Fig. 7.4. This chart is consistent with most industry standards. It has a frequency axis along the bottom, in logarithmic units. The vertical axis is velocity, also in logarithmic units. Displacement and acceleration axis also exist as diagonal lines, also in logarithmic units. Areas have been partitioned off and labeled as "good," "acceptable," "tolerable," "excessive," and "extremely rough." This labeling is for the manager who likes judgment words based on technical data. Any vibration measurement of amplitude and frequency will locate a point on this chart, and a judgment as to severity can be immediately made. The amplitude measurement can be in mils, inches per second, or *g*'s. The lower frequency criteria follow a constant displacement line. Failures at low frequency are usually due to excessive deflections. At higher frequencies, above 200 Hz, the criteria follow a line of constant acceleration. This recognizes acceleration, which is proportional to force, as the factor contributing more to premature failure. In between, where most machinery operates, velocity is the criteria of choice.

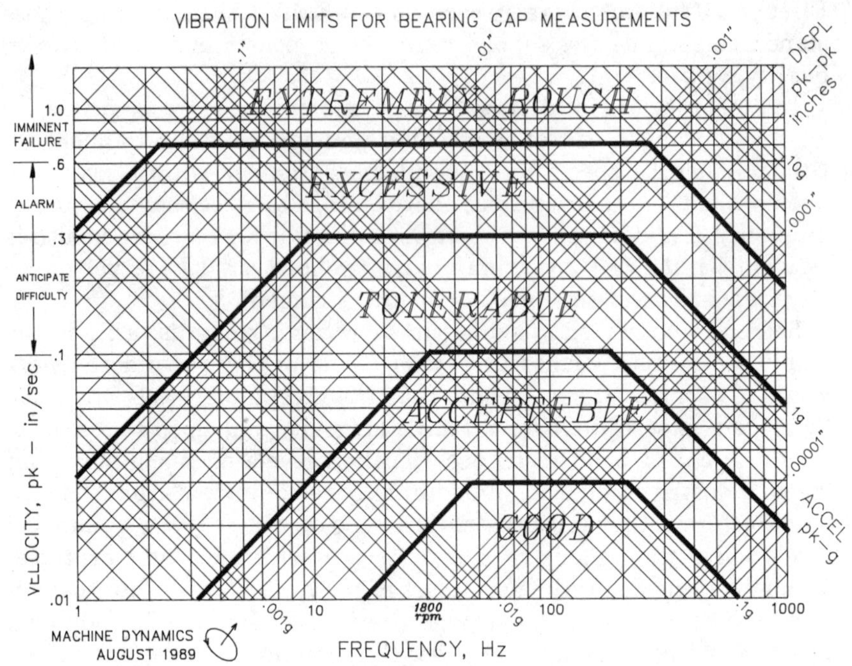

Figure 7.4 Vibration severity judgment chart.

This chart (Fig. 7.4) does not pretend to assume the running speed as the frequency of measurement. A judgment can be made of a vibration at any frequency. The only requirement is that the measurement be made on the bearings, or as close as possible on a rigid member. The vertical axis on the left side contains common velocity criteria of machinery, i.e.,

Less than 0.1 in/sec = Okay.

0.1 to 0.3 in/sec = Anticipate difficulty.

0.3 to 0.6 in/sec = Alarm, or serious concern.

Above 0.6 in/sec = Imminent failure; few machines will operate long above this level.

The interpretation for the verbal criteria is as follows:

"Good" = The best and smoothest machines are in this category.

"Acceptable" = This is okay for normal industrial and commercial applications.

"Tolerable" = This category straddles the typical criteria for building mechanical equipment of 0.2 in/sec depending on the foundation and support stiffness. Equipment in this range may be acceptable if it does not transmit objectionable vibration elsewhere. In terms of machine life, it will probably cause premature wear, but not immediately.

"Excessive" = Too much; many alarms are set at 0.3 in/sec. Vibration levels in this category are uncomfortable to be around.

"Extremely rough" = Serious damage is taking place; do not operate equipment in this category unless you are willing to suffer the consequences of a failure at any time.

This chart introduces simplicity into the present complexity of many vibration standards. These quality judgments need to be tempered with information on the size of the machine and the support stiffness. The criteria are general in nature and are intended to be consistent with industry standards. They are not intended for a stop-run decision. This decision can be made only by the vibration analyst and the production manager as a consensus with experience of previous failures on similar machines.

Finally, Fig. 7.4 can be used to convert amplitudes from one unit to another in lieu of Eqs. (7.1) through (7.3). The only input into this chart is a frequency and an amplitude. Any other amplitude can be read from the intersection point of the two measured values.

As an example of the use of this chart, let's take the vibration level of Number 3 boiler from the first case history in Chapter 5. From Fig. 5.66, the vibration amplitude was 0.8 *g* rms at 58.5 Hz. This is equivalent to 1.13 *g* peak. The intersection of these two points is in the "extremely rough" category on the chart in Fig. 7.4. The velocity is 1.19 in/sec peak. The displacement is 6.5 mils peak to peak. By any measure, this vibration is bad news.

Acceptance Testing

Vibration measurements are performed for one of four general purposes:

- Balancing
- Trending
- Analysis
- Acceptance testing

Balancing measurements will be covered in *Machinery Vibrations: Corrective Methods* (1992). Trending is in the next chapter. Analysis for diagnostic purposes was just covered. Measurements for acceptance testing of new or rebuilt equipment is the subject of this section.

Vibration is a measure of mechanical quality of new manufactured equipment. The standards mentioned in the previous section can be used to specify the acceptable vibration level of a new piece of equipment. The power utilities and the petrochemical industries typically specify acceptable vibration levels for new equipment. Just like any other specification, the equipment must meet the specified vibration level or it is not accepted. Every good trending program starts with baseline vibration measurements to establish the vibration signature when the machine is in a known good running condition. There is no better time to start than when the equipment is new. *New* is not generally synonymous with *good*. More industries are specifying vibration testing in their purchase documents. Included in Appendix C is a sample: "Vibration Specification for Acceptance Testing of New or Rebuilt Machinery."

However, today, vibration specs are not universally accepted by all equipment suppliers. The resistance to vibration specs takes the form of responses like "We don't have instruments to measure this." They don't have instruments because their need has been infrequent. They can always hire a specialist for these measurements just like they hire a test and balance contractor.

"The equipment is guaranteed against defects." Equipment suppliers have devised enough avoidance tactics to make guarantees a useless piece of paper today. Besides, the guarantee does not protect you from lost income or production delays.

"We have no control over assembly or installation factors." The lack of control over installation factors is a valid concern, and the sample specification attempts to compensate the supplier for this.

"This just increases the installation costs." The increase in installation costs is the argument that has allowed vibration specs to be negotiated away. The cost is actually a balancing of up-front cost versus long-term operating costs. The installation contractor has little interest in the latter, but a profit interest in the former. He or she can be expected to take a position that maximizes his or her interest at the expense of others. Likewise, the buyer's representative should insist on minimizing the long-term life cycle cost. A quality-control inspection up front that takes but a few moments is a small insurance premium.

Until vibration specs are more widely accepted, it is recommended to negotiate with the supplier on a vibration criteria for acceptance. Mechanical equipment suppliers or rebuilders that resist vibration data are resisting a quality-control inspection of their product or work. This type of resistance is a normal human reaction. However, those contractors that will still be in business 50 years from now are those that embrace this technology and use it to control their own quality. They will, in the process, gain the respect of their customers, and possibly even the distinction that their product is to be accepted without inspection.

Three machines of exactly the same make and model can have significantly different vibration levels. Variability of a factor of 10 to 1 is not unusual. There is a similar variability between different manufacturers. So the total variation in vibration between the best and worst 10-hp motor can be a ratio of 100 to 1. This is easy to detect with even the simplest of vibration instruments. Even if you don't have any standards, you can compare one machine to another just like it.

Presented in the "Case Histories" section of Chapter 5, "Typical Vibration Problems," are examples of supposedly identical fans. The customers paid the same amount for each fan, but they received different-quality products. The bad news is that those customers without vibration specs and without vibration measuring instruments may be getting the rejects from those customers who do have specs and instruments.

Several years ago, I purchased a rotary phase converter. It vibrated so badly that the concrete floor buzzed at 3600 rpm. It was so loud that the grinder operator 6 ft away could not hear the grinding wheel as it touched the metal workpiece. This is an important sensing feature in precision grinding. The sound level was measured to be 70 dB_A at 3 ft. Some negotiation was carried on with the factory president for several weeks until the unit was finally returned. Not only was I charged shipping in both directions but also a repainting and restocking charge. The factory obviously shipped it to another customer.

Some amount of vibration is normal for all machines. It is the result of manufacturing tolerances and assembly variation. However, a large vibration amplitude compared to similar machines indicates something abnormally wrong. In addition, an increase in vibration after months at a stable level indicates trouble has developed. This is the purpose of trending.

Chapter

8

Trending

This chapter on trending is really a chapter on maintenance management. Statistical process control revolutionized the quality of manufactured goods. The key concepts are understanding the variables and controlling them with periodic monitoring. The variables of machinery condition are all revealed in the vibration signature, which is unique for each machine, much like a fingerprint. In addition, for a specific machine this quantity is not exact, and it varies over time. Vibration monitoring is the periodic measurement of this variable in time intervals, rather than part intervals as in manufacturing. With this distinction, a vibration monitoring program can have the same benefits for operation and maintenance as statistical process control has had for manufacturing. Vibration monitoring is, in fact, statistical process control applied to the care and maintenance of mechanical equipment.

Maintenance is one field of modern technology that has changed dramatically over the past 50 years and that is continuing to evolve because of the increasing size and sophistication of modern machinery and the tools available to attend to it. Modern process industries are more or less in continuous production with high-speed, high-dollar machinery. But even the initial cost of this machinery is modest when compared to the daily loss during an unscheduled shutdown. The risks and consequences of maintenance decisions are greater than the design decisions and purchase decisions, because maintenance decisions are made when the equipment is needed for production and time is of the essence. Maintenance decisions are costly.

The cost to maintain a machine over its lifetime is greater than the initial purchase price, especially if energy consumption is considered. Maintenance is a profit center—able to add dollars to the bottom-line profit by controlling the expense of parts, labor, and unnecessary machine replacements. It is a hidden gold mine within every company. Maintenance can no longer be considered an expense of doing business. It must be viewed as a professional responsibility that must be done well to stay competitive. There are high-technology tools available to do it better. The technology of maintenance engineering is now on an equally sophisticated level as design engineering, and the impact of maintenance actions on the overall company performance are at least as significant.

Managing Maintenance

There are presently three methods of managing a maintenance operation:

1. Run-to-failure
2. Preventive maintenance
3. Predictive maintenance

Run-to-failure is as the name implies, i.e., maintenance only springs into action when a machine breaks down. The majority of maintenance is done with this philosophy. Depending on the size of your operation and the size of your maintenance staff, this method may be appropriate for you. It is also the most expensive on a per-horsepower basis because it requires massive amounts of overtime, spare parts inventory, and generates the highest lost-production rates.

Table 8.1 represents the results of a study conducted for the power utilities. A definition of preventive and predictive maintenance will be given in the next section. Clearly, there is a hidden gold mine in operating cost savings by progressing to a predictive maintenance program using vibration monitoring. The savings are in greater up-time, no longer replacing parts that don't need replacement, and less emergency maintenance.

As a guess, possibly half of Americans manage their homes and their cars with a run-to-failure maintenance program. This type of

TABLE 8.1 Maintenance Costs, $/hp/year*

Run-to-failure	17–18
Preventive	11–13
Predictive	7–8

*1985 U.S. dollars.
SOURCE: Electric Power Research Institute, Palo Alto, Calif.

maintenance management is appropriate where the replacement cost and the adverse consequences are small.

Preventive Versus Predictive Maintenance

Preventive maintenance is a time-based system of parts replacement. This procedure requires that after a period of time has elapsed, parts will be replaced. The idea is that wear parts have a fixed number of cycles to failure. This fixed number of cycles to failure is actually an average. Some will last more cycles, some less. There is some variability in components.

This fixed number of cycles to failure is converted to an operating time. The parts are scheduled for a changeout before this time expires in the hopes of averting a failure. This time to failure does not always consider the load (or stress on components), exposure to chemicals or radiation, temperature effects, mishandling, or the variability in product quality. The result is that the times between changeout gets shorter, because if one premature failure occurs, the maintenance manager begins to think that the time is too long between changeout, and he or she shortens the time. Parts suppliers love this system. It means big business for them. Parts are changed more frequently than necessary.

This time-based preventive maintenance system is equivalent to your going to your cardiac surgeon every 40 years to get a new heart. You know that your old one is going to wear out, so why not just get a new one ahead of time? You wouldn't do this to your body, so why change parts that don't need to be changed on machinery?

The commercial airline industry conducted a study some years ago with the amazing result that overhauling was doing more damage than good. They found that most failures were not time related. In addition, they discovered a strong correlation between failures and recent maintenance work. In other words, the disassembly and handling was creating damage that otherwise would not have occurred.

There is also a better way for industrial machinery. The key is having a good indicator for when serious damage is beginning to occur. For rotating machinery, it is vibration.

Predictive maintenance is letting a machine run as long as it exhibits health. This is equivalent to an annual health physical rather than periodic parts replacement. As long as the body remains healthy, no surgery is done. Similarly, as long as the machine's healthy condition persists, parts are not changed. With a periodic checkup, the machinery can continue to run.

There are three big advantages to a predictive maintenance program:

1. Only machinery that indicates unhealthy conditions is worked on. No effort is wasted working on healthy machines. Maintenance is directed toward known defects.
2. Machines continue to run fine far beyond their regular overhaul periods.
3. Sometimes defective machinery is found shortly after start-up. This is actually a blessing, because a repair can be planned, possibly avoiding a more serious outage later. The machine is probably still under warranty, and the owner avoids the repair costs.

This system of health monitoring of machinery is called "predictive maintenance" because it has the capability to predict failures. Ninety percent of all machinery failures are preceded by a change in vibration. This vibration change is usually detectable months in advance, sometimes a year or more. By detecting the vibration change, then tracking it, it is possible to predict the time to failure.

Compare the following two scenarios.

1. *Without vibration monitoring:* "It seems rough. Something doesn't sound right. We need to overhaul it before winter. It won't last another season."
2. *With vibration monitoring:* "The vibration level has remained steady for 3 years, but began rising 2 months ago. At its present rate of increase, it will probably fail in 4 months. The spectrum indicates that a significant imbalance has developed, but the bearings are okay. We should clean the rotor and inspect it during next month's outage and plan to rebalance it if necessary."

The first scenario will be more expensive because everything will be replaced that can be, in the hopes of replacing the defective part. If the rotor is not rebalanced, then the same problem will still be there on start-up. Money and time would have been wasted on the wrong fix, and the problem would not be corrected. The second scenario provides more information based on sound knowledge. It offers a background history, and suggests the remaining running time. It also suggests the specific actions to be taken to correct the problem.

Trending is the graphical presentation of some variable plotted with time, in this case, vibration amplitude (Fig. 8.1). The vibration level of a machine remains relatively flat when it is healthy, as can be seen in the trend chart in Fig. 8.1. There may be a slight positive slope. When a serious defect develops that is leading to a failure, there is a sharp rise, typically an exponential rise to failure.

By knowing the level at which a failure occurs, it is possible to pre-

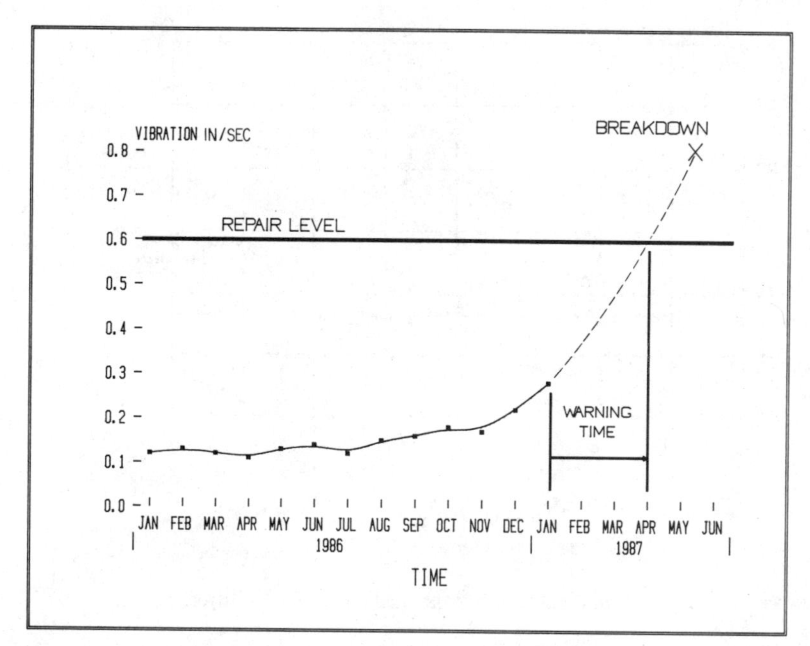

Figure 8.1 Trend chart.

dict the time to failure. The level at which a failure occurs can be from experience with a previous failure on a similar machine, from the vibration standards, or take 0.6 in/sec. Most experts agree that 0.6 in/sec velocity of vibration is very serious on rotating machinery, and a failure is imminent.

The vibration level will vary somewhat between periods. This is normal. A suitable response level is a doubling (2X) or a 6-dB increase in the vibration amplitude. The Canadian navy found that a 6-dB increase is significant; less than 6 dB, leave it alone and continue monitoring. When the level exceeds 6 dB from the baseline data, or a 6-dB change, then the appropriate response is to step up monitoring, and possibly do more diagnostics with a spectrum analyzer and an analyst.

The nuclear power industry regularly applies this technology to their critical reactor coolant pumps and turbines. They have enough historical data to safely predict the level at which failure will occur. With this knowledge and frequent trending, they can predict a failure to within hours.

Figure 8.2 is from actual data at nuclear power facilities.[1] The "Shutdown" and "Notify NRC" (Nuclear Regulatory Commission)

[1]J. H. Maxwell, ASME Paper 88-PVP-11, 1988. (Reprinted with permission.)

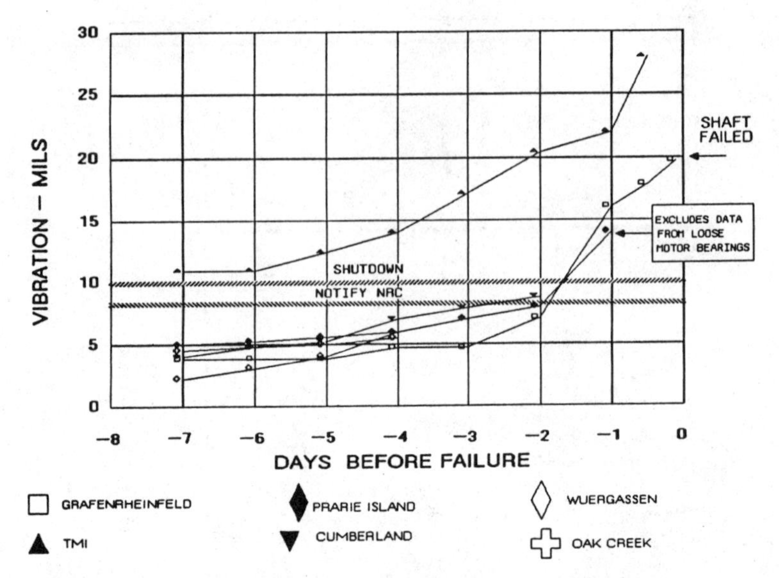

Figure 8.2 Vibration from cracked shafts at nuclear power facilities.

lines are the limits applied at Palo Verde Nuclear Generating Station. The data from Grafenrheinfeld, TMI, and Prairie Island are from reactor coolant pumps. The data from Cumberland, Wuergassen, and Oak Creek are from steam turbines. Both classes of machines tend to operate with about 4 mils vibration. The exception is TMI which started high. Notice in Fig. 8.2 that the machines with cracked shafts were all operated to within 72 h of failure. Most were not run to failure, but the remaining running time was estimated later by examining the crack size, and correlating this to actual failures, like Grafenrheinfeld.

The benefits are being able to operate equipment safely longer and approach its failure point. For power plants and other process industries this can translate into big dollars in the form of revenues that would have been lost if an earlier shutdown was called. In Fig. 8.2, the "Shutdown" limit is higher than the "Notify NRC" line at 8 mils. This is analogous to a production manager who wants to know when a problem is developing but is willing to operate further into the failure cycle if he or she has confidence that a catastrophic failure will not occur. Vibration data has the capability of providing both pieces of information. The trending program provides the initial warning, and the predictive portion can estimate the time to failure.

Predictive maintenance using vibration is statistical process control to the care of machinery. The machinery is characterized with baseline data and trended. Only significant changes are reacted to.

Otherwise, it continues to run. Vibration data can be used to determine overhaul schedules, rather than an elapsed period of time.

The phrase, "If it ain't broke, don't fix it," used to characterize a run-to-failure maintenance philosophy. This same phrase is now used in a predictive system where precious maintenance resources are directed to where they are needed after a diagnostician performs a checkup. In other words, hands and tools are kept off machines that are healthy.

This discussion of preventative maintenance programs is not intended to degrade it. Some parts of preventive maintenance programs are good and should be continued, like oil and filter changes. However, regular parts replacement on a time schedule is wasteful. There is a better way.

Preventive and predictive maintenance programs require an investment in tools and time. The run-to-failure method does not. The maintenance manager must factor this into the equation for his or her overall, long-term costs of doing business. Some dollar figures and manpower requirements for a vibration monitoring program are given in later sections of this chapter.

Screens

A screen is a tool that sifts bulk material and selects what is allowed to pass through. This same concept is desirable in a vibration monitoring program. The data-handling task could become very burdensome if several hundred machines are to be monitored every month. A good screen for this data would allow the "good" machines, or "no-change" machines to pass, and stop and dwell on those machines that have true defects developing. There are several different ways to accomplish this.

One way is to limit the number of monitoring points per machine. One vibration analyst can easily handle 100 points per month. With 1 mounting point per machine, the analyst can handle 100 machines. With 6 mounting points per machine, this same analyst can then handle only 17 machines easily. One effective method of ensuring a good screen is to limit the volume of data that the analyst must interpret. This keeps his or her interest level high and boredom down.

Ninety percent of equipment is in good condition. The vibration monitoring program measures and verifies this and delivers confidence to the owners. The challenge with surveillance monitoring is to identify the 10 percent troublemakers. It is not possible to positively identify all the troublemakers, and it is very probable that false alarms will occur. No screen is perfect, but neither is the theory of vibration monitoring. Eighty percent of machines do follow a normal pattern of stable vibration during most of their lives with an exponential rise toward failure. The important point to convey here is that about 20 percent do not. They fail catastrophically with no warning, or the time from the onset of symptoms to failure is so short that it appears like there was no warning. In some rare instances, the vibration actually decreases immediately prior to failure. When these situations occur, it is not a failure of the screen unless more than 20 percent unpredictables occur.

I remember a 100-hp motor that I reported was a "healthy machine." Four days later a bearing failed, and 2 days after that, the other bearing failed. These instances seem to discredit vibration monitoring, but we should not focus on the exceptions which were unpredictable anyway before vibration monitoring. We should focus on the statistical fact that 80 percent of machine failures are predictable. The present state of the technology is focused on developing better screens. Let's discuss the different types of screens.

The overall vibration level is the oldest screen, and still a very effective one. The decisions to be made when using the overall are:

1. What type of transducer? (accelerometer, velocity, or others)
2. What display to trend in? (acceleration, velocity, or displacement)
3. What frequency range?

The accelerometer is the transducer of choice for machine monitoring, and this choice tends to favor the acceleration display. There is no sense in using a high-frequency transducer when the velocity displays will deemphasize the data. Defects tend to manifest themselves first as high frequencies. Hence, the overall in acceleration is a good indicator.

There is another good reason for favoring the acceleration display with an accelerometer. This concerns some problems with digital integration. Digital integration introduces bogus data at low frequencies, and the overall is not a valid number on some instruments. For these reasons, trending using an accelerometer and digital integration to velocity is not a good idea.

Some companies suggest monitoring vibrations out to 50 orders. I agree with this concept. For an 1800-rpm (30-Hz) machine, this means out to 1500 Hz. The vast majority of useful trending information is in this range. Some instruments can also monitor selective frequency bands within this range. This is also a good idea and brings forth the next concept in screens: monitoring selective frequency bands. This is nothing more than measuring the overall between 20 and 80 Hz, the overall between 80 and 400 Hz, and possibly the overall between 0 and 2000 Hz. This will yield three numbers. The first will be representative of the imbalance and first harmonic of an 1800-rpm machine. The second will encompass bearing frequencies and motor vibrations, and the third band will monitor everything including the high frequencies. This method will yield only three numbers to keep track of that allow some analysis to be done. Other common band monitoring schemes could be 0 to 200 Hz, and 0 to 2000 Hz. These schemes are rather simple, but very effective, in detecting problems.

The most sophisticated type of screen is one that envelopes the frequency domain signature. This screen is always applied with computer software. It applies amplitude limits at every frequency based on a "good" signature, usually the baseline spectrum. This screen is much more effective in detecting small changes, like bearings, in the presence of large dominant signals like imbalance.

As mentioned previously, the time domain contains significant information, especially about transients, that is not available in the frequency domain. Consequently, frequency domain analysis is poor at screening for bearing defects or other metal-to-metal impacts. There are other screening techniques for these. One is the crest factor. This measure is the ratio of the peak amplitude to the root-mean-square amplitude and gives an indication of the wave shape. Bearing analyzers and high-frequency energy detectors also screen for these shock pulses. A third method is to simply view the time domain while at the machine and note the peak amplitudes.

Setting Up a Vibration Monitoring Program

The biggest savings to be gained from a vibration monitoring program is avoiding losses due to unexpected breakdowns. Use this as a criterion for deciding if a vibration monitoring program is for your operation. Critically examine every piece of equipment in your operation and ask yourself, "If this machine broke now, what would the consequences be?" If the consequences are not serious, perhaps an inconvenience for someone, then it may be better to let it run until it breaks, with an occasional greasing.

You may find that a relatively small piece of equipment, like an air compressor, may have serious consequences if it quits unexpectedly. It may provide compressed air for all your pneumatic air cylinders and building controls. The whole plant may be thrown into chaos without compressed air. The appropriate correction may be to back up that air compressor with a second unit.

Some equipment is very large and expensive, and unreasonable to back up. An example is a chiller that provides process chilled water, or perhaps a boiler draft fan. This kind of equipment, unduplicated and critical, is a good candidate for health monitoring.

A vibration monitoring program costs money. There is an initial investment in instrumentation of about $20,000. But this investment is small compared to the manpower required to collect data and analyze it. The running cost in time is the major cost of a vibration monitoring program. Plan on a full-time person for every 200 machines you want to monitor.

These are the types of considerations that need to be made before getting into a vibration monitoring program using vibration signatures. Go through every system in your plant, and consider an unexpected outage for every piece of equipment in that system. Ask yourself:

What are the consequences of an unplanned outage?
Is there a backup?

Then, considering the cost of a predictive maintenance program, ask yourself:

Is it more economical to install a backup?

You may find that a simple building fan is critical to your whole operation if it provides cooling to an area with a large heat rejection need. An example is environmental chambers with built-in refrigeration compressors, or a computer room. You may also discover a simple 10-hp hot water pump may shut down an entire process line and cost $4000 in wasted material in process that must be discarded. In this case, one shutdown would pay for a backup pump.

A useful way to go about this analysis is to review your maintenance history for the past 3 years. This will give you an idea of your troublesome machines and the consequences. You will also get an idea if these breakdowns could have been avoided with vibration monitoring. After this exercise you will know the cost to your maintenance program. In addition to the cost of parts replacement, add the salary cost of your maintenance staff, and the cost of lost product and lost production time.

If, after this, you decide that a vibration monitoring program is for you, there are some steps to implementing a successful program.

1. Get management commitment for a long-term program.

Management must commit financially for the instrument investment, training, and time for data collection. This is a long-term improvement program. Obtain management commitment to allow it to run for at least a year before looking for a payback—better 2 years.

2. Select and train personnel.

Identify one or two people who will have responsibility for this entire program. Ideally, this would be an engineer and a technician teamed-up. Give them ownership of everything. Send them to training seminars. Let them select the instrument to use. Allow them to select the machinery to monitor, and to design the data collection system.

If less than 50 points are monitored, a maintenance engineer can usually handle this alone as a part time responsibility along with other duties. From 50 to 200 points, two part-time people are required; a technician to collect the data and an engineer to analyze it and maintain the system. Greater than 200 points, begin adding full-time people with the job title of *vibration technician.* Greater than 400 points, consider an automated collection and analysis system using a host computer and analysis software.

One caution: Do not attempt any level of vibration monitoring program without engineering support. Engineering support is required to:

- Interface instruments
- Program the system
- Coordinate with equipment manufacturers
- Analysis

3. Select machines to monitor.

Start small with about 20 machines. Pick the most critical machines based on your initial plant survey. You want to show a payback quickly to management, so pick the problem cases to start with.

4. Select and purchase instruments.

Remember to allow the people who will be doing the work to select the instruments they will be working with. The choices are:

- Simple hand-held meter
- Tuneable filter meter
- Portable analyzer and balancer
- FFT spectrum analyzer
- Data logger

The instruments that you specify are mostly dependent on the kind of program you want to have. By this time, you should have an idea of the total number of machines to be monitored and the skills of the people involved.

There are basically three types of vibration monitoring programs:

1. Simple trending with hand-held meters
2. Tuneable filters or FFT analyzer
3. Data loggers with a host computer

Simple trending with hand-held meters is easy to get started and requires few skills (Fig. 8.3). A maintenance person goes to a machine, possibly once a month, attaches the pickup, and gets a reading. The skill involved is where the probe is attached and how it is held. The person also needs to know how to turn the instrument on and perhaps select displacement, velocity, or acceleration on a switch. On some older meters, a range switch may have to be operated to get a needle deflection. The meter is read, and the number is recorded or stored. The single number recorded is an overall value and will show a change when a serious defect has developed. It may not give very much early warning on some machines, since by the time the overall value is affected, the defect may have developed into the later stages of degradation.

This system of monitoring does not allow any balancing to be done or analysis. It is, however, inexpensive to start, requires few skills, and

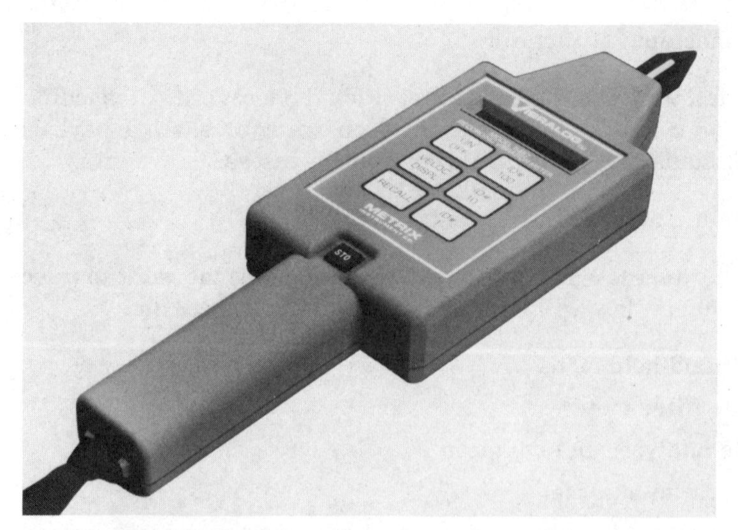

Figure 8.3 Hand-held vibration meter for trending. (*Courtesy Metrix Instrument Company.*)

is better than nothing. When the trending shows that something is going wrong, then the usual thing to do is to call in an expert with an analyzer to diagnose the problem and give you an idea of what to replace before it breaks. You may also want to step up monitoring frequency to weekly or daily. This will give you a better predictor of when to expect a breakdown.

The next type of program is to use a tuneable filter analyzer or FFT analyzers (Fig. 8.4). These instruments allow diagnostics to be done right on the spot. The FFT analyzer is faster for diagnostics. With these instruments and some accessories you can do balancing. Either of these instruments allows you to plot a spectrum; the tuneable filter analyzer usually does it on the spot on a strip of paper, and the FFT analyzer has a memory so you can store traces and plot them later.

Analyzers (tuneable filter and FFT analyzers) are the only instruments available for on-site diagnostics in real time. They can also be used for trending purposes and are powerful in this respect because frequency information is provided. Older analyzers are somewhat heavy, bulky, and cumbersome to carry around all day. Newer analyzers, as in Fig. 8.4, are lighter, more portable, and far more powerful.

To overcome the disadvantages of older analyzers (skills and weight), data loggers have been developed in recent years specifically for vibration monitoring programs (Fig. 8.5). The small and light

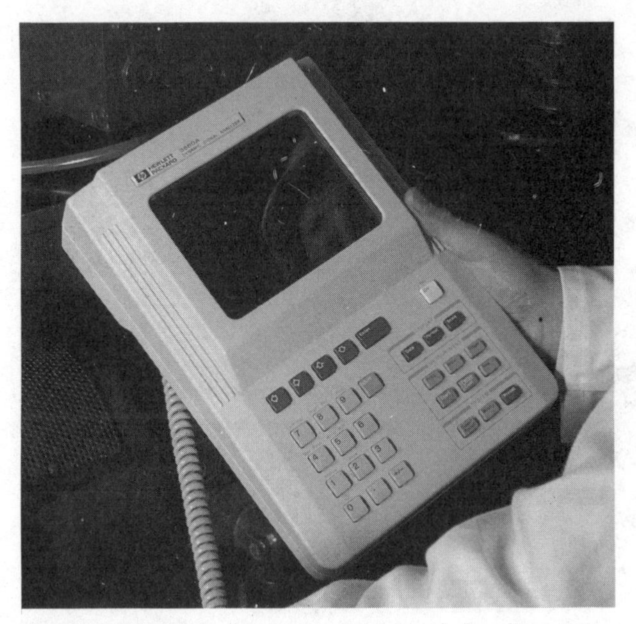

Figure 8.4 Dynamic signal analyzer used for diagnostics. (*Courtesy Hewlett-Packard.*)

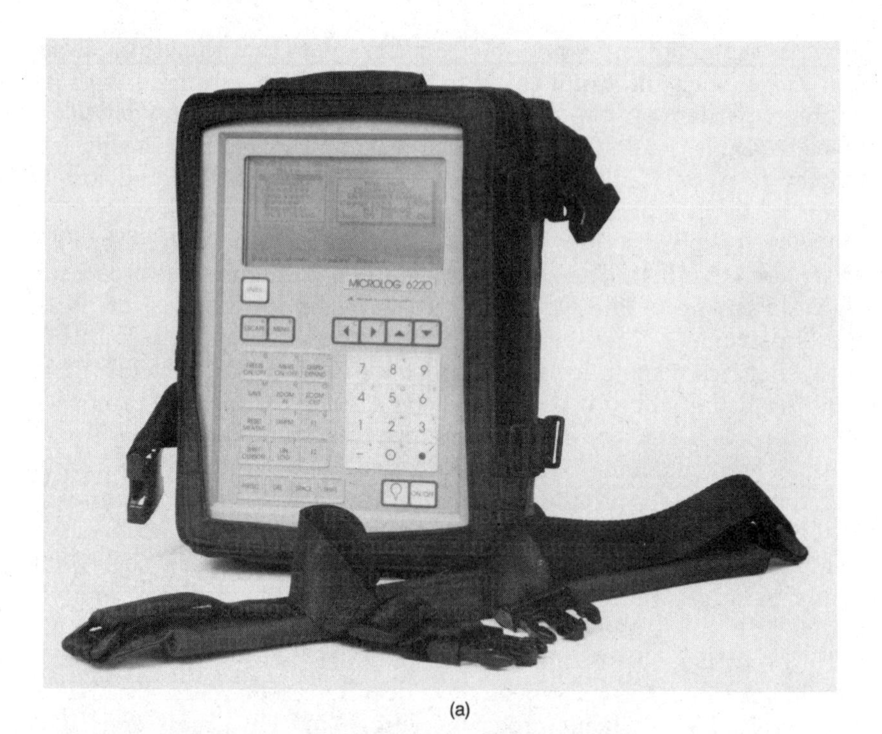

(a)

(b)

Figure 8.5 (*a*) Data logger. (*b*) Analysis of route data with a host computer. (*Courtesy SKF, Palomar Technology.*)

boxes have few buttons and large memories. A typical maintenance person can learn to use one in a few hours. The instrument is carried along a prescribed route, and data is captured and stored in the data logger's memory. At the end of the route, the data logger is returned to an office or laboratory, where the data logger is connected to a host computer. The computer extracts the data, analyzes it, compares it to last month's data, stores it, and issues a warning if necessary. Some data loggers can perform as stand-alone analyzers and even be used to do balancing.

As you can see, to do vibration monitoring with a data logger, a host computer and software are necessary. The computers are usually desktop types. The cost of this kind of system will likely be about $20,000 for the initial instruments, computer, and software. However, for monitoring many machines, like 500 or more, this is the best method.

These instruments along with their computer host represent the most powerful and sophisticated level of a vibration monitoring program. They are not recommended for first-time users just beginning a new program, unless the people have computer skills, machine diagnostic skills, and FFT analyzer skills, or a training budget is allocated for acquisition of these skills. To give you an idea of what instruments to purchase, consider the machines you want to monitor, the people skills available, and the spending allowance you have.

5. Take baseline data.

While your instruments are on order, mark the machines you want to monitor and set up a route. You may want to drill and tap holes, or epoxy on mounting adapters. Some machines may be in hazardous locations or otherwise unsafe to get to. A typical example is a propeller-type cooling-tower fan. In these cases, you can permanently attach a transducer when the equipment is down and run a cable out to a convenient location. When your instruments arrive, you should take about a month to become familiar with them.

The next step is to collect baseline data for all the machinery you initially want to monitor. You should take data at all bearings. Even though you may only monitor one point on a machine, you should collect and store baseline data at all bearings. When problems develop later, this will be invaluable in diagnosis.

Compare this baseline data against industry standards to see how the equipment compares. You can also compare similar machines in your plant. Do some analysis to see what is out of balance, is misaligned, has bad bearings, is cavitating, etc.

6. Start your monitoring program.

A good starting interval is monthly. You can extend this to 2, 3, or 6 months later if things look good. The purpose of this monitoring is to trend the data and look for increases. Remember that a doubling of amplitude (6 dB) represents a significant change and signals the beginning of a breakdown.

Figure 8.6 is a trend chart for a 50-hp motor and pump. After baseline spectrums are recorded, the periodic monitoring takes the form of overall readings. The data is recorded as a row of numbers. Tables of numbers are difficult for humans to interpret, so the data is graphed immediately above it. It is easier to see trends with data in graphical form. A sketch of the machine is useful to show measurement points. It's a good idea to take vibration readings before and after major maintenance. This is a good quality control of the work done.

7. Prepare monthly reports.

Management, who invested in this program, likes to know if it is doing any good. Don't keep them in the dark. If everything is running smoothly, let them know and feel good about it too. This visibility to management is what gives your program credibility. In time, you will

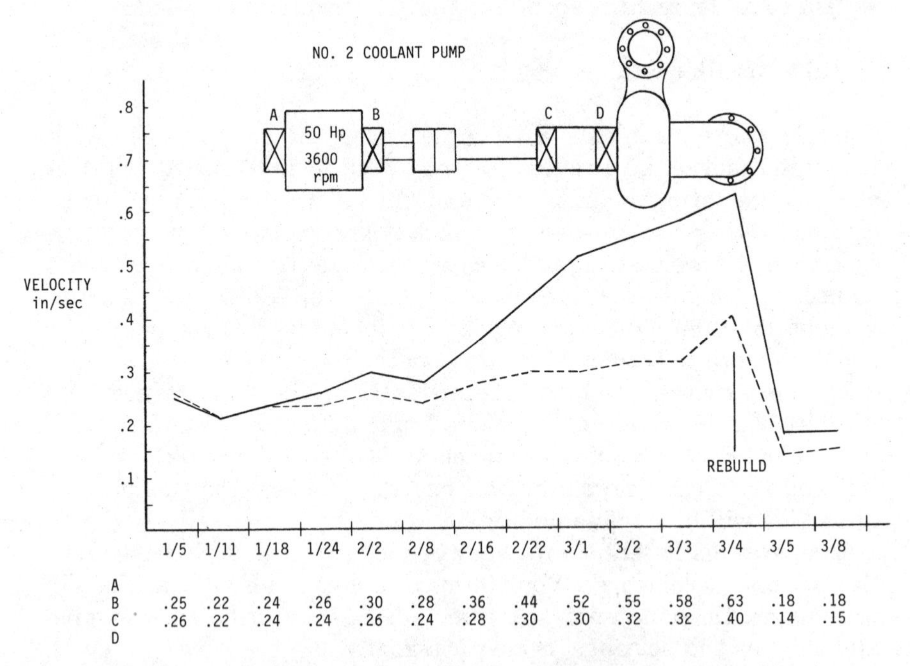

Figure 8.6 Overall vibration readings trended for a 50-hp pump-motor combination.

find management clamoring for vibration reports at the slightest hint of anything going wrong.

The plots in Fig. 8.7 were taken at 6-month intervals in a regular vibration monitoring program. Notice that there is some variability, but there is also consistency as the plots do look similar.

In time, the monitoring program will take the form of a pyramid; as illustrated in Fig. 8.8. At the top of the pyramid are a few machines with serious defects that require close scrutiny. Next are those that require detailed diagnostics because of an upward trend or other irregularity. Below that are the machines that are routinely trended. At the bottom are the large number of good machines that may or may

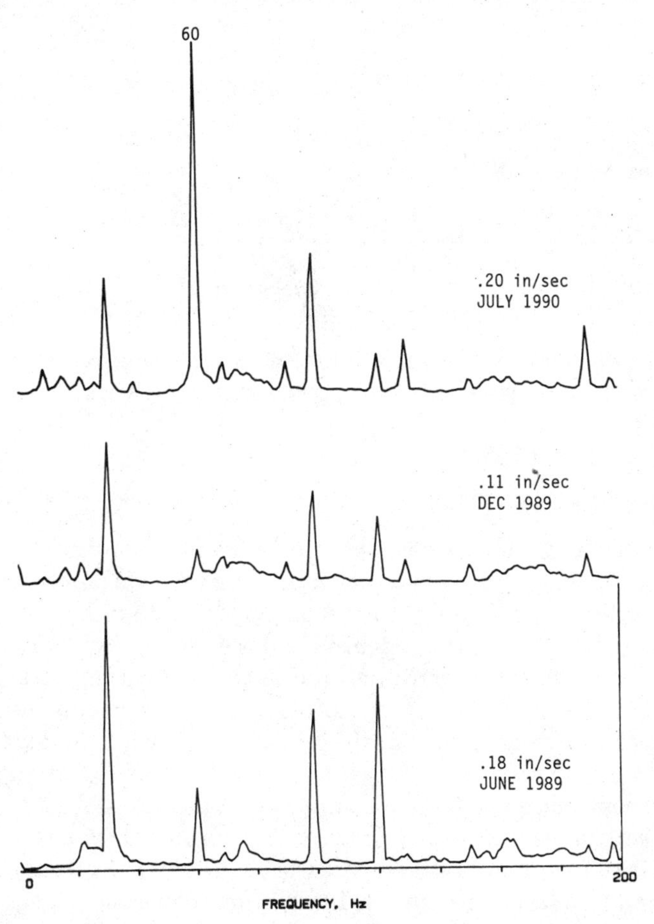

Figure 8.7 Spectrums taken at approximately 6-month intervals from a turbine. The peak at 60 Hz was a growing resonance that was corrected *before* the permanent vibration monitoring system shut down the turbine.

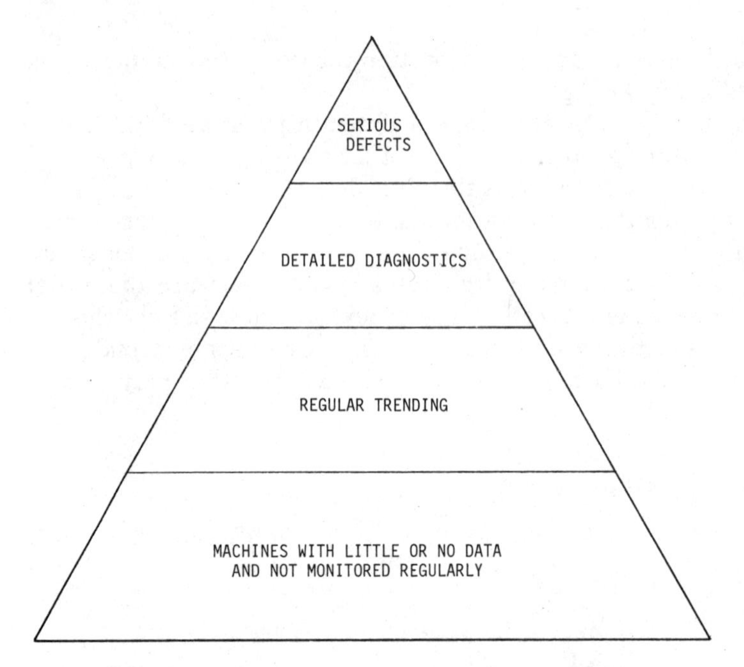

Figure 8.8 Arrangement of machines in a trending program.

not have vibration data on record but are not monitored on a regular basis. This segregation is desirable to focus attention on those machines that need attention.

Permanent Monitoring

Some machines can develop a very serious defect very fast and destroy themselves in a short time. Some examples are high-speed turbines or cooling-tower fans that throw some weight off and go into a serious imbalance condition. To protect the machine, a disconnect relay is installed in the power or control circuit, and the input to this relay is a vibration signal. The output from the vibration transducer can be monitored in the control center with a digital LED display or analog meter (Fig. 8.9).

These control room monitors have a connector to plug into for an oscilloscope or spectrum analyzer to view the vibration signal from the transducer. This is convenient for diagnostic work. On these electronic systems, alarm and trip levels can be set. In addition to disconnecting power, they can also activate a tape recorder to record the vibration data on spin-down. This is invaluable for later diagnostics.

Newer systems have digital waveform recording that can be pretriggered. That is, these systems continuously digitize the input

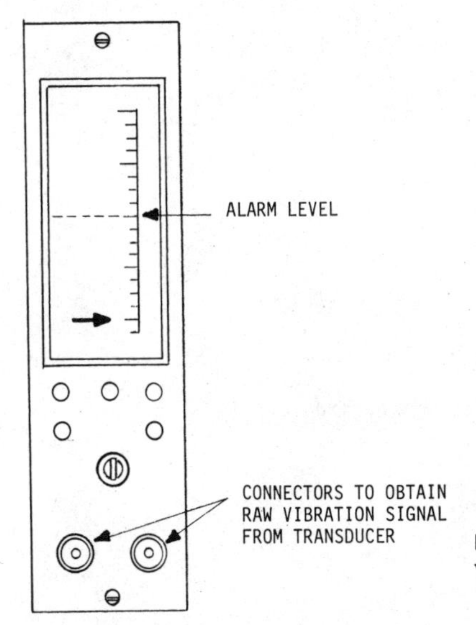

Figure 8.9 Permanent monitor with setpoint for alarm.

waveform and discard the oldest data. When an event occurs, like a trip, the controller can be preprogrammed to retain the data for, say, 1 min prior to the trip. In this way, it can record the events leading up to the trip and the subsequent coast-down. Imbedded in this vibration data is usually the whole story of what went wrong.

Finally, with personal computers and modems, it is possible for an analyst to view the vibration data from a machine in the comfort of his or her home at any time of the day or night.

To conclude this chapter, it is useful to visualize the ideal maintenance strategy. Ideally, the best managed maintenance plan is to never have an unexpected breakdown. The machine is taken out of service just moments before failure, which happens to coincide with a regularly scheduled outage anyway. No additional defects are introduced during maintenance activities, and the cost of this program is nil.

Chapter

9

Acoustics

This chapter is included because there is a fine, almost intangible line dividing acoustical and vibration problems. The difference is mostly in the path. Vibration problems are structureborne oscillations, whereas, acoustical problems are airborne oscillations. The instrumentation used to measure and analyze vibration problems can be used just as well for acoustical problems with the simple exchange of a different transducer. Historically, the two fields have common roots, and today they are both subsets of a larger field of knowledge called simply, *vibrations*. Every vibration analyst finds a need at some time to measure the sound field and work an acoustical problem. This is easily done with the available vibration instruments with the addition of a low-cost microphone.

Acoustics actually is a broad field. It encompasses more than we wish to consider here, specifically, the fields of architectural acoustics, community noise criteria, human hearing exposure, and music, which will not be discussed further here. The field of concern in this chapter is machinery noise. Since machines are installed in buildings and within rooms, these environments need to be considered in conjunction with machines as sources.

Noise must be defined. Sound is a means of communication. Wanted sound communicates information. Unwanted sound is noise. So by definition, *noise* is unwanted sound. This makes noise highly subjective. What may be music to one person may be noise to another. The degree of annoyance cannot be precisely measured. What can be measured is the frequency, amplitude, and phase. This information is sufficient to identify the source and the path and to judge the severity.

The Physics of Sound

Sound is a pressure variation in air. It is a longitudinal compressive wave. Some other moving, vibrating, oscillating, or pulsing object sets the air into motion and sustains that motion. Sound is a form of mechanical energy radiating out from the source. In air, sound travels at 344 m/sec (1129 ft/sec). The intensity drops off with distance in accordance with the inverse square law, i.e., the sound intensity in air decreases by a factor of 4 when the distance doubles. This, of course, assumes free-field conditions with no barriers or other interfering objects.

In steel, sound travels at 5060 m/sec (16,600 ft/sec), and the inverse square law does not apply. Sound travels about 15 times faster in steel than in air. In metal objects, the sound energy does not radiate equally in all directions but is confined within the metal boundaries to some extent. The sound energy is directed along the metal path. In air, sound attenuation is quite rapid. In solids or liquids, the rate of attenuation is much lower. Vibrations in solids or liquids can travel much farther. Ducts, pipes, and almost any continuous rigid component of a building can carry sound energy past the best of air path barriers.

Sound has a source, a path, and a receiver. All three of these must be positively identified to solve noise problems. The source and receiver are usually straightforward. Identifying the path presents the most difficulties, and this is also where most noise control measures are applied.

Sound is a very low level of energy intensity. Sound energy is typically measured in picowatts (10^{-12} W). This extremely small amount of energy is offset by the extreme sensitivity of the human ear. This extraordinary sensitivity of the human ear is comparable to modern electronic sensors. This was, and still is, no doubt, a significant advantage for mammalian survivability. It is also the reason for its annoyance at times. This sensitivity of the human ear can be used to advantage during analysis.

Machinery Noise

Whenever mechanical power is generated, transmitted, or used, a fraction of this power is converted into sound power and radiated into the air. Generally, the fraction of available mechanical power that is converted into acoustical power is very small. However, very little acoustical power is needed to make a sound source audible.

Mechanical equipment generates two kinds of sound; broadband and pure tones. The broadband sound can be desirable or objectionable depending on the frequency content, amplitude, and structureborne versus airborne content. The pure tones are almost always objectionable. Let's discuss the broadband sound first.

Mechanical systems installed to serve buildings must share the interior space with the human occupants. Heating, ventilating, and air conditioning (HVAC) systems create a background level of low-frequency broadband noise. This has been found to be desirable in densely packed buildings. This white noise masks obtrusive sounds (conversation, business machines, and foot traffic) and allows people to work closer to each other without being bothered by each other's presence. Complete silence would be most unpleasant. Humans become disoriented and distressed. Too much quiet is undesirable. People tend to fall asleep. Some broadband background noise is wanted. The design goal for HVAC engineers is to provide enough background noise that is low in level, unobtrusive in quality, and that does not interfere with the building occupants. This goal is breached when the background is no longer steady in level, bland in character, and free of identifiable machinery noises. Humans are more tolerant of low frequencies and broadband noise and less tolerant of high frequencies and pure tones. The usual complaints are audible tones, whines, whistles, hums, and rumbles.

Audible tones are usually pure tones coming through the air delivery system. The tones can be a resonance or a vane passing. It is easiest to investigate for a vane passing first and eliminate this possibility. Airfoil fans are particularly plagued by vane passing tones. The analysis procedure is simply to measure the frequency of the tone in the complaint location. Then measure the speed of the fan serving this area and count the number of blades on it. The fan blades are generating the tones if the speed times the number of blades equals, or is close to, the objectionable frequency. Some slight frequency decrease may be observed, and this is due to damping in the system.

Resonances, which also cause pure tones, may be two kinds: structural or acoustical. Structural resonance is the easier of the two to identify and correct. Simply find the "singing" part and stiffen it. An acoustical resonance is found by matching a length of ductwork with the wavelength of the objectionable tone. The wavelength of sound is calculated with the following formula.

$$\lambda = \frac{1129 \text{ ft/sec}}{f} \tag{9.1}$$

where λ = wavelength, ft
f = frequency, Hz

As with any resonance, a change of speed will both verify and correct it.

Whines may be one person's way to describe a pure tone, or it may be a sheet metal rub or belt rub. For either kind of rub, the on-board human diagnostic instruments are better for analysis. Listen for the character of the noise, and walk around until you can put your finger on it and it stops. Electronic instruments are not very useful here.

Whistles can also be another descriptive term for resonance, or it can be air-flow noise past a diffuser or register. The best, and most permanent, solution is to slow down the air velocity through the register. This can be done by adjusting the damper or increasing the flow area of the register.

Hums are most likely 120 Hz from electrical machines. It is worthwhile here to measure the frequency with a microphone and positively identify the 120 Hz. Common sources for this in occupied spaces are the ballast transformers in fluorescent light fixtures. Other sources are power line transformers near the ceiling or in adjacent equipment rooms.

Rumbles indicate low-frequency structural motion. First, of course, is to identify the source. This can be done with simple turn-off tests. Next, measure vibrations on the structure and investigate the path. Rigid and massive structural members, like columns, will move with low amplitudes, but they can very well transmit this energy to flexible members. The object for this investigation is to find these flexible members in motion. The alternative solutions are to stiffen the flexible members, isolate the machine better, or put more distance between the source and receivers. None of these are pleasant alternatives after the construction is complete. Light barrier panels of gypsum board will do little for rumble-type problems.

Sound Measurement

Sound measurements can be made with electronic instruments or human biological instruments. In some respects the biological instruments are superior. They are superior when identifying the character of the sound.

The objectionable sound can be a pure tone or a multiplicity of tones. The human ear and brain constitute a good spectrum analyzer that can identify pure tones subjectively. When the sound is composed of a multiplicity of tones, possibly harmonics, the human instruments outclass the electronic spectrum analyzer. The spectrum analyzer can measure the amplitude and frequency of each component part, but it has difficulty assigning a fingerprint to it. This is what the human instruments do best. The entire symphony of sound has a characteristic that the human ear and mind can readily identify and remember. For example, middle C from a piano, a violin, and a trumpet all have the primary component of 256 Hz, but differing harmonic content. This is what makes them different to the human. The human is capable of readily distinguishing the difference. This can be used to advantage for machinery noise measurements.

First, a change in harmonic content will be obvious to a person who regularly hears the machine. This is embodied in phrases like "it sounds different" and "something's not right." Second, supposedly identical machines differ in their sound signature, and the human can immediately discern this without expensive instruments and spectrum analysis. And third, a specific machine's sound can be tracked in the presence of other, louder, sounds to the point of origin by focusing on the character of the sound. Spectrum analyzers can also do this, but not so fast nor so inexpensively.

There are other indicators of noise outside the normal range of hearing. At low frequencies and high amplitudes, a pressure fluctuation can be felt, possibly even in the chest cavity. At high frequencies and high amplitudes, pain can be felt in the ear.

Electronic measurement of sound is done with a microphone as a transducer, and a readout instrument. There are two basic microphone types: condenser and electrodynamic. Either one can be used; however, the condenser microphone is favored for machinery noise investigations because of its frequency response. Electrodynamic microphones start to lose sensitivity around 80 Hz and are down considerably at 50 Hz. Condenser microphones, on the other hand, remain linear below 50 Hz and some are even linear down to 20 Hz. Condenser microphones are also more sensitive than electrodynamic types because of preamplifiers. For noise measurements where spectrum analysis is to be done, the calibration of the microphone is unimportant. When amplitudes are to be measured as with a sound-level meter, then the amplitude calibration of the instrumentation is important.

The human ear does not respond uniformly to sounds at all frequencies. We are deaf to low-frequency sounds and most sensitive to sounds around 2000 Hz. This is illustrated in Fig. 9.1. To compensate for this, the A weighting of a sound-level meter weights a signal in a manner which approximates the human ear's response. The LIN weighting is no weighting at all. The raw signal from the microphone passes through unmodified. Therefore, to make amplitude measurements approximating human sensitivity, the A weighting is used. When doing investigative work using frequency analysis, it is undesirable to modify any part of the spectrum, so the LIN weighting is appropriate.

The amplitude scale for sound work is a logarithmic scale. It ranges from 0 dB, which is the threshold of hearing, to 140 dB, which is near a jet aircraft engine. Office areas typically measure around 60 dB. Three decibels is the minimum change that humans can perceive; i.e.,

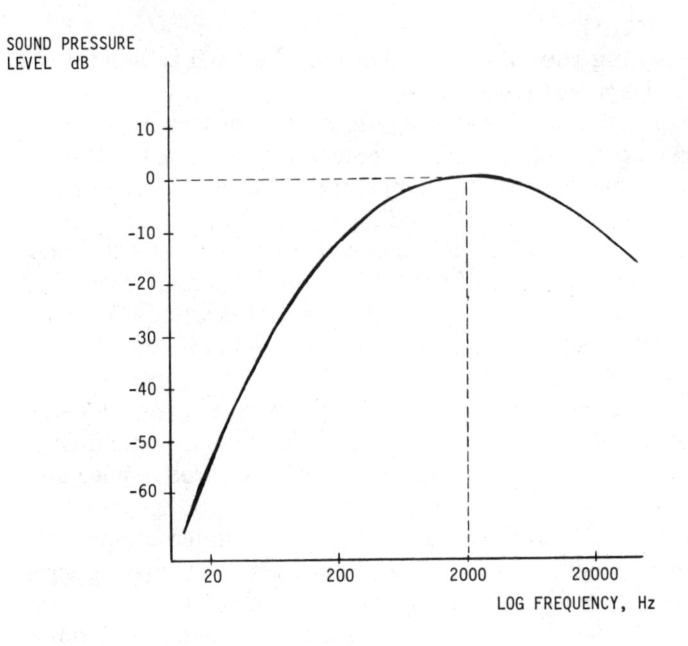

Figure 9.1 Human hearing response.

a change in sound-pressure level of less than 3 dB will go unnoticed. Ten decibels of change is required before the sound subjectively appears to be twice as loud.

To obtain frequency information using a sound-level meter, octave analysis is done with filters. An octave is a doubling of frequency. It gets its name from the musical scale where one octave covers eight notes of the diatonic musical scale. For example, the 1-KHz octave band filter passes the frequencies from 707 to 1414 Hz. Remember that filters are not perfect and the frequency limits are not sharp cutoffs—there is some slope to the sides, and some signals outside this band will sneak in. To get finer resolution of frequency analysis, ⅓-octave bands are available on some filter sets. In ⅓-octave bands, the highest frequency of the passband is 1.26 times the lowest frequency. The same 1-kHz passband in a ⅓-octave filter will have frequency limits of 891 to 1120 Hz. A significant amount of acoustical data is still measured and reported in octave and ⅓-octave filtered measurements, but these measurements are crude compared to a spectrum analyzer that can perform narrowband frequency analysis. A 400-line analyzer set to a frequency span of 2000 Hz will have bandpass filters set at 5 Hz. The 1000-bin filter in this analyzer will have lower- and upper-cutoff frequencies of 997.5 and 1002.5 Hz. With zooming, this can get much narrower. The superiority of spectrum analyzers to obtain precise frequency information should be evident.

Buildings

Machines installed in buildings typically occupy spaces that are not habitated by humans. The objective of noise control is to contain the vibratory energy, both structureborne and airborne, within that space. This space may be a mechanical room, a closet, or the space above the ceiling. For simplicity, let's call any space occupied by noise-generating machinery a *room*.

The acoustic waves emanating from the machine travel to the extreme boundaries of the room in all directions. If an opening is found, then the waves will pass through. Acoustical energy can go wherever air can go. At the boundaries, some of the acoustical energy is absorbed, some is transmitted through the barrier, and some is reflected back into the room. That energy which is absorbed and reflected is of no consequence to inhabitants outside the machine room. It is that portion that is transmitted through or around the room barriers that becomes troublesome.

The acoustical energy impinging on the room barriers sets these partitions into motion. The acoustical energy becomes structureborne at that point and adds to the vibration that is directly coupled to the building at the machine mounting. Acoustical energy is very small, but when acting over a large area, like a wall or ceiling, it can generate significant motion, especially if the partition is light and flexible. The partition then couples to the air on the other side and retransmits the energy acoustically. The partition acts like a big diaphragm. This may be difficult to visualize, but a vibration transducer on a flexible wall panel will easily prove that it is in motion, and the frequency content on the wall and in the air are probably similar.

There are two general fixes for such acoustical transmission across barriers. One is to make the barrier more massive and stiff such that it can't move. This means replacing a gypsum board partition with a concrete masonry unit wall. The barrier resists motion because of its inertia. It takes more energy to move a more massive barrier. Therefore, sound transmission through a barrier depends directly on the mass of the barrier. This is the reason for using soft sheet lead on partitions. Lead increases the partition's mass and also has high internal damping. The other fix is to build enclosures around the machinery such that the number of air-to-partition interfaces increases, and so do the losses in acoustic energy. The choice depends to some extent on the frequency content of the noise. More enclosures of light materials are effective for high-frequency airborne noise above 500 Hz. These enclosures become increasingly less effective at lower frequencies until they are almost useless below 100 Hz. For these low frequencies, massive partition construction is the best cure.

Neither of these acoustical fixes will solve structureborne vibration that enters the structure at the machine mounting. Improved isolation at the machine is the best remedy. It is useful to measure the frequency content of the structureborne vibration that is getting through. High frequencies (above 60 Hz) are best handled by cork, shock pulses by rubber, low frequencies (less than 60 Hz) by mechanical springs, and very low frequencies (less than 10 Hz) by air springs.

At times, a directional barrier on one side only is all that is required. An example would be blocking the machine noise transmission to a residential area across the street. In general, the barrier object must be larger than 1 wavelength in order to significantly disturb the sound. For example, at 100 Hz the wavelength is, according to Eq. (9.1):

$$\lambda = \frac{1129}{100} = 11.3 \text{ ft}$$

At higher frequencies, the wavelength is smaller.

At 1,000 Hz

$$\lambda = \frac{1129}{1000} = 1.1 \text{ ft}$$

So higher frequencies can be blocked with smaller barriers. In addition, sound absorption is readily achieved at high frequencies. These two properties of high frequencies, small barriers and ready absorption, make high frequencies easy to handle. These same properties make low frequencies difficult to block.

The major problem with machinery noise is that it is of low frequency, typically less than 500 Hz. This makes it difficult to block as an airborne vibration.

A sound-level meter without filters has limited usefulness in solving noise problems, because it provides amplitude information only. The amplitude is of least importance in analysis. A sound-level meter can be used to measure the sound amplitude in the complaint location before and after, for documentation purposes. For analysis, the frequency is of paramount importance. The frequency will identify the source and the path. The offending frequency will also determine the type of acoustical treatment. The frequency is needed to specify isolators and absorbers.

Case Histories of Acoustic Problems

Fan blade resonance

This case history is of a vaneaxial fan with resonances on the stationary blades. The complaint was a whine throughout a two-story office building. The whine was worst in an executive office on the second floor. Since the problem was acoustical, the first step was to measure the sound field at the complaint location. Figure 9.2 is a spectrum of the sound on the executive's desk on the second floor.

The peak at 115 Hz is probably from the ballast transformers in the fluorescent light fixtures. The real problem was the two peaks at 335 and 355 Hz. To verify this, a person was dispatched to turn off the building supply fan in that quadrant. As the fan speed slowed down, the two peaks at 335 and 355 Hz dwindled down to nothing. The executive declared that the whine was gone. When the fan was restarted, these two peaks immediately reappeared and the whine returned.

The supply fan was a 20-hp vaneaxial fan powered by a variable-speed drive. The whining was found to be absent below 40 percent speed, and intermittent above 50 percent speed. It was worse at 44.4 percent speed. This was consistent. Whenever the speed was changed, the strong whine always reappeared at exactly 44.4 percent speed. A vibration, or sound, that is speed sensitive is a strong indication of resonance.

Resonance bump tests were performed on the outside, accessible parts of the suspect fan, but none was found that matched the 335- or 355-Hz frequencies. The fan interior was made accessible by removing the inlet diffuser, and resonance bump tests were performed on the ro-

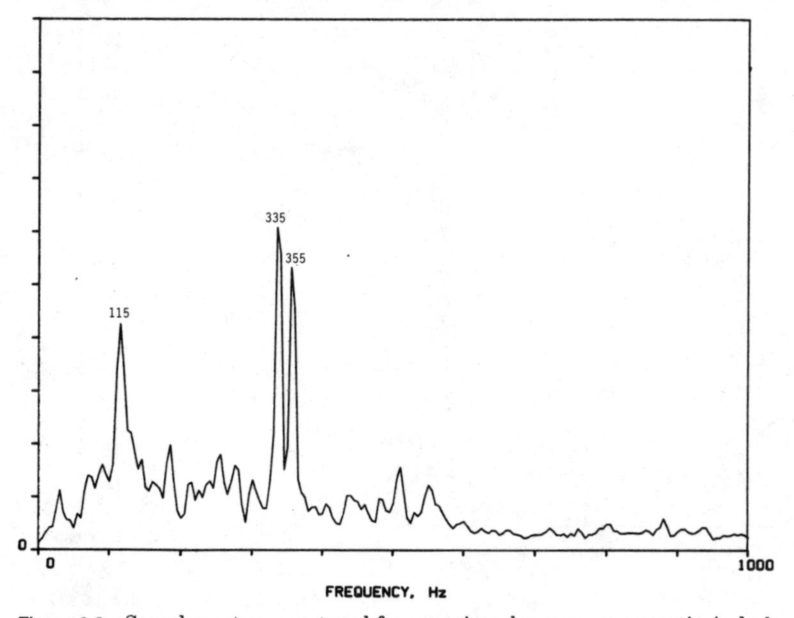

Figure 9.2 Sound spectrum captured from a microphone on an executive's desk.

tating vanes, hub, motor, and stationary vanes. One stationary vane rung loud and clear at 335 Hz. This looked promising. For further testing, an accelerometer was mounted to this stationary vane, and the cable was brought out through a hole drilled in the side of the fan. The fan was run like this, and Fig. 9.3 is the spectrum from this accelerometer with the fan operating at 44.4 percent speed (362 rpm). The 490-Hz vibration was not investigated because it was not part of the acoustic problem. The vibration amplitude was 2.5 *g* at 335 Hz at an unsupported corner of the stationary vane (Fig. 9.4). This level of vibration was serious and would most likely have resulted in cracking in time. But the complaint was noise. It was clear then that this unsupported corner was excited into resonance by the motor which was being driven by a variable-speed drive. The airstream picked up this vibration and carried it to the building occupied spaces via the supply ductwork.

Two solutions were offered for this problem. One was to simply cut off this unsupported corner. This would have affected the air-flow pattern and would have to be approved by the fan manufacturer. The second solution offered was to stiffen this unsupported corner. To test the stiffening idea, a wedge was made of wood and driven between the motor and this corner. Figure 9.5 is a vibration spectrum from the same spot as Fig. 9.3, but now with a wedge in place. This obviously took

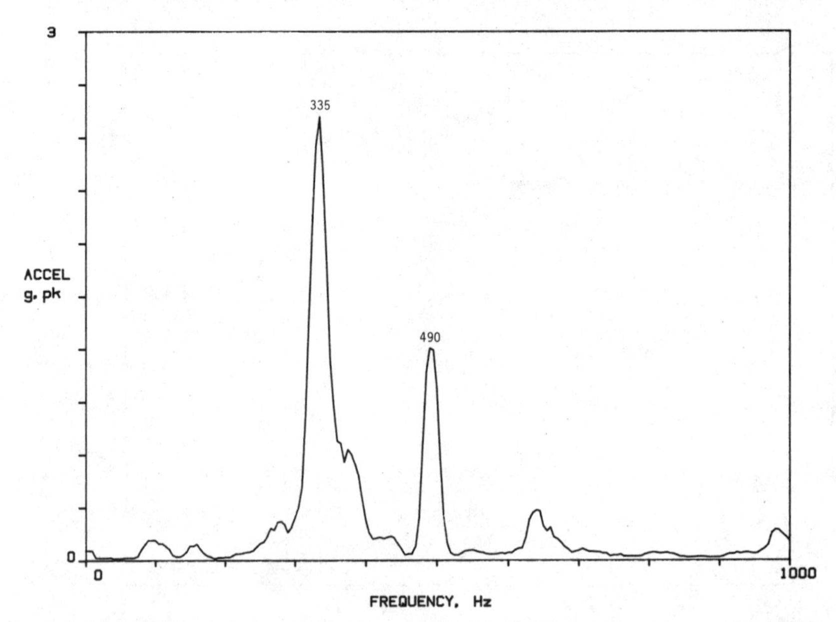

Figure 9.3 Vibration on a stationary vane of a vaneaxial fan. The fan speed was 362 rpm (44.4 percent of full speed).

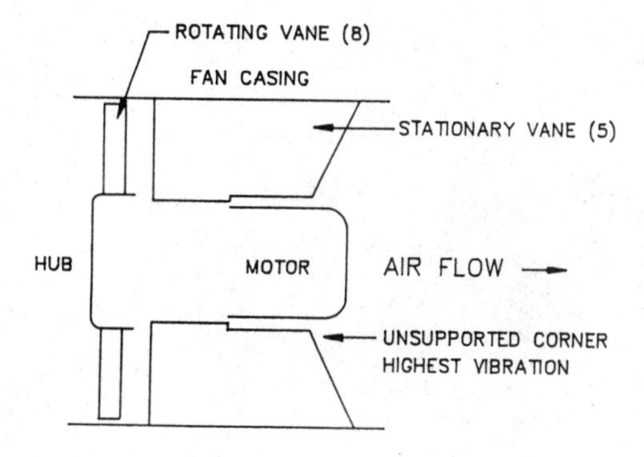

Figure 9.4 Internal arrangement of a vaneaxial fan showing the unsupported corner of the stationary vanes.

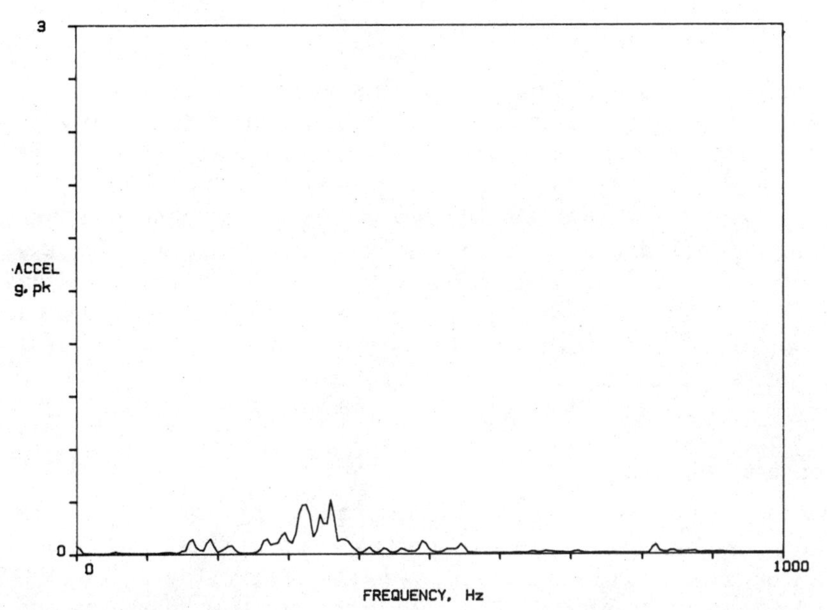

Figure 9.5 Vibration on the same stationary vane as Fig. 9.3 but now with a stiffening wedge in place.

care of this one blade. The 490-Hz vibration also went away. Other blades, slightly different in size or thickness, could be producing the 355-Hz tone.

A permanent solution was designed. A steel ring was welded to all of the unsupported corners to tie them together (Fig. 9.6). This solved the acoustic problem such that the fan could then operate through its entire speed range with no audible tones.

This case history illustrates the need for the fan manufacturer to do vibration testing. It would have been less expensive, and less damag-

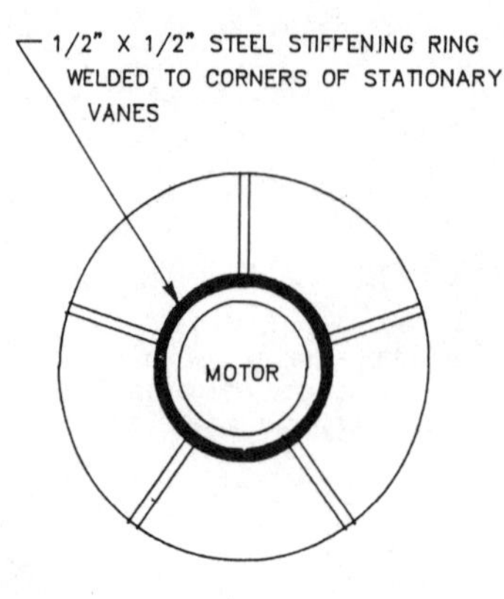

Figure 9.6 Modification of vane-axial fan to tie together all the unsupported corners of the stationary vanes.

ing to his reputation, to have found this resonance and corrected it at the factory. Even if it is a constant-speed machine, the possibility of driving it at a variable speed sometime in the future should be considered. To conserve energy, it is popular to add variable-speed drives to machines that were designed to run at a constant speed. Operating and maintenance professionals should consider the possibility of structural resonances when retrofitting an existing machine to run at different speeds. On variable-speed machines, vibration and acoustic testing should be specified to apply throughout the entire speed range.

Acoustical resonance

This case history is of an acoustical resonance that disguised itself as a vane passing tone. The complaint was also a pure tone throughout a two-story medical building. The worst location was in the hallway under an air supply diffuser. The sound spectrum was measured and plotted, and is shown in Fig. 9.7. There is a pure tone at 193 Hz. The supply fan was an airfoil-type centrifugal fan with nine blades. It rotated at a constant speed of 1290 rpm. Simple multiplication proves that this is a vane passing frequency:

$$1290\,\frac{\text{cycles}}{\text{min}} \times \frac{1\text{ min}}{60\text{ sec}} \times \frac{9\text{ blades}}{\text{cycle}} = 193.5\text{ Hz}$$

Many possible alternatives were considered, all of them expensive. Some of the fixes were to upgrade the existing sound attenuator, acoustically line the ductwork, change the fan, and install an active cancellation system. No one thought of changing the fan speed.

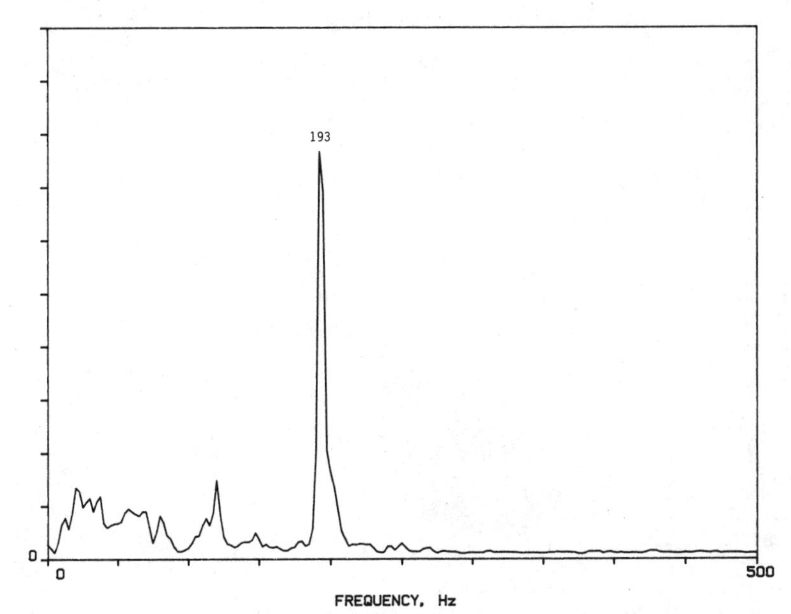

Figure 9.7 Sound spectrum in the hallway, under an air supply diffuser, of a medical building. The 193-Hz pure tone is a vane passing frequency.

Two months later, as the expensive alternatives were being investigated, a test-and-balance contractor changed the fan speed to 1445 rpm by putting a smaller pulley on the fan. The tone disappeared completely. The new vane passing frequency would have been 217 Hz. Figure 9.8 is the new sound spectrum in the same location in the hallway at the higher fan speed. There is nothing at 217 Hz. The vane passing tone should have increased in frequency with an increase in fan speed. The fact that it totally disappeared meant that there was an acoustic resonance.

A length of ductwork was tuned to resonate acoustically at 193 Hz. This duct picked up the weak vane passing tone from the fan and amplified it. When the speed of the fan was increased slightly, the vane passing increased to a frequency that the duct was no longer sensitive to.

I wish all problems were so easily corrected. Figure 9.9 is a plot of speed changes at another similar problem. This second problem was also identified as a vane passing frequency on an airfoil fan.

With the experience gained from the first problem, where a small speed change made a big difference, it was hoped that a speed change would correct this one also. Eight different speeds were tried, from 881

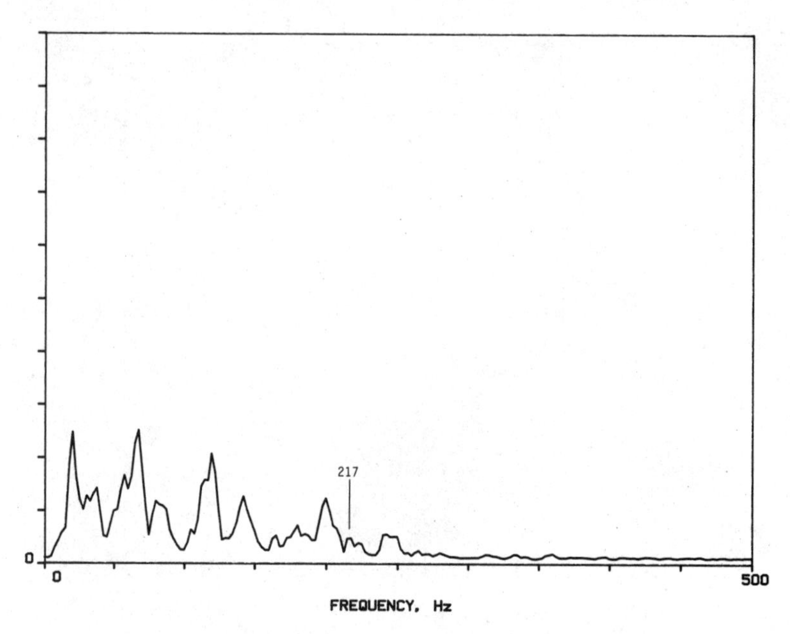

Figure 9.8 New sound spectrum after increasing the fan speed to 1445 rpm.

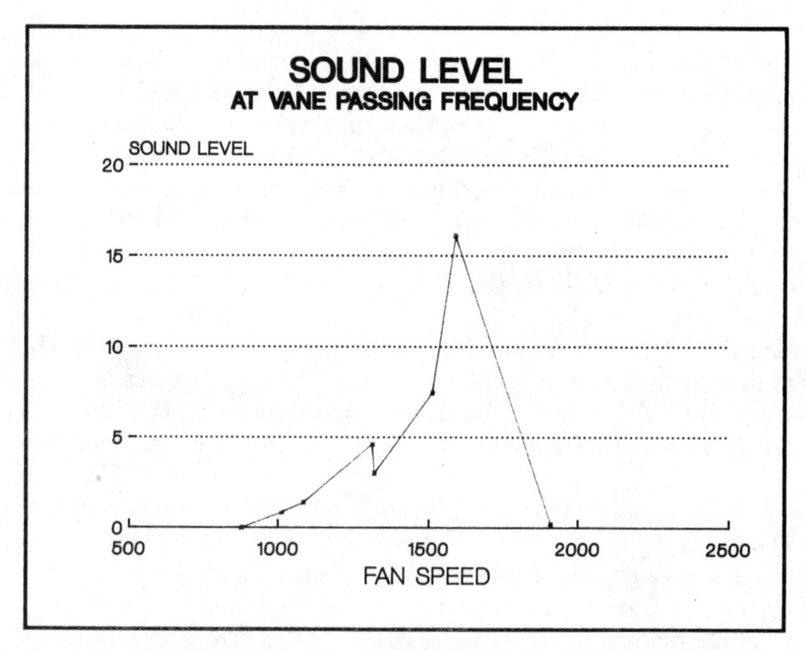

Figure 9.9 A plot of sound-level amplitudes of vane passing tones as the fan speed was changed. This represents a broad resonance diagram.

to 1910 rpm. The vane passing tone did finally drop away at the extreme speeds, but the owner did not want to operate this fan so far away from the nominal speed of 1600 rpm. As can be seen from Fig. 9.9, this speed of 1600 rpm is very close to the peak of this resonance diagram and the worse place to operate in terms of noise. Unfortunately, this resonance diagram is rather broad and a slight speed change did not solve the problem, as in the first case. In this second problem, the owner modified the duct system, in addition to changing the fan to a nonairfoil type.

Chapter

10

Final Remarks

Thus far, this book has only lightly touched on the practical and professional aspects of vibration measurement and analysis. This final chapter will expose the remaining practical considerations to acquiring vibration data and some suggestions for handling it in a professional manner.

Mechanical equipment rooms are generally not laboratory-clean environments. It sometimes seems like a contradiction to be transporting a laboratory-grade electronic instrument into this environment. The first practical aspect to vibration measurements is getting your instruments to the machine and back out undamaged. This may also include cross-country transportation. A rugged instrument case and/or travel case is helpful in maintaining the life of your instruments. Field-portable electronic instruments will never achieve the life that a laboratory bench instrument can enjoy. Plan for significant repair and replacement cost.

The second practical consideration is gaining access to the machine and to its bearings. You may know your instruments and your own skill well enough to estimate 30 min to acquire the vibration data, but have you considered things like:

- Is a ladder necessary?
- Is the machine on the roof exposed to the weather? (And is it snowing?)
- Is 120-VAC power available?
- Is there sufficient light to work in?
- Will coveralls be necessary?
- What tools will be needed to remove panels or covers?

The 30 min could lengthen to 4 h if these access problems need to be solved.

The third practical consideration is retaining the data until it can be presented to the owner. Eventually, the significant data will need to be transferred to paper. Modern instruments have plotters on board or digital memories to store it. A bookkeeping system needs to be devised to keep track of which data goes with which end of the machine. A short pencil is better than a long memory. These are some of the practical aspects of acquiring vibration data. Others follow, along with professional manners.

Safety Precautions

While taking vibration measurements, you will be working around rotating equipment. It may be partially disassembled with safety covers removed. A few safety precautions are in order.

1. Stay out of the machine's "line of fire" as much as possible. *Line of fire* is an infantry term, and every infantryman learns to stay out of the enemies' line of fire. Parts flying off of a rotating machine will fly off on a tangent. Machines do "throw" couplings, and temporary balancing weights can fly off.
2. Tuck in all loose clothing and roll up your sleeves so that it cannot get entangled in rotating equipment. Be especially careful when covers are removed from couplings or belts.
3. Have someone familiar with the system shut down equipment. This ensures that the correct switch is thrown and proper coordination is done with the building occupants or users. If the switch is not in the immediate vicinity where you have visual contact with it and control over its use, be sure that the switch is locked out and/or tagged. This is critical for your safety.
4. Don't inadvertently reconfigure controls. The piece of equipment under test for vibration is usually operating and in use. It is supplying air, water, energy, or something else to the building or process. Care must be taken when working around this equipment and its controls not to inadvertently reconfigure it. No controls should be reconfigured (switches, valves, air pressure) without the user's knowledge and consent.
5. Beware of bumping hazards. Move slowly, but deliberately, around mechanical equipment to avoid bumping hazards to your head. Wear a hard hat and ear plugs where appropriate.
6. Watch your step. Sometimes, when taking vibration measurements, cables are on the floor or spanning open spaces between

transducers, instruments, and power sources. Do not step on these cables or tug at them. It is a good idea to secure cables with duct tape wherever practical. Your instrument may also be on the floor. Watch your step and work slowly but deliberately to avoid crushing cables or tripping.

7. Be aware of stroboscopic illusions. Be always alert when the machinery is rotating. Avoid the temptation to touch it when freezing its motion with a strobe. Turn off the stroboscope when not in use.
8. Be aware of high voltages. You will be working around high-voltage, and many times, high-current wiring. Be observant for, and avoid, high-voltage hazards such as exposed wires, frayed insulation, damaged conduit, and improper grounding.
9. Everything that goes in must come out. Tools, instruments, and portable radios have become entangled in machine rotating parts resulting in complete destruction of both. That is a very embarrassing situation. To avoid this embarrassment, have a system to take everything out with you. Some suggestions are a checklist, hand rotate the machine to check for freedom before energizing, and shine your flashlight back in and visually check before closing the door.
10. Limit the number of starts on large motors. The current surge on starting, especially under load, overheats the windings. Time is required for the windings to recover thermally. Generally, two starts under load per hour is safe. If more are required, then the motor manufacturer should be consulted.
11. Don't get burned. Test the surface temperature of all pipes, valves, motors, heat exchangers, etc., before grabbing hold.

Approaching a Vibration Problem

In App. B is a procedure that I use in my company when approaching a vibration problem. I do not follow this procedure step by step every time, but I know it well. Usually the problem is obvious after performing a few steps of this. I keep this procedure in my case in reserve. Occasionally, when I come across that difficult problem that is not obvious, I pull out the procedure and work it step by step.

You may encounter a person, typically in the maintenance organization, who does not agree with your diagnosis. This person is usually very close to the machine and knows it better than anyone else. It has been my experience that this person's intuition is probably right. His or her intimate knowledge of how the machine responds, sounds, and feels is sometimes more valuable than the most sophisticated electronic instruments. Get to know this person and get him or her on your side. Give respect and credibility to his or her position. Try to find reasons that your data supports, rather than refutes, his or her position. There are two reasons for doing this.

First, you want him or her on your side. In the event of a conflict, management, who may have an incomplete understanding of vibration analysis, is likely to side with the technician. His or her fix may be the wrong one, but then after trying it, you will both be closer to the right one. It benefits everyone concerned if the vibration analyst remains connected to the problem until a successful conclusion is achieved.

Second, teach this person everything you know about vibration analysis in the time you have together. The objective is to have him or her embrace this technology. By mating his or her intuitive "feel" for machinery with modern instruments and analysis, this person will be a powerful force for the operation after you are gone.

Do's

You will now have specialized knowledge of machinery condition. This knowledge of machine health should be guarded with confidentiality just like a doctor protects information of a patient's condition. A normal human response is to associate a defect with the last person who worked on the machine. This does not establish a cause-and-effect relationship and should not even be suggested. Information about a machine's unhealthy condition should only be shared with those who have a need to know. Use this information for good. Remember that defects are no one's fault, and everyone wants to do a good job. Let's not dwell on any negatives but rather move forward toward healthier machinery.

Here are a few do's:

- Do respond to situations.
- Do teach your fellow workers.
- Do document returns in writing—especially on how catastrophes were avoided.
- Do enlist the cooperation of others.
- Do inform others as soon as a problem appears.
- Do report results to managers. Focus on the facts, with analysis and recommendations, avoiding opinions.

The continuity of life in our modern society depends absolutely on the continuity of the supply of water, energy, transportation, communication, and waste removal. A disruption in any of these services has serious consequences. We are past the point of return to a rural agricultural society to still maintain our standard of living. The machinery of our civilization must keep turning. It is maintenance professionals, like yourselves, who keep the machinery working. There is no one else to do it. Let's perform this task smarter, not harder, and always strive for excellence.

I wish you success and satisfaction in solving your vibration problems.

Epilogue

With all of the damaging environmental factors and human errors in design, installation, and maintenance of machinery, it is a wonder that it works as long or as well as it does. The fact is that it *does* work. The good news is that it can work much longer. In fact, if the problems of wear and corrosion can be solved, there is no reason why machinery cannot function indefinitely. These are the two most significant problems in mechanical engineering. They have been the most significant problems for engineers for centuries. Even with our modern computer technology and aerospace travel, wear and corrosion still remain the most fundamental mechanical engineering problems. These problems have been in the academic and research arenas for some time. Great progress has been made in delaying these degrading effects but not in cancelling them. With so much effort expended for so long, the probability of a solution forthcoming from these arenas diminishes with every sunset. A fresh outlook is required. I believe that the solution will come from the maintenance arenas where people struggle with these problems on a daily basis. Dynamic vibration holds the key to understanding the forces in machinery.

By taking up the serious study of vibrations, humans have embarked on a journey of knowledge that is guaranteed to produce a deeper understanding of our dynamic universe. For centuries, the engineering of machines and structures was based on static analysis, i.e., the strength of materials. Safety factors were added to account for dynamic, real-world conditions. With a proper understanding of dynamic behavior, these safety factors can be shrunk considerably. It is also anticipated that with regular vibration monitoring, new rules for design of machines and structures will be fed back to the design engineers from the maintenance professions. It can also be foreseen that with a better understanding of our dynamic universe, the static laws of science will need to be modified.

Specifically, Newton's second and third laws require assumptions about rigid bodies that neither absorb nor amplify motions. Real bodies are made of elastic materials that deflect. These deflections, in an oscillating environment, can absorb or amplify motion. Also, the transmission of force is not instantaneous. Forces travel through materials at the speed of sound through that material, and conditions can change in that short time interval. Finally, resonance makes Newton's second law invalid as a frequency-independent expression. The motion attained with a given force is clearly dependent on frequency of oscillation, not just the inertia and the input force. The universe is an oscillating one. All physical phenomena have a frequency component that should be factored into the equation. My suspicion is that this frequency component was assumed to be negligible in the macroscopic world view. By considering the dynamic factor of oscillation, this technology of vibration promises to open new windows into the physical universe and possibly solve some stubborn problems that have remained unsolved due to a limited perspective.

Appendix

Tuneable Bandpass Filter

This appendix contains Fig. A.1, an electrical schematic diagram for an active tuneable bandpass filter that is suitable for low-frequency analysis and balancing. This filter was made with less than $50 worth of parts. In addition to filtering, it provides a gain of 10.7 (amplification).

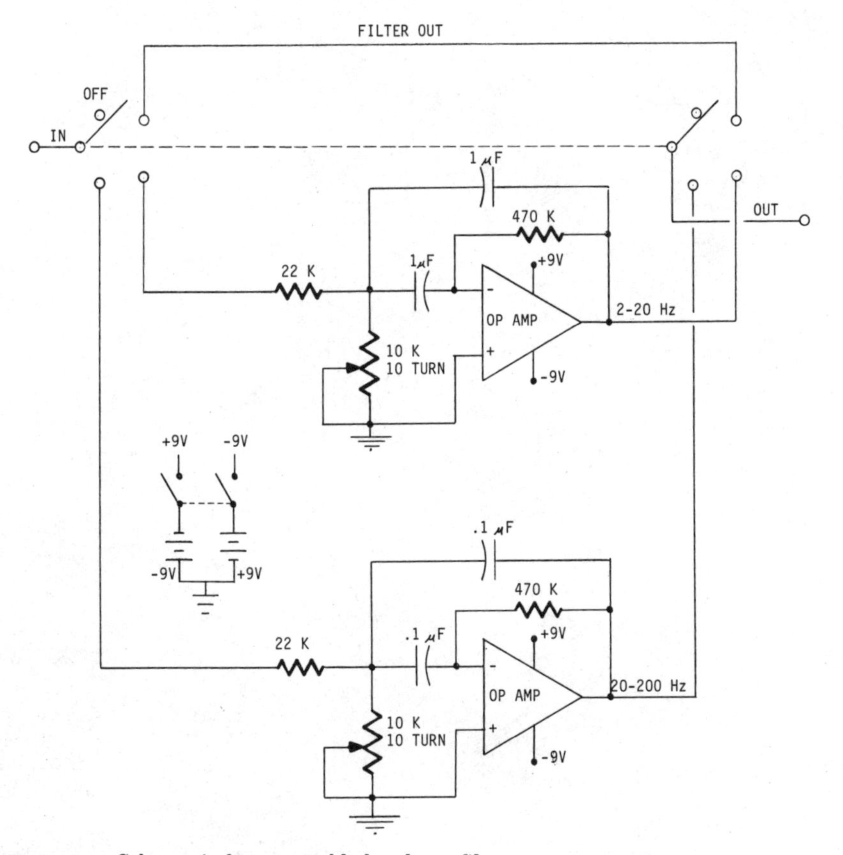

Figure A.1 Schematic for a tuneable bandpass filter.

Appendix

B

Vibration Analysis Procedure

1. Gather known facts. Have a quick meeting with those closest to the machine to learn something of its history. Try to find out what is not wrong with the machine. What has been the trend?

2. At the machine, don't let the vibration scare you. Use a step-by-step analysis. Remember that vibration problems are:

40 percent unbalance
30 percent misalignment
20 percent resonance
10 percent the rest

Make a sketch, or take a picture.

3. Look, listen, and feel. Hold a coin or probe in your hand and "hand feel" the vibration at the usual points for measuring the vertical, horizontal, and axial vibration at the bearing caps and housings. Mentally record which ones seem to be the roughest. Continue by feeling the attached typical items such as pipes, base, pedestals, valve stems, and ductwork. Walk around the floor area reaching out about 4 or 5 yards. Especially note if the floor vibration is considerably greater away from the machine (may indicate resonance in the floor).

4. If the complaint is building vibration, go to the actual location in the building and measure the frequency and amplitude on the floor or other part of the structure. Move around on the same floor to see if it decreases or increases. Go up or down to different levels and record the vibration amplitude at the offending frequency. For an acoustical problem, measure the sound field with a microphone.

5. Go back to the machine and repeat your probing around. This

time with the analyzer and accelerometer, record the amplitude and frequency at the worst points. Start at 20 KHz and step down.

6. Measure 1X-rpm vibration at all bearings in the horizontal, vertical, and axial direction. Also record phase at each point. Consider mapping the deflection-mode shape to find out how the machine is shaking.

7. From the above, determine unbalance, misalignment, or possible resonance.

8. Check similar machines in the area for comparative data.

9. Tighten all bolts. Check for soft foot.

10. Shut down equipment. Look at background vibration. For a building problem, go back to the complaint area and measure floor vibration with the machine off.

11. Check for structural resonances.

12. At start-up, view the run-up plot, looking for resonances.

13. Change load. A decrease in vibration at higher load could indicate looseness. Change valve positions to examine for cavitation.

14. Record data.

15. Now go to a quiet location with your data, and the machine's history. List all the possible sources that could cause vibration at that specific frequency. Analyze the phase readings to visualize how the machine is shaking rather than just how much.

16. Determine what is not the problem.

17. Ask yourself, "What else could cause the same symptoms?" and "What are the real probabilities for this to really have gone wrong?"

18. You can never be 100 percent sure. List the possible causes in order of priority.

19. Eliminate the possible causes by corrective methods.

20. If the problem remains, then the true cause has not yet been found. Gather additional data and do further analysis. Ask yourself, "What forces within the machine are acting in this direction at this frequency?"

Appendix

C

Vibration Specifications for Acceptance Testing of New or Rebuilt Machinery

1.0 General

The purpose of this specification is to guarantee to the buyer that mechanical equipment is in reasonably good health initially.

It is recognized that all manufactured goods contain variability. No two items are exactly alike. Vibration is a test for the dynamic characteristics of machinery. It is an overall measure of mechanical quality, and a good yardstick to compare similar machines. It is also recognized that the foundation and installation factors affect the vibration level and are also variable. The intention of this specification is to provide confidence to the buyer that the equipment is statistically normal based on industry standards of vibration and at the same time to limit the liability of the manufacturer to only the manufactured machine and relieve the manufacturer of foundation or installation factors.

2.0 Scope

2.1. This specification applies to normal industrial and commercial construction. Sensitive spaces and processes may require greater attention and control of vibration.

2.2. This specification covers measurement and analysis only. Corrective methods are additional work, the extent of which is unknown until after the measurement and analysis is completed. The responsibility of correction is to be negotiated between the buyer and supplier.

2.3. The responsibility of the supplier is to deliver equipment that is statistically normal and meets industry specifications for vibration. It is also the responsibility of the supplier to prove this to the buyer.

2.4. All equipment of ½ hp or larger shall be tested.

2.5. The equipment shall be tested under normal operating conditions under normal load. The intention here is to avoid idle tests. The equipment shall be fully functional, under normal control, supply energy or fluids to the system it is intended to serve, be installed in its final operating location, and under load.

2.6. A professional regularly engaged in vibration analysis shall supervise the measurements and interpret the results.

2.7. The equipment shall be tested in accordance with industry standards, some of which are listed in Section 6.0. In the absence of appropriate industry standards, then the requirements of Section 3.0 apply.

3.0 Requirements

3.1. Measurements shall be taken on the bearings in the horizontal, vertical, and axial directions. If the bearings are inaccessible, then the measurement location shall be a rigid part of the machine as close as practical to the bearing that faithfully follows the bearing motion. Flexible or resonant panels are to be avoided.

3.2. The vibration amplitude at all frequencies from 1 Hz to 50X running speed shall be continuously monitored and recorded for a minimum of 30 sec. Summation averaging on an FFT spectrum analyzer is acceptable. If the amplitude at any frequency fluctuates wildly, i.e., more than 50 percent, then the peak amplitude shall be recorded for this 30-sec monitoring period.

3.3. During this test, the machine shall have normal freedom of movement. The vibration isolators shall be functioning normally with no binding.

3.4. The equipment vibration amplitude shall not exceed 0.2 in/sec peak at all frequencies on rotating equipment. Reciprocating equipment shall not exceed 0.4 in/sec peak.

3.5. Variable-speed machines shall be tested for compliance throughout their entire speed range.

3.6. It is intended that this measurement for compliance be made at the machine's final installed location. However, the supplier may take measurements at the factory under similar foundation conditions. If the difference between factory measurements and final installed location measurements increased by more than 50 percent, then it can be presumed that the variability is in the foundation and the supplier is not responsible. It is the buyer's responsibility to supply a stable foundation.

3.7. If multiple machines are being purchased of the same make and model, then no single machine shall display a measured vibration amplitude at any frequency greater than 2 standard deviations from the norm for this group at that frequency. The intention of this is to weed out the 5 percent troublemakers.

3.8. A walkthrough shall be conducted throughout the facility listening and feeling for vibration and noise. The purpose is to detect objectionable tones and building transmission problems. Objectionable tones emanating from the machine are the supplier's responsibility. Building transmission problems are the buyer's responsibility.

4.0 Instruments

4.1. The vibration measuring instruments shall be capable of narrowband frequency analysis from 1 Hz to 50X running speed. They shall be capable of summation averaging for a time period of 30 sec minimum at all frequencies specified.

4.2. The entire instrumentation chain, from transducer to readout instrument, shall be calibrated in accordance with ANSI S2.2–1959 (R 1990).

4.3. The supplier is responsible for providing instrumentation and for making these measurements to prove the mechanical integrity of the equipment.

5.0 Documentation

5.1. Hard-copy plots of the vibration amplitude versus frequency of each measurement point shall be supplied to the buyer. This establishes the machine health condition at the time of ownership transfer and protects both the supplier and the buyer. These plots serve as baseline data if future problems develop. (*See Figures C.1 and C.2.)*

5.2. The shaft speeds shall be identified on these plots.

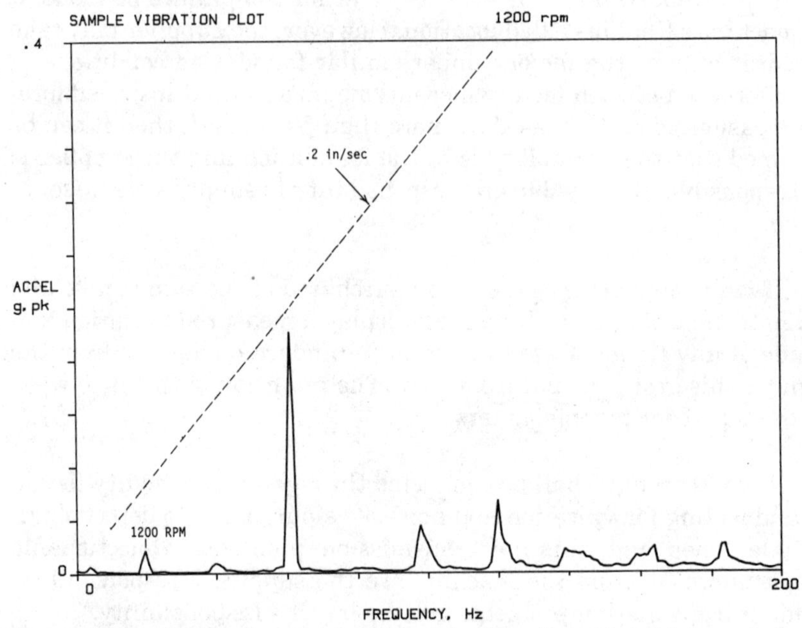

Figure C.1 Sample vibration plot from a new building fan with 0.2 in/sec velocity limit drawn as a dashed line: 0- to 200-Hz spectrum.

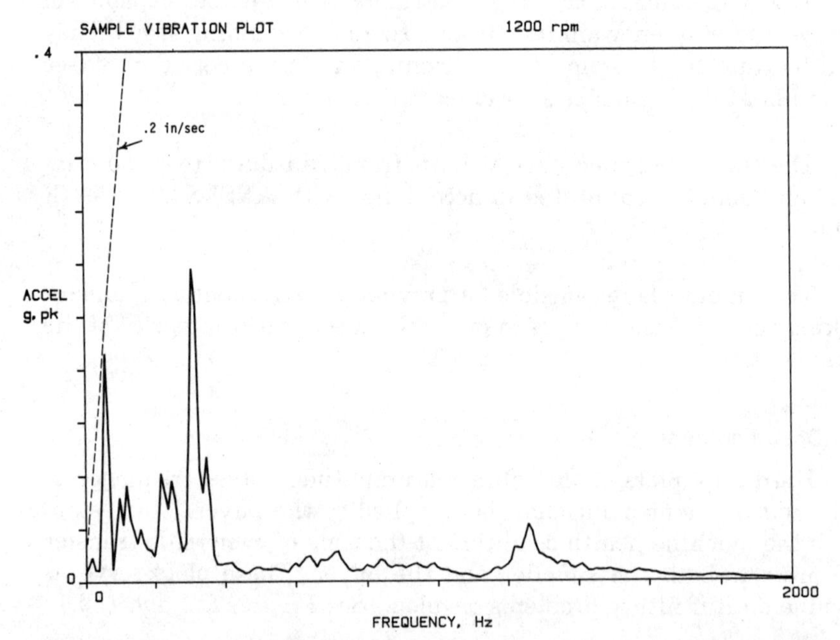

Figure C.2 Sample vibration plot from a new building fan with 0.2 in/sec velocity limit drawn as a dashed line: 0- to 2000-Hz spectrum.

5.3. A report, to accompany the plots, shall list noncompliant measurements, identify the vibration sources, and recommend corrective actions. Any other unusual conditions observed shall be reported.

5.4. The vibration amplitude can be plotted in acceleration, velocity, or displacement; however, the limit of 0.2 in/sec shall be clearly identified at each frequency of interest.

6.0 Industry standards

There are many standards for specific types of equipment that specify measurement methods different than those presented in this specification. These specific standards also vary from country to country. The purpose for providing this list of standards is for the general buyer of mechanical equipment where no specific standard applies.

NEMA MG1-12.05, Dynamic Balance of Motor, June 1978.
NEMA MG1-12.06, Method of Measuring the Motor Vibration, June 1978.
NEMA MG1-20.52, Balance of Machines, October 1977.
NEMA MG1-20.53, Method of Measuring the Motor Vibration, October 1977.
API 541, Recommended Practice for Form-wound Squirrel-cage Induction Motors, December 1972.
Hydraulic Institute, Acceptable Field Vibration Limits for Horizontal Pumps, 14th edition, Centrifugal Pumps Applications.
Compressed Air and Gas Institute, In-Service Vibration Standards for Centrifugal Compressors, 1963.
ISO Draft Standard, Determination of Mechanical Vibration of Gear Units During Acceptance Testing, ISO/TC-60/WG-9-N55, 1987.
ANSI S2.41–1985, The Measurement and Evaluation of Vibration Severity of Large Rotating Machines, in situ, operating at speeds from 10 to 200 rps.

Appendix

D

Conversion Formulas

$$1 \text{ oz} = 28.35 \text{ g}$$

$$\text{dB} = 20 \log_{10} \frac{V}{V_{\text{ref}}}$$

where V_{ref} = 1 V rms in most FFT spectrum analyzers

$1\ g = 32.2 \text{ ft/sec}^2$

$= 386 \text{ in/sec}^2$

$= 9.8 \text{ m/sec}^2$

$= 980 \text{ cm/sec}^2$

Centrifugal force due to imbalance:

$$F = mr\omega^2$$

where m = mass

r = radius

ω = rotational speed, rad/sec

in English units:

$$F(\text{lb}_f) = 1.77\left(\frac{\text{rpm}}{1000}\right)^2 \text{oz-in}$$

in SI units:

$$F(\text{N}) = 1.1\left(\frac{\text{rpm}}{10}\right)^2 \text{kg-m}$$

$$A = 0.1022 f^2 D$$

$$V = 61.7 \frac{A}{f}$$

$$D = 9.78 \frac{A}{f^2}$$

where A = peak acceleration, g
V = peak velocity, in/sec
D = peak displacement, in
f = frequency, Hz

in/sec = 0.0254 m/sec

mil = 25.4 μm

pound force = 4.448 N

1 cycle = 360° = 6.283 rad

hertz = 60 rpm

rms = 0.707 peak

Glossary

ac/dc coupling A selection on the front panel of modern readout instruments. The ac-coupling position switches a capacitor into the input conductor. This is used for piezoelectric accelerometers to remove the power supply bias voltage. The dc-coupling position removes this capacitor from the input line. This is used for transducers that have an output at zero frequency (such as piezoresistive accelerometers and proximity probes), and it is desired to make measurements below 1 Hz.

acceleration The time rate of change of velocity.

accelerometer Transducer whose voltage output is proportional to acceleration.

A/D converter Analog-to-digital converter. An electronic device that measures the voltage of a waveform at intervals and converts the visual presentation of this waveform to a list of numbers for digital processing.

Aeolian Pertaining to wind.

alias False signals in the frequency domain caused by a measuring rate for digitizing that is too slow.

alignment The condition of proper orientation of machine drive components such that vibratory forces unnecessary to power transmission are minimized.

amplitude The maximum departure of motion measured from the mean position to an extreme. Units are peak, peak to peak, and root mean square.

antialias filter A low-pass filter designed to block frequencies greater than one-half the measuring rate.

attenuate To make small.

axial In the direction parallel to the shaft centerline.

balance A condition where the rotating centerline between bearings coincides with the line that defines the center of mass distribution.

balancing A procedure for adjusting the mass distribution of a rotor by adding or removing weight, with the goal of achieving less vibration amplitude at rotational speed.

bandwidth The difference between the upper- and lower-cutoff frequencies, at which the signal is attenuated by 3 dB.

baseline spectrum A vibration spectrum captured from a machine when it is presumed to be in good running condition. Future spectrums are compared to this baseline when looking for changes during analysis.

beats The alternating rise and fall of vibration amplitude caused by two sources vibrating at close to the same speeds.

bin One spectral line in the frequency display of an FFT analyzer. The bin can be viewed as a bandpass filter with a bandwidth that is adjusted by the frequency span selected.

blade passing frequency See vane passing frequency.

block size The number of voltage measurements in a time block of data sampled by an FFT spectrum analyzer. This number is usually 512 for a 200-line frequency display, 1024 for a 400-line frequency display, etc.

Bodé plot A plot of amplitude versus frequency and phase verses frequency with all data points at 1X running speed. The data is admitted through a tracking filter as the machine changes speed.

brinnelling A permanent surface deformation caused by high static forces or high vibratory forces in a single spot, or more commonly a combination of the two, that results in local yielding at the points of contact.

calibration A test to verify the accuracy and repeatability of measurement instruments. For vibration, a transducer is subjected to a known motion, usually on a shaker table, and the output readings are verified or adjusted.

calibration constant The sensitivity of a transducer within its linear range, expressed as a ratio of millivolts per vibration amplitude. Typical units are millivolts per *g*, millivolts per inches per second, and millivolts per mil.

cantilever A beam supported at one end only.

carrier frequency (*see* modulation).

center of gravity The point representing the average position of matter in a body.

charge amplifier An electronic amplifier used to convert the high impedance of a piezoelectric accelerometer to low impedance for acceptance by common readout instruments.

critical speed The rotational speed corresponding to a natural frequency of the rotor-shaft-bearing system, above which the rotor is considered flexible.

damping Progressive diminishing in amplitude of oscillation.

dB Decibel, $\frac{1}{10}$ of a bel. A logarithmic unit of amplitude for noise or vibration measurement.

dc coupling (*see* ac/dc coupling).

differentiation A mathematical process of converting displacement to velocity, and velocity to acceleration. In an FFT spectrum analyzer, a single differentiation is represented by multiplication by $j\omega$.

dynamic range The difference between the highest voltage level that will overload the instrument and the lowest voltage level that is detectable. Dynamic range is usually expressed in decibels, typically 60 to 90 dB for modern instruments.

dynamic stiffness The apparent stiffness of a spring member under vibration or shock loading. This apparent stiffness is frequency dependent.

eddy current A circulating current induced in a conductive material by a changing electromagnetic field.

EU Engineering units. These units are selected by the user of an FFT spectrum analyzer in conjunction with the transducer sensitivity, to display the motion output directly in physical units of mils, inches per second, or *g*'s, which then become the EUs.

external sampling The rate of measurement in the digitizing process is controlled by a multiplied tachometer signal. The result is a stationary display of vibration as the speed changes. Useful for analyzing variable-speed machines.

external trigger The beginning of each time block is controlled by an external signal, typically a point on the rotating system that is sensed as it passes by a fixed detector (Keyphasor).

fast Fourier transform (FFT) A computer technique to calculate the frequency components of a time waveform from the digitized voltage measurements. The result is a display of amplitude versus frequency, and phase versus frequency.

fatigue The tendency for a material to crack under repeated deflection.

FEA (*see* finite element analysis).

FFT (*see* fast Fourier transform).

filter A device (mechanical or electronic) that passes or rejects information within specific frequency limits.

finite element analysis (FEA) A modeling technique for predicting dynamic behavior before any hardware is built.

flattop window A manipulation that provides the best amplitude accuracy.

flexible coupling A device, connecting the shafts of the driver and driven machines, that is capable of being bent and at the same time strong enough to transfer the power. *Semiflexible coupling* is a better term.

flexible rotor A rotor that deforms significantly at running speed. This term is used for rotors that operate close to or above their first critical speed.

forced vibration Oscillation occurring at the frequency of a driving force input.

free running A term used to describe the data acquisition of an analyzer, where the instrument acquires the next time block when it is finished processing the previous time block and is ready for more data. It continuously operates this way, updating the display.

free vibration Oscillation occurring at a natural frequency, after an initial force input.

frequency The number of complete oscillations per unit of time; units are hertz, revolutions per minute, and orders.

frequency domain Vibration represented as a graph of amplitude versus frequency.

FTF Fundamental train frequency. The speed of rotation of the cage assembly in a roller element bearing.

g Acceleration due to gravity; equal to 980 cm/sec^2 (32.2 ft/sec^2) on the surface of the earth.

gear-mesh frequency The speed of rotation multiplied by the number of teeth of the gear on that shaft. The mating gear has the same gear-mesh frequency because its different speed multiplied by its different number of teeth results in the same number.

GPIB General-purpose interface bus, or IEEE-488 standard bus. Used for computer interfacing to electronic instruments.

ground loop Circulating current between two or more connections to electrical ground. This signal can be detected and displayed by electronic instruments. These signals are generally not associated with the variable to be measured and represent noise in the measuring system.

Hanning window A manipulation that reduces leakage and provides a good compromise between frequency resolution and amplitude accuracy.

harmonics Vibration frequencies which are integral multiples of the fundamental.

heavy spot The imaginary vector sum of the nonuniform mass distribution within a rotating body. The corrective procedure of balancing places a weight to compensate for the forces caused by the heavy spot.

high-pass filter A filter that passes all frequencies above a lower-cutoff frequency.

HVAC Heating, ventilating, and air conditioning.

Hz Hertz, one cycle per second.

imbalance (or unbalance) Unequal weight distribution within a body that results in a periodic force (as measured at a stationary point) when the body is rotated.

inertia A property of matter by which it resists a change in motion, including resistance to change from a position of rest.

integration A mathematical process of converting acceleration to velocity, and velocity to displacement. In an FFT spectrum analyzer, a single integration is represented by division by $j\omega$; a double integration by division by $G - \omega^2$.

isolation Diminishing the transfer of vibration amplitude by the judicious selection of barrier materials and barrier configuration. One hundred percent isolation is not possible unless there is no physical contact and a perfect vacuum surrounds the object to be isolated.

journal That part of the shaft that rides in a plain cylindrical bearing.

Keyphasor[1] A sensing device, or signal, that detects the passage of a point on the rotor. The sensing device may be a magnetic, capacitive, eddy current, or photoelectric probe. The signal is used as a trigger for the external trigger input of other electronic instruments, such as FFT spectrum analyzers.

leakage An error introduced into the FFT process caused by using finite-length time blocks that do not match at the ends. Its effects are a smearing of the frequency lines at lower amplitudes. This error is minimized by the use of window functions.

linearity In dynamic systems, a property of proportionality where the force input and motion output maintain the same ratios when either is changed. Examples of nonlinearity are severe strain and hitting mechanical stops.

low-pass filter A filter that passes all frequencies below an upper-cutoff frequency.

mechanical impedance An apparent opposition to the transfer of vibration amplitude that is frequency dependent.

micrometer One millionth of a meter. One micrometer (also called micron) equals 0.04 mils.

mils Thousandths of an inch (0.001 in). One mil equals 25.4 μm.

modal analysis The study of mode shapes with the goal of finding the locations of maximum deflection to know where to apply stiffeners.

mode shape The shape that a part takes in oscillation at a specific frequency.

---

[1]Keyphasor is a Bently Nevada Corporation trade name.

modulation The mixing of two signals that causes amplitude or frequency variation. The higher frequency is usually called the *carrier,* and its amplitude or frequency is changed by another periodic event. In vibration, this means that two vibrations arrive at the transducer location. When separated into discrete frequencies by the FFT process, the result is a component at the carrier frequency, and adjacent components, or sidebands, spaced at the frequency of the modulating signal. Modulation implies that the two vibrations are related physically.

natural frequency The frequency that a part, or system, will oscillate at if excited with an impulse.

nodes Points or lines where very little motion takes place during a condition of resonance.

noise Unwanted signals. Signals entering the measuring system that do not represent the variable being measured.

noise floor The lower sensitivity limit of an electronic measuring instrument, expressed in microvolts (10^{-6} V).

octave The interval between two frequencies where the higher frequency is two times the lower frequency.

orders A unit of frequency unique to rotating machinery where the first order is equal to rotational speed.

period The time required for one complete cycle of oscillation.

phase A time delay, expressed in degrees of rotation, between two transducers.

phasor A rotating vector.

piezoelectric The property of certain crystals to generate an electric charge when the crystal is mechanically strained.

pressure waves Motion which considers the body as elastic where particles move very small distances relative to the center and transfer energy of motion to neighboring particles which carry the energy as a traveling wave. This wave travels to the extremes of the body, at the speed of sound through that material, where it is reflected and coupled to the adjacent material.

prime mover The driver machine that converts electrical or chemical energy into mechanical motion.

radial In a direction perpendicular to the shaft centerline.

random A variable whose value at some future time cannot be predicted with confidence.

rectangular window Sometimes called *uniform window.* This is actually no manipulation at all and provides no protection against leakage. It is used to measure frequency most accurately and for transient events.

resonance A condition of amplification of vibratory motion when the driving force input oscillates at the natural frequency of a physical part.

rigid rotor A rotor that does not deform significantly at running speed. A rotor whose parts do not take up motion relative to each other, i.e., all points move in the same direction at the same instant of time.

rms Root mean square. Equal to 0.707 times the peak.

rotor Rotating part.

rpm Revolutions per minute or cycles per minute.

running speed The speed of rotation of a machine expressed in revolutions per minute or hertz.

seismic Referenced to inertial space.

shock An impact, and the resulting transient pulse.

sidebands Spectral lines of equal spacing on both sides of a common signal. The sidebands result from the mixing of two signals, the common signal is of one frequency and the spacing is the frequency of the second signal.

signature The unique vibration characteristics of a machine. This signature changes as physical conditions or force events change within the machine or on its support system.

signature analysis The interpretation of the electric signal from a transducer to determine the significant physical forces at work, or to look for changes, in a machine.

spectrum analyzer An instrument which displays dynamic data (vibration) in the frequency domain.

stiffness The property of elastic deflection under a force load.

time domain Vibration represented as a graph of amplitude versus time, as usually seen on an oscilloscope. A trace of the actual voltage output from a transducer.

torsional vibrations The oscillatory twisting of a shaft about its longitudinal axis.

trigger An event used as a timing device to initiate measurement.

tuneable bandpass filter An adjustable filter with upper- and lower-cutoff frequencies, where the center frequency can be adjusted.

uniform window (*see* rectangular window).

vane passing frequency The number of vanes, or blades, times the rotational speed.

vector A quantity having both magnitude and direction.

vibration An oscillatory motion.

vortices Whirlpools of fluid flow.

white noise Random vibration characterized by a uniform distribution of energy across the frequency spectrum.

whole body motion Motion which considers the body as a rigid mass where all particles move in the same direction at the same time.

window A manipulation of data, especially at the extremes of a time block, to minimize leakage.

Bibliography

ANSI S2.2–1959 (R 1990). American National Standard Methods for the Calibration of Shock and Vibration Pickups.

ANSI S2.11–1961 (R 1986). American National Standard for the Selection of Calibrations and Tests for Electrical Transducers Used for Measuring Shock and Vibration.

ANSI S2.10–1971 (R 1990). American National Standard Methods for Analysis and Presentation of Shock and Vibration Data.

ANSI S2.17–1980 (R 1986). American National Standard—Techniques of Machinery Vibration Measurement.

ANSI S2.7–1982 (R 1986) (ASA 42). American National Standard—Balancing Terminology.

ANSI S2.38–1982 (R 1990). American National Standard—Field Balancing Equipment—Description and Evaluation.

ANSI S2.42–1982 (R 1990). American National Standard—Procedures for Balancing Flexible Rotors.

ANSI S2.40–1984 (R 1990). American National Standard—Mechanical Vibration of Rotating and Reciprocating Machinery—Requirements for Instruments for Measuring Vibration Severity.

ANSI S2.43–1984 (R 1990). American National Standard—Criteria for Evaluating Flexible Rotor Balance.

ANSI S2.41–1985 (R 1990). American National Standard—Mechanical Vibration of Large Rotating Machines With Speed Range from 10 to 200 rev/s—Measurement and Evaluation of Vibration Severity in situ.

ANSI S2.60–1987. American National Standard—Balancing Machines—Enclosures and Other Safety Measures.

ANSI S2.19–1989 (ASA 86). American National Standard—Mechanical Vibration—Balance Quality Requirements of Rigid Rotors, Part I: Determination of Permissible Residual Unbalance.

ANSI S2.61–1989. American National Standard Guide to the Mechanical Mounting of Accelerometers.

API 670. American Petroleum Institute. Non-contact Vibration and Axial Position Monitoring System.

API 678. American Petroleum Institute. Accelerometer Based Vibration Monitoring System.

API 541. American Petroleum Institute. Maximum Permissible Vibration for Form-wound Squirrel-cage Induction Motors.

Compressed Air & Gas Institute. In Service Vibration Standards for Centrifugal Compressors.

Bently Nevada Seminar. *Machine Protection Seminar,* Bently Nevada, Minden, Nevada.

Michael P. Blake and William S. Mitchell. *Vibration and Acoustic Measurement Handbook.* Spartan Books, 1972.

Heinz P. Bloch and Fred K. Geitner. *Machinery Failure Analysis and Troubleshooting.* Gulf Publishing Company, 1986.

Simon Braun. *MSA—Mechanical Signature Analysis.* American Society of Mechanical Engineers (ASME), 1983.

Jens T. Broch, *Mechanical Vibration and Shock Measurements,* 2d ed., 3rd impression, Brüel & Kjær, Denmark, April 1984.
Brüel & Kjær Seminar. *Modern Techniques of Machine Vibration Analysis.* Brüel & Kjær, Nærum, Denmark.
Brüel & Kjær, *Vibration Analysis of Machinery.* Brüel & Kjær, Nærum, Denmark.
Brüel & Kjær, *Digital Signal Analysis Using Digital Filters and FFT Techniques.* Brüel & Kjær, Nærum, Denmark, January 1985.
Ralph T. Buscarello. *Practical Solutions to Machinery and Maintenance Vibration Problems.* Update International, Lakewood, Colo.
Edgar J. Gunter. *Selected Papers on Field Balancing of Rotating Machinery—Advanced Theory and Techniques.* Vibration Institute, Willowbrook, Ill., May 1983.
Cyril M. Harris and Charles E. Crede. *Shock and Vibration Handbook,* 2d ed., McGraw-Hill, New York, 1976.
Cyril M. Harris. *Shock and Vibration Handbook,* 3rd ed., McGraw-Hill, New York, 1988.
J. P. Den Hartog. *Mechanical Vibrations,* 4th ed., McGraw-Hill, New York, 1956.
J. P. Den Hartog. *Mechanical Vibrations,* 4th ed., Dover Publications, Mineola, N.Y., 1985.
Hewlett-Packard Application Note 243-1. *Effective Machinery Maintenance Using Vibration Analysis.* Hewlett-Packard Company, Palo Alto, Calif., 1983.
Hewlett-Packard Application Note 243. *The Fundamentals of Signal Analysis.* Hewlett-Packard Company, Palo Alto, Calif., July 1982.
Robert S. Jones. *Noise and Vibration Control in Buildings.* McGraw-Hill, New York, 1984.
J. M. Juran. *Quality Control Handbook,* 3rd ed., McGraw-Hill, New York, 1979.
Mechanical Technology Incorporated (MTI). June 1987 Seminar, *Rotating Machinery, Vibration Analysis and Diagnostic Techniques.* MTI, Latham, N.Y., 1973.
John S. Mitchell. *An Introduction to Machinery Analysis and Monitoring.* PennWell Publishing, Tulsa, Okla., 1981.
John Piotrowski. *Shaft Alignment Handbook.* Marcel Dekker, New York, 1986.
Singiresu S. Rao. *Mechanical Vibrations.* Addison-Wesley, Reading, Mass., 1986.
J. D. Smith. *Gears and Their Vibration: A Basic Approach to Understanding Gear Noise.* Marcel Dekker, New York, 1983.
W. Soedel. *Vibration of Shells and Plates.* Marcel Dekker, New York, 1981.
Vibration Institute Seminar Proceedings. *Balancing of Rotating Machinery—Houston, Texas.* Vibration Institute, Willowbrook, Ill., February 26–28, 1980.
Vibration Institute Course. *Machinery Vibration Analysis I Course—Tempe, Arizona.* Vibration Institute, Willowbrook, Ill., March 1–4, 1988.
"The Balance Quality of Rigid Rotors," ISO DR 1940.
AD-A171 031, P. D. McFadden. "Examination of a Technique for the Early Detection of Failure in Gears by Signal Processing of the Time Domain Average of the Meshing Vibration." Commonwealth of Australia, 1986, ARL-AERO-PROP-TM-434.
AD-A182-572, P. D. McFadden. "Interpolation Techniques for the Time Domain Averaging of Vibration Data with Application to Helicopter Gearbox Monitoring." Commonwealth of Australia, 1986, ARL-AERO-PROP-TM-437.
AD-A173 851, P. D. McFadden. "Analysis of the Vibration of the Input Bevel Pinion in Ran Wessex Helicopter Main Rotor Gearbox WAK 143 Prior to Failure." Commonwealth of Australia, 1985, ARL-AERO-PROP-R-169.
AD-A155 196, P. D. McFadden. "Proposal for Modifications to the Wessex Helicopter Main Rotor Gearbox Vibration Monitoring Program." Commonwealth of Australia, 1985, ARL-AERO-PROP-TM-422.
Hydraulic Institute Application Standards B-74-1, 1967.
S2.5. American Standard Recommendations for Specifying the Performance of Vibration Machines.
J. I. Taylor. "Identification of Bearing Defects by Spectral Analysis." ASME Publication 79-DET-14, 1979.
Harvey L. Balderston. "The Detection of Incipient Failure in Bearings." *Materials Evaluation,* Vol. 27, No. 6, June 1969, pp. 121–128.

Index